Geometry of Mechanics

Advanced Textbooks in Mathematics

Print ISSN: 2059-769X
Online ISSN: 2059-7703

The *Advanced Textbooks in Mathematics* explores important topics for postgraduate students in pure and applied mathematics. Subjects covered within this textbook series cover key fields which appear on MSc, MRes, PhD and other multidisciplinary postgraduate courses which involve mathematics.

Written by senior academics and lecturers recognised for their teaching skills, these textbooks offer a precise, introductory approach to advanced mathematical theories and concepts, including probability theory, statistics and computational methods.

Published

Geometry of Mechanics
  by Miguel C Muñoz-Lecanda and Narciso Román-Roy

Classical and Modern Optimization
  by Guillaume Carlier

An Introduction to Machine Learning in Quantitative Finance
  by Hao Ni, Xin Dong, Jinsong Zheng and Guangxi Yu

Conformal Maps and Geometry
  by Dmitry Beliaev

Crowds in Equations: An Introduction to the Microscopic Modeling of Crowds
  by Bertrand Maury and Sylvain Faure

Mathematics of Planet Earth: A Primer
  by Jochen Bröcker, Ben Calderhead, Davoud Cheraghi, Colin Cotter,
  Darryl Holm, Tobias Kuna, Beatrice Pelloni, Ted Shepherd and Hilary Weller
  edited by Dan Crisan

Periods and Special Functions in Transcendence
  by Paula B Tretkoff

The Wigner Transform
  by Maurice de Gosson

Advanced Textbooks in Mathematics

Geometry of Mechanics

Miguel C Muñoz-Lecanda
Narciso Román-Roy

Universitat Politècnica de Catalunya · BarcelonaTech, Spain

W **World Scientific**

NEW JERSEY · LONDON · SINGAPORE · BEIJING · SHANGHAI · HONG KONG · TAIPEI · CHENNAI · TOKYO

Published by

World Scientific Publishing Europe Ltd.

57 Shelton Street, Covent Garden, London WC2H 9HE

Head office: 5 Toh Tuck Link, Singapore 596224

USA office: 27 Warren Street, Suite 401-402, Hackensack, NJ 07601

Library of Congress Cataloging-in-Publication Data

Names: Muñoz Lecanda, Miguel, author. | Román-Roy, Narciso, author.
Title: Geometry of mechanics / Miguel C Muñoz-Lecanda, Narciso Román-Roy.
Description: New Jersey : World Scientific, [2026] | Series: Advanced textbooks in mathematics,
    2059-769X | Includes bibliographical references and index.
Identifiers: LCCN 2024058423 | ISBN 9781800616639 (hardcover) |
    ISBN 9781800616790 (paperback) | ISBN 9781800616646 (ebook) |
    ISBN 9781800616653 (ebook other)
Subjects: LCSH: Mechanics, Analytic. | Symplectic geometry.
Classification: LCC QA805 .M88 2026 | DDC 531.01/51612--dc23/eng/20250221
LC record available at https://lccn.loc.gov/2024058423

British Library Cataloguing-in-Publication Data
A catalogue record for this book is available from the British Library.

For any available supplementary material, please visit
https://www.worldscientific.com/worldscibooks/10.1142/Q0490#t=suppl

Typeset by The Manila Typesetting Company

To our respective families for their constant support and understanding over the years.

To our collaborator and friend Arturo Echeverría-Enríquez (in memoriam).

To the coauthor MCML, who passed away just as we completed the manuscript.

Contents

Preface

The aim of this book is to study the geometry underlying mechanics and its application to describe autonomous and nonautonomous conservative dynamical systems, as well as dissipative dynamical systems. We use different geometric descriptions to study the main properties and characteristics of these systems, such as their Lagrangian, Hamiltonian and unified formalisms, their symmetries, and the variational principles, among others. The study is done mainly for the regular case, although some comments and explanations about singular systems are also included.

This book is intended for readers who have completed at least the first courses of a degree in mathematics or physics and who have a basic knowledge of differential geometry of smooth manifolds and analytical mechanics.

This text includes the extended content of a course on Advanced Mathematical Models of Physics that the authors taught for several years at the Faculty of Mathematics and Statistics (FME) of the Universitat Politècnica de Catalunya · BarcelonaTech (UPC).

MC Muñoz-Lecanda and N Román-Roy

About the Authors

Miguel C Muñoz-Lecanda (1946–2023) was a founding member of the Geometry, Mechanics and Control Network. He joined the Department of Applied Mathematics at the Universitat Autònoma de Barcelona in 1984 and was made Full Professor of Applied Mathematics in 1992. Over the course of his career, he published more than 80 papers in scientific journals, proceedings and books. Professor Muñoz-Lecanda delivered many undergraduate and graduate courses in the Faculty of Sciences at both the Universitat Autònoma de Barcelona and the Universitat de Barcelona, as well as at both the Barcelona School of Telecommunications Engineering and the Faculty of Mathematics and Statistics at the Universitat Politècnica de Catalunya · BarcelonaTech. His research interests were mathematical physics, applied differential geometry, theoretical physics, and applied mathematics, in general; and geometric mechanics, geometric classical field theory, geometric control theory, and geometric quantization in particular.

Narciso Román-Roy (born in 1957) is a founding member of the Geometry, Mechanics and Control Network. He has been a Full Professor in Mathematics at the Universitat Politècnica de Catalunya · BarcelonaTech since 1988 and has published around 115 papers in scientific journals, proceedings, and books. Professor Román-Roy has delivered many undergraduate and graduate courses in the Faculty of Science at the Universitat Autònoma de Barcelona, as well as in the School of Industrial Engineering of Barcelona, the Barcelona School of Telecommunications Engineering, and the Faculty of Mathematics and Statistics at the Universitat Politècnica de Catalunya · BarcelonaTech. His research fields are mathematical physics, applied differential geometry, and theoretical physics, in general; and specifically, geometric mechanics, geometric classical field theory, gravitation, and geometric quantization.

Acknowledgments

We extend our special thanks to Prof. Xavier Gràcia-Sabaté for the many clarifying discussions held on much the contents of this book and Prof. Xavier Rivas-Guijarro for his careful reading of the entire manuscript and his very helpful comments and suggestions.

We also thank Prof. Manuel de León-Rodríguez for his comments and expert advice on the final editing and publication process of this manuscript.

Finally, we want to thank all our colleagues from the *Geometry, Dynamics, and Field Theory Network* for their enriching collaboration over all these years.

We acknowledge the financial support from the projects PID2021-125515NB-C21, funded by MICIU/AEI/10.13039/501100011033/FEDER, UE, and RED2022-134301-T of AEI, as well as from the Ministry of Research and Universities of the Catalan Government, project 2021 SGR 00603 *Geometry of Manifolds and Applications, GEOMVAP.*

MINISTERIO
DE CIENCIA, INNOVACIÓN
Y UNIVERSIDADES

Cofinanciado por
la Unión Europea

AGENCIA
ESTATAL DE
INVESTIGACIÓN

Notation and Terminology

$C^\infty(M)$	Smooth functions (of class C^∞) on a manifold M.
$\Omega^k(M)$	Differentiable k-forms on a manifold M.
$Z^k(M)$	Differentiable closed k-forms on a manifold M.
$\mathfrak{X}(M)$	Vector fields on a manifold M.
$\mathcal{T}_k^m(M)$	Tensor fields of type (k, m) on a manifold M.
Diff (M)	Diffeomorphisms of the manifold M.
$\mathfrak{X}^{V(\pi)}(M)$	π-vertical vector fields on a bundle $\pi \colon M \longrightarrow N$.
$\mathfrak{X}(M, \pi)$	Vector fields along a map π on a manifold M.
$\Gamma(\pi)$	Sections of the projection π on a bundle $\pi \colon M \longrightarrow N$.
d	Exterior differential.
$\beta(X) \equiv \langle X \vert \beta \rangle$	Natural pairing between differential 1-form β and vector fields X.
$i(X)\beta$	Contraction of a differential k-form β with a vector field X.
$L(X)\beta$	Lie derivative of a differential form β by a vector field X.
$X(f) \equiv L(X)f$	Action of a vector field X (Lie derivative) on a function f.
TQ , $\tau_Q \colon TQ \longrightarrow Q$	Tangent bundle of a manifold Q, canonical projection.
T^*Q , $\pi_Q \colon T^*Q \longrightarrow Q$	Cotangent bundle of a manifold Q, canonical projection.

$(q^i, v^i), (q^i, p_i)$	Natural coordinates on TQ and T^*Q.
$\Omega \in \Omega^2(M)$, (M, Ω)	(Pre)symplectic form, (pre)symplectic manifold.
$\Theta \in \Omega^1(M)$	(Pre)symplectic potential ($\Omega = d\Theta$).
$\flat_\Omega \colon TM \longrightarrow T^*M$	Canonical isomorphisms on a symplectic manifold (M, Ω).
$\sharp_\Omega = \flat_\Omega^{-1} \colon T^*M \longrightarrow TM$	
(M, Ω, α) , (M, Ω, h)	Hamiltonian system on a symplectic manifold.
$\{f, g\}$	Poisson bracket between the functions f and g.
$\mathfrak{X}_{lh}(M)$, $\mathfrak{X}_H(M)$	Local/global Hamiltonian vector fields on a symplectic manifold.
$X_h \in \mathfrak{X}(M)$	Hamiltonian vector field of $f \in C^\infty(M)$ on a symplectic manifold.
G , $\mathbf{g}$	Lie group, Lie algebra.
$\xi_X \in \mathfrak{X}(M)$	Fundamental vector field of $\Phi \colon G \longrightarrow \mathrm{Diff}(M)$ associated with $X \in \mathbf{g}$.
$\mathrm{j}^* \colon \mathbf{g} \longrightarrow C^\infty(M)$	Comomentum map associated with an action $\Phi \colon G \longrightarrow \mathrm{Diff}(M)$.
$\mathrm{J} \colon M \longrightarrow \mathbf{g}^*$	Momentum map associated with an action $\Phi \colon G \longrightarrow \mathrm{Diff}(M)$.
$\Delta \in \mathfrak{X}(M)$	Liouville (dilation) vector field on a vector bundle $\pi \colon M \longrightarrow N$.
$T\phi \colon TM \longrightarrow TN$	Canonical lift of a map $\phi \colon M \longrightarrow N$ to the tangent bundles.
$Z^V \in \mathfrak{X}(TQ)$	Vertical lift of $Z \in \mathfrak{X}(Q)$ to the tangent bundle TQ.
$Z^C \in \mathfrak{X}(TQ)$	Complete canonical lift of $Z \in \mathfrak{X}(Q)$ to the tangent bundle TQ.
$\widetilde{\gamma} \colon \mathbb{R} \longrightarrow TQ$	Canonical lift of a curve $\gamma \colon \mathbb{R} \longrightarrow Q$ to the tangent bundle TQ.
$T^*\phi \colon T^*N \longrightarrow T^*M$	Canonical lift of a map $\phi \colon M \longrightarrow N$ to the cotangent bundles.
$Z^* \in \mathfrak{X}(T^*Q)$	Canonical lift of $Z \in \mathfrak{X}(Q)$ to the cotangent bundle T^*Q.
$\mathcal{FL} \colon TQ \longrightarrow T^*Q$	Legendre map defined by $\mathcal{L} \in C^\infty(TQ)$.
$J \in \mathcal{T}_1^1(TQ)$	Canonical endomorphism on TQ.
$\mathcal{L} \in C^\infty(TQ)$, $(TQ, \mathcal{L})$	Lagrangian function, Lagrangian system.

$\Theta_{\mathcal{L}} \in \Omega^1(TQ)$	Lagrangian (Cartan) 1-form associated with $\mathcal{L} \in C^\infty(TQ)$.
$\Omega_{\mathcal{L}} = -d\Theta_{\mathcal{L}} \in \Omega^2(TQ)$	Lagrangian (Cartan) 2-form associated with $\mathcal{L} \in C^\infty(TQ)$.
$X_{\mathcal{L}} \in \mathfrak{X}(TQ)$, $\Gamma_{\mathcal{L}} \in \mathfrak{X}(TQ)$	Lagrangian/Euler–Lagrange vector field of a Lagrangian system.
$\Phi \colon Q \times \mathbb{R}^n \longrightarrow T^*Q$	Complete solution to the Hamilton–Jacobi equation.
$\mathcal{S} \colon Q \times \mathbb{R}^n \longrightarrow \mathbb{R}$	Generating function of a complete solution to the Hamilton–Jacobi equation.
$h \in C^\infty(T^*Q)$, (T^*Q, Ω, h)	Canonical Hamiltonian function, canonical Hamiltonian system.
$\mathcal{W} = TQ \times_Q T^*Q$	Unified (Pontryagin) bundle.
$\varrho_1 \colon \mathcal{W} \longrightarrow TQ$ $\varrho_2 \colon \mathcal{W} \longrightarrow T^*Q$ $\varrho_0 \colon \mathcal{W} \longrightarrow Q$	Canonical projections of the unified bundle $\mathcal{W}$.
$\Theta_{\mathcal{W}} \in \Omega^1(\mathcal{W})$	Canonical 1-form on the unified bundle $\mathcal{W}$.
$\Omega_{\mathcal{W}} = -d\Theta_{\mathcal{W}} \in \Omega^2(\mathcal{W})$	Canonical 2-form on the unified bundle $\mathcal{W}$.
$\mathcal{C} \colon \mathcal{W} \longrightarrow \mathbb{R}$	Coupling function on the unified bundle $\mathcal{W}$.
$\mathcal{H} \colon \mathcal{W} \longrightarrow \mathbb{R}$	Hamiltonian function on the unified bundle $\mathcal{W}$.
$X_{\mathcal{H}} \in \mathfrak{X}(\mathcal{W})$	Dynamical vector field of the system $(\mathcal{W}, \Omega_{\mathcal{W}}, \mathcal{H})$.
$\jmath_0 \colon \mathcal{W}_0 \hookrightarrow \mathcal{W}$	Compatibility submanifold of the system $(\mathcal{W}, \Omega_{\mathcal{W}}, \mathcal{H})$.
(η, ω), (M, η, ω)	(Pre)cosymplectic structure, (pre)cosymplectic manifold.
$R \in \mathfrak{X}(M)$	Reeb vector field on a cosymplectic manifold (M, η, ω).
$\flat_{(\eta,\omega)} \colon TM \longrightarrow T^*M$	Canonical isomorphisms on a cosymplectic manifold (M, η, ω).
$\sharp_{(\eta,\omega)} = \flat_{(\eta,\omega)}^{-1} \colon T^*M \longrightarrow TM$ $X_f \in \mathfrak{X}(M)$	Hamiltonian vector field on a cosymplectic manifold (M, η, ω).
$\operatorname{grad} f \in \mathfrak{X}(M)$	Gradient vector field on a cosymplectic manifold (M, η, ω).
$\mathcal{E}_f \in \mathfrak{X}(M)$	Evolution vector field on a cosymplectic manifold (M, η, ω).
$\rho \colon \mathbb{R} \times Q \longrightarrow \mathbb{R}$	Canonical projection.

$\tau_1 \colon \mathbb{R} \times TQ \to \mathbb{R}$ $\tau_2 \colon \mathbb{R} \times TQ \to TQ$ $\tau_0 \colon \mathbb{R} \times TQ \to Q$ $\tau_{1,0} \colon \mathbb{R} \times TQ \to \mathbb{R} \times Q$	Canonical projections of the bundle $\mathbb{R} \times TQ$.
$\pi_1 \colon \mathbb{R} \times T^*Q \to \mathbb{R}$ $\pi_2 \colon \mathbb{R} \times T^*Q \to T^*Q$ $\pi_0 \colon \mathbb{R} \times T^*Q \to Q$ $\pi_{1,0} \colon \mathbb{R} \times T^*Q \to \mathbb{R} \times Q$	Canonical projections of the bundle $\mathbb{R} \times T^*Q$.
$\boldsymbol{\gamma} \colon \mathbb{R} \longrightarrow \mathbb{R} \times Q$	Canonical lift of a curve $\gamma \colon \mathbb{R} \longrightarrow Q$ to $\mathbb{R} \times Q$.
$\widehat{\mathbf{c}} \colon \mathbb{R} \longrightarrow \mathbb{R} \times TQ$	Canonical lift of a curve $\mathbf{c} \colon \mathbb{R} \longrightarrow \mathbb{R} \times Q$ to $\mathbb{R} \times TQ$.
$\mathcal{J} \in \mathcal{T}_1^1(\mathbb{R} \times TQ)$	Extension of the canonical endomorphism to $\mathbb{R} \times TQ$.
$F\mathcal{L} \colon \mathbb{R} \times TQ \longrightarrow \mathbb{R} \times T^*Q$	Legendre map defined by $\mathcal{L} \in C^\infty(\mathbb{R} \times TQ)$.
$\vartheta_{\mathcal{L}} \in \Omega^1(\mathbb{R} \times TQ)$	Cartan Lagrangian 1-form associated with $\mathcal{L} \in C^\infty(\mathbb{R} \times TQ)$.
$\omega_{\mathcal{L}} = -\mathrm{d}\vartheta_{\mathcal{L}} \in \Omega^1(\mathbb{R} \times TQ)$	Cartan Lagrangian 2-form associated with $\mathcal{L} \in C^\infty(\mathbb{R} \times TQ)$.
$R_{\mathcal{L}} \in \mathfrak{X}(\mathbb{R} \times TQ)$	Reeb vector field on a cosymplectic manifold $(\mathbb{R} \times TQ, \mathrm{d}t, \eta_{\mathcal{L}})$.
$\boldsymbol{\Theta}_{\mathcal{L}} \in \Omega^1(\mathbb{R} \times TQ)$	Poincaré–Cartan 1-form associated with $\mathcal{L} \in C^\infty(\mathbb{R} \times TQ)$.
$\boldsymbol{\Omega}_{\mathcal{L}} = -\mathrm{d}\boldsymbol{\Theta}_{\mathcal{L}} \in \Omega^1(\mathbb{R} \times TQ)$	Poincaré–Cartan 2-form associated with $\mathcal{L} \in C^\infty(\mathbb{R} \times TQ)$.
$\boldsymbol{\Theta}_{\mathrm{h}} \in \Omega^2(\mathbb{R} \times T^*Q)$	Hamilton–Cartan 1-form associated with $\mathrm{h} \in C^\infty(\mathbb{R} \times T^*Q)$.
$\boldsymbol{\Omega}_{\mathrm{h}} = -\mathrm{d}\boldsymbol{\Theta}_{\mathrm{h}} \in \Omega^2(\mathbb{R} \times T^*Q)$	Hamilton–Cartan 2-form associated with $\mathrm{h} \in C^\infty(\mathbb{R} \times T^*Q)$.
$\nu \colon T(\mathbb{R} \times Q) \to \mathbb{R} \times TQ$ $\varpi \colon T(\mathbb{R} \times Q) \to \mathbb{R}$	Canonical projections of the bundle $T(\mathbb{R} \times Q)$.
$pr_1 \colon T^*(\mathbb{R} \times Q) \to T^*\mathbb{R}$ $pr_2 \colon T^*(\mathbb{R} \times Q) \to T^*Q$ $\mu \colon T^*(\mathbb{R} \times Q) \to \mathbb{R} \times TQ$ $u \colon T^*(\mathbb{R} \times Q) \to \mathbb{R}$	Canonical projections of the bundle $T^*(\mathbb{R} \times Q)$.
$\boldsymbol{\Theta}_{\mathcal{L}_{ext}} \in \Omega^1(T(\mathbb{R} \times Q))$	Lagrangian 1-form associated with $\mathcal{L}_{ext} \in C^\infty(T(\mathbb{R} \times Q))$.
$\boldsymbol{\Omega}_{\mathcal{L}_{ext}} = -\mathrm{d}\boldsymbol{\Theta}_{\mathcal{L}_{ext}} \in \Omega^2(T(\mathbb{R} \times Q))$	Lagrangian 2-form associated with $\mathcal{L}_{ext} \in C^\infty(T(\mathbb{R} \times Q))$.
$\boldsymbol{\Theta}_{ext} \in \Omega^1(T^*(\mathbb{R} \times Q))$	Canonical 1-form on $T^*(Q \times \mathbb{R})$.
$\boldsymbol{\Omega}_{ext} = -\mathrm{d}\boldsymbol{\Theta}_{ext} \in \Omega^2(T^*(\mathbb{R} \times Q))$	Canonical 2-form on $T^*(Q \times \mathbb{R})$.
$\mathcal{M} = \mathbb{R} \times TQ \times_Q T^*Q$	Extended unified (Pontryagin) bundle.

$\kappa_1 \colon \mathcal{M} \longrightarrow \mathbb{R} \times TQ$ $\kappa_2 \colon \mathcal{M} \longrightarrow \mathbb{R} \times T^*Q$ $\kappa_0 \colon \mathcal{M} \longrightarrow \mathbb{R} \times Q$	Canonical projections of the extended unified bundle $\mathcal{M}$.
$\Theta_{\mathcal{M}} \in \Omega^1(\mathcal{M})$	Canonical 1-form on the extended unified bundle $\mathcal{M}$.
$\Omega_{\mathcal{M}} = -\mathrm{d}\Theta_{\mathcal{M}} \in \Omega^2(\mathcal{M})$	Canonical 2-form on the extended unified bundle $\mathcal{M}$.
$\mathrm{C} \colon \mathcal{M} \longrightarrow \mathbb{R}$	Coupling function on the unified bundle $\mathcal{M}$.
$\mathrm{H} \colon \mathcal{M} \longrightarrow \mathbb{R}$	Hamiltonian function on the unified bundle $\mathcal{M}$.
$X_{\mathrm{H}} \in \mathfrak{X}(\mathcal{M})$	Dynamical vector field of the system $(\mathcal{M}, \Omega_{\mathcal{M}}, \mathrm{H})$.
$\jmath_0 \colon \mathcal{M}_0 \hookrightarrow \mathcal{M}$	Compatibility submanifold of the system $(\mathcal{M}, \Omega_{\mathcal{M}}, \mathrm{H})$.
$\mathbf{g} \in \mathcal{T}_2(Q)$	Riemannian metric on a manifold Q.
$\mathrm{F} \in \mathfrak{X}(Q)$	Vector force field.
$\omega \in \Omega^1(Q)$	Work differential form.
$(Q, \mathbf{g})$	Riemannian manifold.
$(Q, \mathbf{g}, \mathrm{F})\,,\,(Q, \mathbf{g}, \omega)$	Newtonian system.
∇	Connection on a manifold (Levi-Civita).
∇_X	Covariant derivative by a vector field $X \in \mathfrak{X}(M)$.
$\nabla_t \equiv \nabla_{\gamma'}$	Covariant derivative along a curve $\gamma \colon I \subseteq \mathbb{R} \longrightarrow M$.
$\mathbf{T} \in \mathcal{T}_2^1(M)$	Torsion tensor of a connection.
$R \in \mathcal{T}_3^1(M)$	Curvature tensor of a connection.
$\mathrm{Rie} \in \mathcal{T}_4(M)$	Riemann's curvature tensor of a Riemannian manifold.
$\mathrm{Ric} \in \mathcal{T}_2(M)$	Ricci tensor of a Riemannian manifold.
$\mathrm{S} \in C^\infty(M)$	Scalar curvature of a Riemannian manifold.
$\boldsymbol{\eta}\,,\,(M, \boldsymbol{\eta})$	(Pre)contact form, (pre)contact manifold.
$\mathcal{D}^{\mathrm{C}}\,,\,\mathcal{D}^{\mathrm{R}}$	Contact and Reeb distributions on a contact manifold $(M, \boldsymbol{\eta})$.
$\mathcal{R} \in \mathfrak{X}(M)$	Reeb vector field on a contact manifold $(M, \boldsymbol{\eta})$.
$\flat_{\boldsymbol{\eta}} \colon TM \longrightarrow T^*M$ $\sharp_{\boldsymbol{\eta}} = \flat_{\boldsymbol{\eta}}^{-1} \colon T^*M \longrightarrow TM$	Canonical isomorphisms on a contact manifold $(M, \boldsymbol{\eta})$.
$X_f \in \mathfrak{X}(M)$	Hamiltonian vector field on a contact manifold $(M, \boldsymbol{\eta})$.

$\mathbf{grad}\, f \in \mathfrak{X}(M)$	Gradient vector field on a contact manifold $(M, \boldsymbol{\eta})$.
$\boldsymbol{\varepsilon}_f \in \mathfrak{X}(M)$	Evolution vector field on a contact manifold $(M, \boldsymbol{\eta})$.
$\tau_1 \colon TQ \times \mathbb{R} \to TQ$	Canonical projection of the bundle $TQ \times \mathbb{R}$.
$\mathfrak{F}\mathcal{L} \colon TQ \times \mathbb{R} \longrightarrow T^*Q \times \mathbb{R}$	Legendre map defined by $\mathcal{L} \in C^\infty(TQ \times \mathbb{R})$.
$\theta_\mathcal{L} \in \Omega^1(TQ \times \mathbb{R})$	Cartan Lagrangian 1-form associated with $\mathcal{L} \in C^\infty(TQ \times \mathbb{R})$.
$\omega_\mathcal{L} = -\mathrm{d}\theta_\mathcal{L} \in \Omega^2(TQ \times \mathbb{R})$	Cartan Lagrangian 2-form associated with $\mathcal{L} \in C^\infty(TQ \times \mathbb{R})$.
$\boldsymbol{\eta}_\mathcal{L} = \mathrm{d}s - \theta_\mathcal{L} \in \Omega^1(TQ \times \mathbb{R})$	(Pre)contact Lagrangian form associated with $\mathcal{L} \in C^\infty(TQ \times \mathbb{R})$.
$\mathcal{R}_\mathcal{L} \in \mathfrak{X}(TQ \times \mathbb{R})$	Reeb vector field on the (pre)contact manifold $(TQ \times \mathbb{R}, \boldsymbol{\eta}_\mathcal{L})$.
$\mathfrak{M} = TQ \times_Q T^*Q \times \mathbb{R}$	Extended precontact unified (or Pontryagin) bundle.
$\boldsymbol{\eta}_\mathfrak{M} \in \Omega^1(\mathfrak{M})$	Canonical precontact form on the precontact unified bundle $\mathfrak{M}$.
$\mathfrak{C} \colon \mathfrak{M} \longrightarrow \mathbb{R}$	Coupling function on the unified bundle $\mathfrak{M}$.
$\mathbf{H} \in C^\infty(\mathfrak{M})$	Hamiltonian function on the precontact unified bundle $\mathfrak{M}$.
$X_\mathbf{H} \in \mathfrak{X}(\mathfrak{M})$	Dynamical vector field of the system $(\mathfrak{M}, \Omega_\mathfrak{M}, \mathbf{H})$.
$\jmath_0 \colon \mathfrak{M}_0 \hookrightarrow \mathfrak{M}$	Compatibility submanifold of the system $(\mathfrak{M}, \Omega_\mathfrak{M}, \mathbf{H})$.

Chapter 1

Introduction

The study of mechanics advanced substantially during the 18th and 19th centuries with the emergence of the so-called *analytical* or *rational mechanics*. Many notable mathematicians contributed to its development, such as *L. Euler, W.R. Hamilton, C.G.J. Jacobi, J.L. Lagrange, A.M. Legendre,* and *S.D. Poisson.* Their techniques are based on the application of variational methods to obtain the equations of motion of dynamical systems and their subsequent development, and are widely collected in classical treaties, such as [161, 165, 180, 225, 234, 237, 361] or from a more modern perspective, [12, 47, 99, 148, 222, 246, 318, 332, 364].

Although the dynamical equations of analytical mechanics had been well established by *J.L. Lagrange* and others, the connection between the variational methods and mechanics was established by the *minimum action principles* [33, 34, 159, 173, 234, 246]. Thus, the *Hamilton principle* leads to the Euler–Lagrange equations and the *Lagrangian formalism* of mechanics. Introducing the *Legendre transformation* and the "momentum coordinates," the dynamical equations become the *Hamilton equations*, originating in this way the *canonical* or *Hamiltonian formalism* of mechanics. As in the Lagrangian case, these Hamiltonian equations can also be derived from the *Hamilton–Jacobi variational principle*.

In the second half of the 20th century, an extensive group of mathematicians and physicists used the techniques of differential geometry to formalise analytical mechanics and physics in general, and studied the properties of dynamical systems intrinsically, giving rise to what is known today as *geometric mechanics*. Among the most relevant books and contributions, we can cite [1, 2, 11, 69, 91, 178, 208, 209, 217, 218, 227, 247, 249, 250, 255, 265, 291, 303, 318, 327, 358], in addition to many others.

In the geometric formulation, differentiable manifolds that model the phase spaces of dynamical systems are endowed with different kinds of geometric structures, which are used to study the properties of these systems. Thus, for *conservative systems*, the *symplectic manifolds* are used to describe *autonomous* or *time-independent* dynamical systems [1, 7, 11, 91, 138, 247, 249, 250, 265, 276, 327, 346, 347, 358]; meanwhile, for the *nonautonomous* or *time-dependent* case, other types of manifolds can be considered, such as *jet bundles*, *contact*, and *cosymplectic manifolds* [1, 7, 54, 89, 106, 109, 154, 255, 275, 317]. In particular, for variational dynamical systems, that is, those of *Lagrangian type*, their phase spaces have the geometric structure of a tangent bundle [105, 108, 164, 178, 227, 337, 338] or a cotangent bundle in the dual canonical Hamiltonian formalism [1, 11, 91, 249, 337, 339]. These formalisms can also be described in a single unified framework [322, 323]. Furthermore, for those dynamical systems of *mechanical type*, their configuration spaces are endowed with a (semi)Riemannian metric [1, 11, 46, 91, 179, 291], which is used to construct their Lagrangian functions in a canonical way. Finally, symplectic mechanics has also been used to describe *nonholonomic systems*, that is, systems subjected to constraints depending on positions and velocities [11, 100, 127, 131, 165, 248, 282, 344], and *vakonomic systems*, which consist of considering those systems but modifying the variational principle using only the variations allowed by the constraints [11, 34, 159, 273].

In addition, there are other more general geometric structures that are used to model some special types of dynamical systems. These are *Poisson* and *Jacobi manifolds* [249, 256] or *Lie algebroids* [78, 130, 188, 192, 270, 272, 360]. Finally, *dissipative* or *nonconservative dynamical systems* are modelled using *contact manifolds* [38, 40, 93, 114, 117, 141, 169, 224, 235]. In all these cases, there are deep connections between the differential equations describing the dynamics and the underlying geometric structures.

Another very important fact that characterises the behaviour of dynamical systems is the existence of symmetries. The interest in studying symmetries is a consequence of the well-known fact that their existence is closely related to that of conserved quantities, and the main result comes from the work of *Emmy Noether* [283] (see [229] for a review of her results). In turn, the existence of conserved quantities facilitates the integration of the dynamical equations, applying the suitable reduction procedures. Concerning this, it is interesting to mention the fundamental *Arnold–Liouville Theorem* for *integrable systems* [11] (see also [19, 277]), and also those by *J.E. Marsden* and *A. Weinstein* on the problem of the *(symplectic) reduc-*

tion by symmetries [266] (see also [263, 265, 267, 292–294] and the references therein).

The physical systems for which these geometric approaches were initially developed had the characteristic of being *regular* [1]. The dynamical systems are of this kind in classical analytical mechanics. Nevertheless, with the advent of relativity theory and the development of field theories in modern physics, in general, many systems appear for which that property does not hold, that is, they are *singular systems*. For autonomous systems, this feature of being regular or not manifests in the fact that the underlying geometric structure is *symplectic* or *presymplectic*, respectively [60, 63, 72, 73, 184–186, 196, 258, 280, 281, 324].

The aim of this book is to undertake a detailed exposition of the geometry underlying dynamical systems and how it is applied to describe their different kinds: conservative autonomous and nonautonomous dynamical systems in general, and Newtonian systems in particular, as well as dissipative systems. We use different geometric descriptions to study the main properties and characteristics of these systems, such as their Lagrangian, Hamiltonian and unified formalisms, their symmetries, and the variational principles. The study is done mainly for the regular case, although comments and explanations about singular systems are also introduced throughout the exposition and in some proposed problems.

The organisation of the book is as follows:

The first chapter of the exposition, Chapter 2, is devoted to presenting the most general geometric framework for autonomous mechanics, that is, *symplectic manifolds* and their structures and characteristics (*Poisson brackets, symplectomorphisms*, etc.), as well as the fundamental notion of a *Hamiltonian dynamical system*. A general description of *symmetries, conserved quantities*, and the *Noether Theorem* is also provided for this general situation.

In Chapter 3, we continue the exposition on *symplectic mechanics*, studying, in particular, autonomous mechanical systems of variational type, which are characterised by means of time-independent Lagrangian functions. The phase spaces of these kinds of systems are represented by tangent and cotangent bundles of manifolds representing the configuration spaces of the systems; so, the canonical geometric structures of these bundles are introduced first. Starting from the Lagrangian function and using the

---

[1] For Lagrangian systems, locally, this means that the Hessian matrix with respect to the generalised velocities is regular everywhere.

canonical structures of the tangent bundle, we can define a (pre)symplectic form and the *Legendre map*, which allow us to establish the Lagrangian formalism, the associated canonical Hamiltonian formalism for these systems, and the equivalence between both formalisms. We also state the foundations of a new geometric setting of the *Hamilton–Jacobi theory* for Hamiltonian systems. In addition, we give a description of the so-called Skinner–Rusk unified formalism, which combines the two previous formalisms into a single framework. The study of symmetries, conserved quantities, and Noether's Theorem is also performed for Lagrangian systems in both formalisms, which lead to the introduction of some new types of symmetries. As these dynamical systems are of variational type, a section is devoted to introducing the variational formulation and deriving the dynamical equations from the corresponding variational principle. Finally, two of the most classic examples in mechanics, the *harmonic oscillator* and the *Kepler problem*, are analysed.

Chapter 4 is devoted to developing one of the most interesting and generic geometric formulations for describing *nonautonomous dynamical systems*, both the Lagrangian and Hamiltonian formalisms. It is based on using *cosymplectic manifolds*, which are presented in the first section of this chapter. Then, the dynamics and symmetries for these time-dependent systems are studied and the formalism is used to describe the classical examples introduced in the previous chapter when external time-dependent forces act on the oscillator or for variable mass systems in the case of the Kepler problem. In this chapter, we also give a brief presentation of two other very common formulations of nonautonomous mechanics, namely, the *contact* and the *extended symplectic formulations*, and show their equivalence with the cosymplectic picture.

Next, Chapter 5 deals with the study of those dynamical systems whose configuration spaces are endowed with additional geometric structures, such as a *metric*. First, we review the foundations of connections in manifolds and Riemannian geometry, which are the geometric structures needed to develop what is called *Newtonian mechanics*. In particular, the existence of a metric allows us to construct what are known as *mechanical Lagrangians*. We analyse their different types, namely, *conservative* and *coupled systems*, and systems with *holonomic* and *nonholonomic constraints*, stating their dynamics and variational formulation. As a particular situation, the case of *(nonautonomous) Newtonian systems* is displayed along this exposition.

Chapter 6 is an introduction to the study of *autonomous dissipative systems* using *contact geometry*. After reviewing the foundations of contact manifolds, we establish the generic concepts about *contact Hamiltonian systems* and develop the Lagrangian, Hamiltonian, and unified Lagrangian–Hamiltonian formalisms for these kinds of systems. Next, we study symmetries and the concept of *dissipated quantities* for contact Hamiltonian and Lagrangian systems, establishing the so-called dissipation theorems, and show how to associate dissipated and conserved quantities with these symmetries. As mentioned in the above chapters, the examples of the damped harmonic oscillator and the Kepler problem with friction are analysed in this context.

The book ends with an appendix covering various contents on other geometric structures that appear throughout the exposition, in particular, tangent and cotangent bundles, and Lie groups and Lie algebras.

Furthermore, a collection of proposed exercises is displayed at the end of each chapter, some of which contain results complementary to those presented in the chapters.

Along the exposition, all manifolds are supposed to be real, second countable, and C^∞. All the maps and the structures are smooth. The summation criterion for repeated crossed indices is adopted.

Chapter 2

Symplectic Mechanics (I): Autonomous Hamiltonian Dynamical Systems

The general geometric framework for describing autonomous mechanical systems (with a finite number of degrees of freedom) uses some particular types of differentiable manifolds to model the phase spaces of these systems: *symplectic* and *presymplectic manifolds*. This formulation is known as *symplectic mechanics*. Taking these kinds of manifolds as phase spaces, we obtain a very general setting for studying dynamical systems, from which other descriptions, such as the Lagrangian and the Hamiltonian formalisms of Lagrangian systems, can be analysed as specific cases[1].

The symplectic description of (autonomous) Hamiltonian systems has been explained in many works and books (see, for instance, [1, 11, 138, 209, 231, 249, 250, 265, 327, 358] and the references quoted therein).

In this chapter, after reviewing the fundamental concepts of symplectic geometry, we present the symplectic description of the autonomous Hamiltonian systems and introduce several types of symmetries and their associated conserved quantities from a geometric perspective.

2.1   Notions on Symplectic and Presymplectic Geometry

In this section, we state the fundamental concepts and properties of symplectic (and presymplectic) manifolds[2] (see, for instance, [1, 3, 50, 208, 230, 249, 358]).

---

[1]This general formulation allows us to describe dynamical systems that are not of Lagrangian type, for instance, the system of a classical spin particle [327].

[2]The word *symplectic* comes from the Greek word "$\sigma \upsilon \mu \pi \lambda \varepsilon \kappa \tau \iota \kappa o \varsigma$" and was introduced by *H. Weyl*. Although the structure of the symplectic manifolds had been implicitly considered above, it was not until the decade of 1950 when the symplectic geometry appeared as a differentiated branch of differential geometry, and *A. Lichnerowicz* was the first to use the term *symplectic manifold*.

2.1.1   *Symplectic and presymplectic vector spaces*

Let $\mathbf{E}$ be a finite dimensional real vector space.

> **Definition 2.1.** A **symplectic inner product**, or a **linear symplectic structure**, on $\mathbf{E}$ is a nondegenerate skew-symmetric bilinear function ω on $\mathbf{E}$. The pair $(\mathbf{E}, \omega)$ is said to be a **symplectic vector space**.

Nondegeneracy means that if $\omega(\mathbf{x}, \mathbf{y}) = \mathbf{0}$, for every $\mathbf{y} \in \mathbf{E}$, then $\mathbf{x} = 0$. Let $\mathbf{e}_1, \ldots, \mathbf{e}_r$ be a basis of $\mathbf{E}$ and $\boldsymbol{\alpha}^1, \ldots, \boldsymbol{\alpha}^r$ its dual basis. If $\omega_{ij} = \omega(\mathbf{e}_i, \mathbf{e}_j)$, then the expression of ω on these bases is

$$\omega = \omega_{ij}\, \boldsymbol{\alpha}^i \otimes \boldsymbol{\alpha}^j,$$

being $(\omega) = (\omega_{ij})$, the matrix of ω in these bases.

The linear mapping $\omega^\flat : \mathbf{E} \longrightarrow \mathbf{E}^*$ is defined by

$$(\omega^\flat(\mathbf{x}))(\mathbf{y}) = \langle \omega^\flat(\mathbf{x}), \mathbf{y}\rangle = \omega(\mathbf{x}, \mathbf{y}),$$

and its matrix relative to these bases is (ω_{ij}), since $\omega(\mathbf{e}_i) = \omega_{ij}\boldsymbol{\alpha}^j$. Due to the skew symmetry of the bilinear mapping ω, we have that the matrix (ω_{ij}) is skew symmetric. The nondegenerate implies that the mapping $\omega^\flat$ is one-to-one, and therefore an isomorphism, and that (ω_{ij}) is regular. As a consequence, the dimension of $\mathbf{E}$ is even, that is, $r = 2n$, because $(\omega) = -(\omega)^t$ (the transpose matrix) and $\det(\omega) = \det(\omega)^t = (-1)^r \det(\omega)$.

A basis $\mathbf{a}_1, \ldots, \mathbf{a}_n, \mathbf{a}_{n+1}, \ldots, \mathbf{a}_{2n}$ of $\mathbf{E}$ is called **symplectic** if

$$\begin{aligned}
\omega(\mathbf{a}_i, \mathbf{a}_j) &= 0; & i, j &= 1, \ldots, n, \\
\omega(\mathbf{a}_i, \mathbf{a}_j) &= 0; & i, j &= n+1, \ldots, 2n, \\
\omega(\mathbf{a}_i, \mathbf{a}_{n+j}) &= \delta_{ij}; & i, j &= 1, \ldots, n.
\end{aligned}$$

The existence of a symplectic basis is given by the following lemma.

> **Lemma 2.1.** *Let $(\mathbf{E}, \omega)$ be a $2n$ dimensional symplectic vector space.*
>
> *(1) There exists a symplectic basis $(\mathbf{e}_k)$ $(k = 1, \ldots, 2n)$ on $\mathbf{E}$.*
>
> *(2) If $(\boldsymbol{\alpha}^k)$ is the corresponding dual basis on $\mathbf{E}^*$, then $\omega = \displaystyle\sum_{i=1}^{n} \boldsymbol{\alpha}^i \wedge \boldsymbol{\alpha}^{i+n}$*
>
> *or equivalently, the matrix of ω in this basis is*
>
> $$\begin{pmatrix} 0_n & I_n \\ -I_n & 0_n \end{pmatrix},$$
>
> *where I_n denotes the identity matrix of order n and 0_n is the zero square matrix $n \times n$.*

Proof. (By induction on the dimension of $\mathbf{E}$). Let $\mathbf{e}_1, \mathbf{e}_{n+1} \in \mathbf{E}$ with $\omega(\mathbf{e}_1, \mathbf{e}_{n+1}) \neq 0$. We can choose two such vectors except where $\omega = 0$. Dividing $\mathbf{e}_1$ by a scalar, we have $\omega(\mathbf{e}_1, \mathbf{e}_{n+1}) = 1$. Then, on the plane $\mathbf{P}_1$ spanned by $\mathbf{e}_1, \mathbf{e}_{n+1}$, the matrix of ω is

$$\begin{pmatrix} 0 & 1 \\ -1 & 0 \end{pmatrix}.$$

Let $\mathbf{E}_1$ be the ω-orthogonal complement of $\mathbf{P}_1$ in $\mathbf{E}$, that is:

$$\mathbf{E}_1 = \{\mathbf{a} \in \mathbf{E} \,|\, \omega(\mathbf{a}, \mathbf{a}_1) = 0, \forall \mathbf{a}_1 \in \mathbf{P}_1\}.$$

Observe that $\mathbf{E}_1 \cap \mathbf{P}_1 = \{0\}$ and $\mathbf{E}_1 + \mathbf{P}_1 = \mathbf{E}$. If $\mathbf{a} \in \mathbf{E}$, we have

$$\mathbf{a} - \omega(\mathbf{a}, \mathbf{e}_{n+1})\mathbf{e}_1 + \omega(\mathbf{a}, \mathbf{e}_1)\mathbf{e}_{n+1} \in \mathbf{E}_1.$$

Then, $\mathbf{E}_1 \oplus \mathbf{P}_1 = \mathbf{E}$ and we can repeat the process on $\mathbf{E}_1$, with a dimension less than $2n$, and we have finished the proof.

The result on the expression of ω as $\omega = \sum_{i=1}^{n} \alpha^i \wedge \alpha^{i+n}$ is straightforward.

$\square$

Remark 2.1.

(1) The nondegeneracy condition is equivalent to stating that $\omega^n = \overbrace{\omega \wedge \ldots \wedge \omega}^{n} \equiv \bigwedge^{n} \omega$ is a volume form on $\mathbf{E}$.

(2) If we remove the nondegeneracy condition, then the mapping $\omega^\flat$ is not one-to-one. In this case, we need to add the basis $\mathbf{u}_1, \ldots, \mathbf{u}_h$ of $\ker \omega^\flat$ to the given basis for the decomposition of $\mathbf{E}$. Then, the corresponding matrix is

$$\begin{pmatrix} 0_n & I_n & 0_{n \times h} \\ -I_n & 0_n & 0_{n \times h} \\ 0_{h \times n} & 0_{h \times n} & 0_h \end{pmatrix},$$

where $2n + h$ is the dimension of $\mathbf{E}$ and $0_{n \times h}$ denotes the zero matrix of n rows and h columns. We say that $2n$ is the rank of ω.

2.1.2 *Subspaces of a symplectic linear space*

Associated with a linear symplectic structure, we have a notion of orthogonality in parallel to the ideas related to the Euclidean scalar product, but the results are very different. In the following, we develop these ideas that are necessary to understand our later description of Lagrangian and Hamiltonian systems.

Let $(\mathbf{E}, \omega)$ be a $2n$-dimensional symplectic vector space and $\mathbf{F} \subset \mathbf{E}$ a linear subspace. The ω-**orthogonal complement** of $\mathbf{F}$ is defined as

$$\mathbf{F}^{\perp} = \{\mathbf{u} \in \mathbf{E} \mid \omega(\mathbf{u}, \mathbf{u}') = 0, \text{ for every } \mathbf{u}' \in \mathbf{F}\}.$$

In general, observe that $\mathbf{F} \cap \mathbf{F}^{\perp} \neq 0$ (e.g., if $\mathbf{F} = \text{span}\{\mathbf{u}\}$, then $\mathbf{F} \subset \mathbf{F}^{\perp}$).

Definition 2.2. Let $\mathbf{F}$ be a subspace of a linear symplectic $2n$-dimensional vector space $(\mathbf{E}, \omega)$.

(1) $\mathbf{F}$ is *isotropic* if $\mathbf{F} \subset \mathbf{F}^{\perp}$, that is, $\omega(\mathbf{u}, \mathbf{u}') = 0$, for every $\mathbf{u}, \mathbf{u}' \in \mathbf{F}$.
(2) $\mathbf{F}$ is *coisotropic* if $\mathbf{F} \supset \mathbf{F}^{\perp}$, that is, if $\omega(\mathbf{u}, \mathbf{u}') = 0$, for every $\mathbf{u}' \in \mathbf{F}$, then $\mathbf{u} \in \mathbf{F}$.
(3) $\mathbf{F}$ is **Lagrangian** if $\mathbf{F}$ is isotropic and there exists an isotropic complement $\mathbf{F}'$, that is, an isotropic subspace $\mathbf{F}' \subset \mathbf{E}$ such that $\mathbf{F} \oplus \mathbf{F}' = \mathbf{E}$.
(4) $\mathbf{F}$ is *symplectic* if ω restricted to $\mathbf{F}$ is nondegenerate, that is, $(\mathbf{F}, \omega_{\mathbf{F}} = \omega|_{\mathbf{F}})$ is a symplectic vector space.

These notions have several associated properties that we resume in the following two propositions, with some proofs and indications.

Proposition 2.1. *Let $(\mathbf{E}, \omega)$ be a symplectic vector space and $\mathbf{F}, \mathbf{G}$ subspaces of $\mathbf{E}$.*

(1) If $\mathbf{F} \subset \mathbf{G}$, then $\mathbf{F}^{\perp} \supset \mathbf{G}^{\perp}$.
(2) $\mathbf{F}^{\perp} \cap \mathbf{G}^{\perp} = (\mathbf{F} + \mathbf{G})^{\perp}$.
(3) $\dim \mathbf{E} = \dim \mathbf{F} + \dim \mathbf{F}^{\perp}$.
(4) $\mathbf{F} = \mathbf{F}^{\perp\perp}$.
(5) $(\mathbf{F} \cap \mathbf{G})^{\perp} = \mathbf{F}^{\perp} + \mathbf{G}^{\perp}$.

Proof. Items (1) and (2) are straightforward.

(3) Consider the linear map $\omega^{\flat} : \mathbf{E} \longrightarrow \mathbf{E}^{*}$. We observed that if $\mathbf{u} \in \mathbf{F}$, then $\omega^{\flat}(\mathbf{u}) \in \mathbf{E}^{*}$ annihilates the subspace $\mathbf{F}^{\perp}$. Hence, the restricted linear map $\omega_{\mathbf{F}}^{\flat} : \mathbf{F} \longrightarrow \mathbf{E}^{*}$ induces another one $\hat{\omega}_{\mathbf{F}}^{\flat} : \mathbf{F} \longrightarrow (\mathbf{E}/\mathbf{F}^{\perp})^{*}$. This last map is injective, but not necessarily surjective. Thus, we have

$$\dim \mathbf{F} \leq \dim (\mathbf{E}/\mathbf{F}^{\perp})^{*} = \dim (\mathbf{E}/\mathbf{F}^{\perp}) = \dim \mathbf{E} - \dim \mathbf{F}.$$

Conversely, consider the map $\mathbf{F} \xrightarrow{\omega^{\flat}} \mathbf{E}^{*} \xrightarrow{j} \mathbf{F}^{*}$, where j is the restriction from $\mathbf{E}$ to $\mathbf{F}$. Let $\bar{\omega}_{\mathbf{F}}^{\flat} = \omega^{\flat} \circ j$ and observe that $\ker \bar{\omega}_{\mathbf{F}}^{\flat} = \mathbf{F}^{\perp}$.

Hence,

$$\dim \mathbf{F} = \dim \mathbf{F}^* \geq \dim(\text{img}\,\bar{\omega}_{\mathbf{F}}^{\flat}) = \dim \mathbf{E} - \dim \mathbf{F}^* = \dim \mathbf{E} - \dim \mathbf{F}.$$

From both expressions, we obtain the desired result.

(4) It is clear that $\mathbf{F} \subset \mathbf{F}^{\perp\perp}$. If we apply the previous result to $\mathbf{F}$ and $\mathbf{F}^{\perp}$, we have

$$\dim \mathbf{E} = \dim \mathbf{F} + \dim \mathbf{F}^{\perp} = \dim \mathbf{F}^{\perp} + \dim \mathbf{F}^{\perp\perp};$$

that is, $\dim \mathbf{F} = \dim \mathbf{F}^{\perp\perp}$, hence $\mathbf{F} = \mathbf{F}^{\perp\perp}$ as the desired result.

(5) From (2) and (4), we have

$$(\mathbf{F} \cap \mathbf{G})^{\perp} \overset{4.}{=} (\mathbf{F}^{\perp\perp} \cap \mathbf{G}^{\perp\perp})^{\perp} \overset{2.}{=} (\mathbf{F}^{\perp} \cap \mathbf{G}^{\perp})^{\perp\perp} \overset{4.}{=} \mathbf{F}^{\perp} \cap \mathbf{G}^{\perp}. \qquad \square$$

The next proposition gives alternative definitions for a subspace to be Lagrangian.

Proposition 2.2. *Let $(\mathbf{E}, \omega)$ be a symplectic vector space $(\mathbf{E}, \omega)$ and $\mathbf{F}$ a subspace of $\mathbf{E}$. The following statements are equivalent:*

(1) $\mathbf{F}$ is Lagrangian.

(2) $\mathbf{F} = \mathbf{F}^{\perp}$.

(3) $\mathbf{F}$ is isotropic and $\dim \mathbf{F} = \frac{1}{2}\dim \mathbf{E}$.

Proof. $(1 \implies 2)$. As $\mathbf{F}$ is Lagrangian, it is isotropic, hence $\mathbf{F} \subset \mathbf{F}^{\perp}$, and there exists another isotropic subspace $\mathbf{F}'$ with $\mathbf{F} \oplus \mathbf{F}' = \mathbf{E}$.

Suppose now that $\mathbf{u} \in \mathbf{F}^{\perp}$, and then, put $\mathbf{u} = \mathbf{a} + \mathbf{b}$ with $\mathbf{a} \in \mathbf{F}$, $\mathbf{b} \in \mathbf{F}'$. We prove that $\mathbf{b} = 0$. In fact, $\mathbf{b} \in \mathbf{F}'^{\perp}$ since $\mathbf{F}'$ is isotropic. We also prove that $\mathbf{b} = \mathbf{u} - \mathbf{a} \in \mathbf{F}^{\perp}$ since $\mathbf{u}, \mathbf{a} \in \mathbf{F}$ and $\mathbf{F}$ is isotropic. Then

$$\mathbf{b} \in \mathbf{F}'^{\perp} \cap \mathbf{F}^{\perp} = (\mathbf{F}' + \mathbf{F})^{\perp} = \mathbf{E}^{\perp} = \{0\},$$

because ω is nondegenerate. Then, we have $\mathbf{u} = \mathbf{a} \in \mathbf{F}$ and $\mathbf{F}^{\perp} \subset \mathbf{F}$.

$(2 \implies 3)$. The hypothesis and item (3) of the above proposition give the result.

$(3 \implies 2)$. From item (3) and the above proposition, we obtain $\dim \mathbf{F} = \dim \mathbf{F}^{\perp}$, and $\mathbf{F}$ being isotropic, we have $\mathbf{F} = \mathbf{F}^{\perp}$, that is, statement (2).

$(2 \implies 1)$. We have $\mathbf{F} = \mathbf{F}^{\perp}$, hence $\mathbf{F}$ is isotropic. We need to construct a subspace $\mathbf{F}'$. Let $\mathbf{a}_1 \notin \mathbf{F}$ and $\mathbf{F}_1 = \text{span}\{\mathbf{a}_1\}$. Then $\mathbf{F} \cap \mathbf{F}_1 = \{0\}$, hence $\mathbf{F}^{\perp} + \mathbf{F}_1^{\perp} = \mathbf{F} \oplus \mathbf{F}_1^{\perp} = \mathbf{E}$, that is, $\mathbf{F} \oplus \mathbf{F}_1^{\perp} = \mathbf{E}$ from the above proposition. Now, let $\mathbf{a}_2 \notin \mathbf{F} + \mathbf{F}_1$, $\mathbf{a}_2 \in \mathbf{F}_1^{\perp}$. We have two alternatives:

(1) This vector $\mathbf{a}_2$ does not exist. In this case, $\mathbf{F}_1^{\perp} \subset \mathbf{F} + \mathbf{F}_1$, hence $\mathbf{F} + \mathbf{F}_1 = \mathbf{E}$ and

$$\mathbf{F}_1^{\perp} = (\text{span}\{\mathbf{a}_1\})^{\perp} \supset \text{span}\{\mathbf{a}_1\} = \mathbf{F}_1,$$

then $\mathbf{F}_1$ is isotropic, and we can take $\mathbf{F}' = \mathbf{F}_1$.

(2) There exists such $\mathbf{a}_2$. Let $\mathbf{F}_2 = \mathbf{F}_1 + \text{span}\{\mathbf{a}_2\}$. Following the same procedure as above, we have $\mathbf{F} \cap \mathbf{F}_2 = \{0\}$, hence $\mathbf{F}^{\perp} + \mathbf{F}_2^{\perp} = \mathbf{F} \oplus \mathbf{F}_2^{\perp} = \mathbf{E}$, that is, $\mathbf{F} \oplus \mathbf{F}_2^{\perp} = \mathbf{E}$. However,

$$\mathbf{F}_2^{\perp} = (\text{span}\{\mathbf{a}_1, \mathbf{a}_2\})^{\perp} \supset \text{span}\{\mathbf{a}_1, \mathbf{a}_2\} = \mathbf{F}_2,$$

hence $\mathbf{F}_2$ is isotropic, and we can take $\mathbf{F}' = \mathbf{F}_2$.

Inductively, we can continue, and as $\dim \mathbf{E}$ is finite, we arrive at some $\mathbf{F}_k$ such that $\mathbf{F}' = \mathbf{F}_k$, and we have finished the proof. □

This last proposition states that a Lagrangian subspace is a *maximal isotropic subspace*.

2.1.3 *Symplectic and presymplectic manifolds*

Considering the results of the above section, first we define:

Definition 2.3.

(a) Let M be a differentiable manifold. A **symplectic form** in M is a differential 2-form $\Omega \in \Omega^2(M)$ such that:

   (1) It is closed: $d\Omega = 0$, (we write $\Omega \in Z^2(M)$).

   (2) It is nondegenerate at every point of M, that is, Ω_{p} is a linear symplectic structure in $T_{\mathrm{p}}M$ for every $\mathrm{p} \in M$.

If the form Ω is closed but degenerate, it is said to be a **presymplectic form**, and if Ω is nondegenerate but not closed, it is an **almost-symplectic form**.

(b) A **symplectic** (respectively, **presymplectic**) **manifold** is a pair (M, Ω) where M is a differentiable manifold and Ω is a symplectic (respectively, presymplectic) form.

If the symplectic (respectively, presymplectic) form is exact, that is, there exists $\Theta \in \Omega^1(M)$ such that $d\Theta = \Omega$, then Ω is an **exact symplectic** (respectively, **exact presymplectic**) form, and (M, Ω) is an **exact symplectic** (respectively, **exact presymplectic**) man-

ifold. The form Θ is called a **symplectic** (respectively, **presymplectic**) **potential**.

Remark 2.2.

(1) As a consequence of the second condition of the definition, a nondegenerate differential 2-form can only be defined in manifolds of even dimension, thus we write $\dim M = 2n$.

(2) According to *Poincaré's Lemma*, every closed form is locally exact; if Ω is a symplectic or a presymplectic form, for every point $p \in M$, there is an open neighbourhood $U \subset M$, $p \in U$, and $\vartheta \in \Omega^1(U)$ such that $\Omega \mid_U = d\vartheta$. Every 1-form ϑ satisfying this condition is called a *local symplectic (or presymplectic) potential*.

Observe that if $\vartheta \in \Omega^1(U)$ and $\vartheta' \in \Omega^1(U')$ are two different symplectic (or presymplectic) potentials, then $\vartheta' = \vartheta + df$ in $U \cap U'$ for some $f \in C^\infty(U \cap U')$.

(3) In the case of a presymplectic form, the dimension of $\ker \omega^\flat_p$, for $p \in M$, may depend on the chosen point $p \in M$. Usually, we demand that this dimension not depend on the chosen point and say that the presymplectic form is *regular*.

The following theorem describes the local structure of the symplectic manifolds [110].

> **Theorem 2.1 (Darboux).** *Let (M, Ω) be a $2n$-dimensional symplectic manifold. For every point $p \in M$, there exists an open neighbourhood $U \subset M$, $p \in U$, which is the domain of a local chart $(U; x^i, y_i)_{i=1\ldots n}$, such that Ω has the expression*
>
> $$\Omega \mid_U = dx^i \wedge dy_i.$$
>
> *These local charts are called* **symplectic charts** *and their coordinates are the* **canonical coordinates** *or* **Darboux coordinates** *of the symplectic manifold in this chart.*

Proof. (This proof is taken from [1]. For other different proofs, see, for example, [44, 108, 143]).

The proof is organised into several parts:

(1) As this is a local result, we can suppose that M is $\mathbb{R}^{2n}$ and $p = 0$.

(2) Let Ω be a symplectic form in $\mathbb{R}^{2n}$ and $\Omega_0 = \Omega(p)$ the constant symplectic form in $\mathbb{R}^{2n}$ equal to Ω at p.

It is enough to prove that there exists a neighbourhood U of p and a diffeomorphism $\phi : U \longrightarrow U$ such that $\phi^* \Omega_0 = \Omega$ on U. This is true since given the above Lemma 2.1, we can choose a global coordinate system in $\mathbb{R}^{2n}$ such that the symplectic form Ω_0 takes the expression we need.

(3) Consider the 2-form $\omega_t \equiv \Omega + t(\Omega_0 - \Omega)$ in $\mathbb{R}^{2n}$ for $t \in [0,1]$. We have:

(a) $\omega_0 = \Omega$, $\omega_1 = \Omega_0$, $d\,\omega_t = 0$, for every $t \in [0,1]$.

(b) $\omega_t(p) = \Omega(p) + t(\Omega_0(p) - \Omega(p)) = \Omega(p) = \Omega_0(p)$ is nondegenerate.

(c) As the interval $[0,1]$ is compact, there exists an open set $U_0 \subset U$ with $p \in U_0$ such that $\omega_t|_{U_0}$ is nondegenerate for every $t \in [0,1]$.

(d) With $\Omega_0 - \Omega$ being closed, we can suppose that U_0 is a ball with centre in p such that there exists $\alpha \in \Omega^1(U_0)$ with $\Omega_0 - \Omega = d\,\alpha$, by the Poincaré Lemma.

(4) Let $X_t \in \mathfrak{X}(U_0)$ be defined by $i(X_t)\omega_t = -\alpha$. This time-dependent vector field is well defined because ω_t is nondegenerate. Observe that $X_t(p) = 0$ since $\alpha(p) = 0$.

(5) Let $F_{t,s}$ be the time-dependent flux of X_t, satisfying $F_{t,s} \circ F_{s,r} = F_{t,r}$ and $F_{t,t} = I$, and we can consider it to be defined in U_0, or reduce it if necessary. Let $F_t = F_{t,0}$ be the associated diffeomorphic flux with $F_{t+h} = F_{t+h,t} \circ F_{t,0}$. Then, we have:

$$\frac{d}{dt} F_t^* \omega_t = \lim_{h \longrightarrow 0} \frac{F_{t+h}^* \omega_t - F_t^* \omega_t}{h} + F_t^*(\Omega_0 - \Omega)$$
$$= F_t^*(L(X_t)\omega_t) + F_t^*(\Omega_0 - \Omega)$$
$$= F_t^*(d\, i(X_t)\omega_t + \Omega_0 - \Omega) = F_t^*(0) = 0,$$

that is $F_t^* \omega_t$ is constant, hence $F_1^* \omega_1 = F_0^* \omega_0$. Then,

$$F_1^* \Omega_0 = F_1^* \omega_1 = F_0^* \omega_0 = \Omega,$$

and F_1 is the diffeomorphism transforming the constant symplectic form Ω_0 into our given symplectic form Ω. $\qquad \square$

Remark 2.3. For presymplectic manifolds, there is a similar result (see, for instance, [108, 143]).

Finally, as a straightforward consequence of the definition, we have:

Proposition 2.3 (Liouville). *Every symplectic manifold is an oriented manifold.*

Proof. Using the symplectic form, we can define the volume form $\Omega^n :=$ $\overset{n}{\bigwedge} \Omega \in \Omega^{2n}(M)$, which is called the **Liouville volume form** on M.   □

Example 2.1. Some examples of symplectic manifolds are as follows:

(1) The cotangent bundle T^*Q of a manifold Q is an example of symplectic manifold, which is also the canonical model of these kinds of manifolds (see Theorem 3.1 for all the details). Here, we do a short survey: We have a canonical 1-form $\Theta \in \Omega^1(T^*Q)$ which, in natural coordinates (q^i, p_i) of T^*Q, is $\Theta = p_i dq^i$. Then $\Omega = -d\Theta = dq^i \wedge dp_i$ is a symplectic form. Observe that the natural coordinates of T^*Q are Darboux coordinates for Ω.

(2) $\mathbb{R}^{2n}$ with the Cartesian coordinate system, $(x^1, \ldots, x^n, y_1, \ldots, y_n)$, and the 2-form $\Omega = dx^i \wedge dy_i$ is a symplectic manifold.

(3) The two-sphere $S^2 \subset \mathbb{R}^3$ as a Riemannian submanifold with the induced area 2-form is a symplectic manifold.

(4) Another relevant example, which will be used later, is given by the following:

Proposition 2.4. *Let (M_1, Ω_1), (M_2, Ω_2) be symplectic manifolds with* $\dim M_1 = \dim M_2$, *and* $\pi_j \colon M_1 \times M_2 \longrightarrow M_j$, $j = 1, 2$ *be the natural projections. Then, the product manifold $(M_1 \times M_2, \Omega_{12} = \pi_1^*\Omega_1 - \pi_2^*\Omega_2)$ is a symplectic manifold.*

2.1.4 *Submanifolds of a symplectic manifold*

As in the case of linear symplectic structures, in a symplectic manifold, there exist some interesting kinds of submanifolds. They are associated with the idea of orthogonality in the tangent space at every point with respect to the induced linear symplectic structure at this point. Lagrangian submanifolds play an important role in the study of dynamics of Hamiltonian systems.

Definition 2.4. Let (M, Ω) be a symplectic manifold and $j : L \longrightarrow M$ an immersion.

(1) L is an ***isotropic immersed submanifold*** of (M, Ω) if $T_p j(T_p L) \subset T_{j(p)} M$ is an isotropic subspace of $(T_{j(p)} M, \Omega_{j(p)})$ as a linear symplectic space for every $p \in L$.

(2) L is a ***coisotropic immersed submanifold*** of (M, Ω) if $T_p j(T_p L) \subset T_{j(p)} M$ is a coisotropic subspace of $(T_{j(p)} M, \Omega_{j(p)})$ as a linear symplectic space for every $p \in L$.

(3) L is a ***symplectic immersed submanifold*** of (M, Ω) if $T_p j(T_p L) \subset T_{j(p)} M$ is a symplectic subspace of $(T_{j(p)} M, \Omega_{j(p)})$ as a linear symplectic space for every $p \in L$.

(4) L is a ***Lagrangian submanifold*** if it is isotropic and $\dim L = \frac{1}{2} \dim M$.

This terminology is also used for subbundles of TM over submanifolds of M.

Remark 2.4.

(1) A submanifold $j : L \longrightarrow M$ is isotropic if, and only if, $j^* \Omega = 0$.

(2) From the above study about linear Lagrangian subspaces, if $L \subset M$ is Lagrangian, then $\dim L = \frac{1}{2} \dim M$ and $(T_p L)^\perp = T_p L$.

Example 2.2. Examples of Lagrangian submanifolds are as follows:

(1) We know that $\mathbb{R}^{2n}$ with elements denoted by $(x, y) \in \mathbb{R}^{2n}$ and coordinates $(x, y) = (x^i, y_i)$ has a natural symplectic form given by $\omega = dx^i \wedge dy_i$. Examples of Lagrangian submanifolds are the following:

$$L_1 = \{(x, y) | x = 0\}, \quad L_2 = \{(x, y) | y = 0\}, \quad L_3 = \{(x, y) | x = y\}.$$

(2) Taking the cotangent bundle of a manifold Q, we have a symplectic manifold, $(T^* Q, \omega)$. Then Lagrangian submanifolds are the fibres of the bundle, that is, $q = $ constant, or the section zero, that is, the manifold Q as a submanifold of $T^* Q$.

(3) In the symplectic manifold $(T^* Q, \omega)$, let $\alpha : Q \longrightarrow T^* Q$ be a differential form and $N = \{(q, p) \in T^* Q \mid p = \alpha(q)\}$ the graph of α. Then N is a Lagrangian submanifold of $T^* Q$ if, and only if, α is a closed form. In fact, we have

$$d\alpha = d(\alpha^* \theta) = \alpha^* d\theta = -\alpha^* d\omega,$$

by the properties defining the canonical forms in T^*Q. For a detailed account of this information, please refer to the corresponding sections in the following chapters.

(4) As a particular case, given a function $f : Q \longrightarrow \mathbb{R}$, the graph of df is a Lagrangian submanifold of T^*Q.

2.1.5 Canonical isomorphism: Hamiltonian vector fields

The fact that a symplectic form is nondegenerate has important consequences. One of the main consequences is the following: every differential form $\Omega \in \Omega^k(M)$ defines a linear map

$$
\begin{aligned}
\flat_\Omega : \quad TM &\longrightarrow \wedge^{k-1}T^*M \\
(p, X_p) &\mapsto (p, i(X_p)\Omega_p)
\end{aligned}
,
$$

and its natural extension (which we denote with the same notation)

$$
\begin{aligned}
\flat_\Omega : \mathfrak{X}(M) &\longrightarrow \Omega^{k-1}(M) \\
X &\mapsto i(X)\Omega
\end{aligned}
.
$$

The inverse of this isomorphism is denoted as $\sharp_\Omega := \flat_\Omega^{-1}$.

Given a differentiable manifold M and a form $\Omega \in \Omega^2(M)$, it is obvious that Ω is nondegenerate (symplectic) if, and only if, $\flat_\Omega$ is an isomorphism between TM and T^*M. Then:

Definition 2.5. If (M, Ω) is a symplectic manifold, then the map $\flat_\Omega$ is called the **canonical isomorphism** induced by Ω.

Given a symplectic manifold (M, Ω), every function $f \in C^\infty(M)$ has associated a unique vector field $X_f \in \mathfrak{X}(M)$ by means of the map

$$
\sharp_\Omega \circ d : C^\infty(M) \xrightarrow{\ d\ } \Omega^1(M) \xrightarrow{\ \sharp_\Omega\ } \mathfrak{X}(M);
$$

that is, defined as $X_f := \sharp_\Omega(df)$, or equivalently, given implicitly by

$$
i(X_f)\Omega := df. \tag{2.1}
$$

Remark 2.5. Observe that the map $\sharp_\Omega \circ d$ is not surjective. This means that although the canonical isomorphism allows us to associate with every vector field X, a differential 1-form $i(X)\Omega$, it is not always possible to associate a function since, in order to do this, the form will need to be exact, but that form is not even closed, in general. Neither is the map $\sharp_\Omega \circ d$ injective, since functions differing in an additive constant have the same vector field associated with this map.

Considering this comment, we have this:

Definition 2.6. Let (M, Ω) be a symplectic manifold. A vector field $X \in \mathfrak{X}(M)$ is a *(global) Hamiltonian vector field* if $i(X)\Omega$ is an exact form. In this case, the function $f \in C^\infty(M)$ such that $i(X)\Omega = df$ is called a *(global) Hamiltonian function* of the vector field X.

  The set of global Hamiltonian vector fields in M is denoted as $\mathfrak{X}_H(M)$.

Considering the comment before Eq. (2.1), every function $f \in C^\infty(M)$ is a Hamiltonian function of a global Hamiltonian vector field X_f.

Nevertheless, the requirement in this definition is too restrictive, and for the physical interest, it is sufficient to have this definition:

Definition 2.7. Let (M, Ω) be a symplectic manifold. A vector field $X \in \mathfrak{X}(M)$ is a *local Hamiltonian vector field* if $i(X)\Omega$ is a closed form.

  In this case, for every point $p \in M$, Poincaré's Lemma assures the existence of a neighbourhood $U \subset M$, $p \in U$, and a function $f \in C^\infty(U)$ such that $i(X)\Omega = df$ in U. This function is called a *local Hamiltonian function* of the vector field X in U.

  The set of local Hamiltonian vector fields in M is denoted as $\mathfrak{X}_{lh}(M)$.

Remark 2.6.

(1) Clearly, $\mathfrak{X}_H(M) \subset \mathfrak{X}_{lh}(M)$. Thus, all we state for local Hamiltonian vector fields holds for global Hamiltonian vector fields.

(2) The above definitions are also valid for presymplectic manifolds. The difference is that, in this case, map $\flat_\Omega$ is not an isomorphism because it is not injective, thus not exhaustive, and hence, not every function in the manifold is associated with a Hamiltonian vector field.

Remember that a curve $c \colon I \subseteq \mathbb{R} \longrightarrow M$ is an *integral curve* of a vector field $X \in \mathfrak{X}(M)$ if $\dot{c}(t) = (X \circ c)(t)$ for $t \in I$, where $\dot{c}(t)$ denotes the derivative of c at t (i.e., the *tangent vector* of the curve at $c(t)$). Let $\tilde{c} \colon I \subseteq \mathbb{R} \longrightarrow TM$ be the canonical lift of c to the tangent bundle TM, that is, $\tilde{c}(t) = (c(t), \dot{c}(t))$, for $t \in I$ (see Definition A.6). Then, considering these definitions and Eq. (2.11), it is straightforward to prove that:

Theorem 2.2. *A vector field $X \in \mathfrak{X}(M)$ in a symplectic manifold (M, Ω) is the (local) Hamiltonian vector field corresponding to the func-*

> *tion* $f \in C^\infty(M)$, *that is,* $X = X_f$, *if, and only if, the integral curves*
> $c: I \subset \mathbb{R} \longrightarrow M$ *of* X *are the solutions to the equation*
>
> $$i(\tilde{c})(\Omega \circ c) = df \circ c . \tag{2.2}$$
>
> *Equation* (2.2) *is the* **Hamilton equation** *for the integral curves of* X.

Remember that Eq. (2.2) is a straightforward consequence of Eq. (2.1), the definitions of integral curve of a vector field, and the contraction $i(\tilde{c})(\Omega \circ c)$ (see Remark A.1).

Local expressions: If $(U; x^i, y_i)$ is a symplectic chart, we have

$$X_f \mid_U = A^i \frac{\partial}{\partial x^i} + B_i \frac{\partial}{\partial y_i},$$

$$df \mid_U = \frac{\partial f}{\partial x^i} dx^i + \frac{\partial f}{\partial y_i} dy_i;$$

then, for X_f, the solution to Eq. (2.1), we have

$$0 = (i(X_f)\Omega - df) \mid_U = \left(-B_i - \frac{\partial f}{\partial x^i}\right) dx^i + \left(A^i - \frac{\partial f}{\partial y_i}\right) dy_i;$$

that is,

$$A^i = \frac{\partial f}{\partial y_i}, \quad B_i = -\frac{\partial f}{\partial x^i}, \tag{2.3}$$

and then

$$X_f \mid_U = \frac{\partial f}{\partial y_i} \frac{\partial}{\partial x^i} - \frac{\partial f}{\partial x^i} \frac{\partial}{\partial y_i}; \tag{2.4}$$

therefore, the integral curves $c(t) = (x^i(t), y_i(t))$ of X_f are the solutions to the system of first-order differential equations

$$\frac{dy_i}{dt} = -\frac{\partial f}{\partial x^i}(c(t)), \quad \frac{dx^i}{dt} = \frac{\partial f}{\partial y_i}(c(t)). \tag{2.5}$$

Equations (2.3) and (2.5) are the local expressions of Eqs. (2.1) and (2.2), respectively, and are called the **Hamilton equations** of the (local) Hamiltonian vector field X_f and its integral curves, respectively.

An important result which is used later is:

> **Lemma 2.2.** *Let* (M, Ω) *be a symplectic manifold. For every point* $p \in M$, *there exist vector fields* $X_j \in \mathfrak{X}_{lh}(M)$, $j = 1, \ldots, 2n$, *such that* $\{X_j(p)\}$ *is a basis of* $T_p M$.

Proof. This Lemma can be proven using symplectic charts, since the local coordinate vector fields $\dfrac{\partial}{\partial x^i}, \dfrac{\partial}{\partial y_i}$ are locally Hamiltonian vector fields associated with the Hamiltonian functions y_i and $-x^i$, respectively. $\qquad\square$

This means that a local Hamiltonian vector field locally expands the tangent bundle of M.

2.1.6 Invariant forms

The properties of Hamiltonian vector fields are closely related to the properties of the symplectic form. This relationship was first established by studying *integral invariants* of mechanics [79, 150, 165]. We go on to explore this relationship, but we first introduce the following concept:

Definition 2.8. Let M be a differentiable manifold and $X \in \mathfrak{X}(M)$. A form $\beta \in \Omega^p(M)$ is an **absolute invariant form** for X if $\mathrm{L}(X)\beta = 0$.

Remark 2.7. Remembering the interpretation of the Lie derivative to be an absolute invariant form for X means that β is invariant along the integral curves of X and if F_t denotes the flux of the vector field X, this is equivalent to the demand that $F_t^*\beta = \beta$ for every t.

Now, we can state the following result, which is usually taken as an alternative definition of a Hamiltonian vector field.

Theorem 2.3. *Let (M, Ω) be a symplectic (respectively, presymplectic) manifold. The vector field $X \in \mathfrak{X}(M)$ is a local Hamiltonian vector field if, and only if, Ω is an absolute invariant form for X.*

Proof. As Ω is a closed form, we have

$$\mathrm{L}(X)\Omega = i(X)\mathrm{d}\Omega + \mathrm{d}\,i(X)\Omega = \mathrm{d}\,i(X)\Omega = 0$$
$$\iff i(X)\Omega \in Z^1(M) \iff X \in \mathfrak{X}_{lh}(M).$$

$\qquad\square$

Remark 2.8. This result relates the Hamiltonian vector fields by the fact that Ω is closed. However, it is less precise than Definition 2.7, since it does not allow us to distinguish the global Hamiltonian vector fields inside the set of the local Hamiltonian ones.

From this theorem, we deduce:

Theorem 2.4 (Liouville). *Let* (M, Ω) *be a symplectic manifold and* Ω^n *the Liouville volume form in* M. *Then* $L(X)\Omega^n = 0$ *for every* $X \in \mathfrak{X}_{lh}(M)$.

Proof. It is straightforward since, as the Lie derivative is a derivation,

$$L(X)\Omega^n = n\,(L(X)\Omega) \wedge \overbrace{\Omega \wedge \ldots \wedge \Omega}^{(n-1 \; times)} = 0.$$

$\square$

Proposition 2.5. *Let* (M, Ω) *be a symplectic (respectively, presymplectic) manifold. The set* $\mathfrak{X}_{lh}(M)$ *is closed for the Lie bracket of vector fields and it is a real Lie algebra.*

Proof. We have to prove that $[X, Y] \in \mathfrak{X}_{lh}(M)$ for every $X, Y \in \mathfrak{X}_{lh}(M)$. Considering the relationship

$$i([X, Y])\Omega = L(X)\,i(Y)\Omega - i(Y)\,L(X)\Omega,$$

and Theorem 2.3, if $i(Y)\Omega|_U = df$, on an open set U, we have

$$i([X, Y])\Omega = L(X)\,i(Y)\Omega|_U = L(X)df = d\,L(X)f,$$

that is, in U, the local Hamiltonian function for $[X, Y]$ is $L(X)f$.

Linearity, skew symmetry, and the Jacobi identity of the Lie bracket complete the proof. $\square$

So, given a symplectic manifold (M, Ω), we have proven that the symplectic and the Liouville volume forms are invariant by the local Hamiltonian vector fields, that is, by the groups of local diffeomorphisms generated by their fluxes. Actually, there is a more general property stating that some geometrical structures, such as those defined by a volume or a symplectic form on a differentiable manifold, are determined by their automorphism groups (i.e., the groups of volume-preserving and symplectic diffeomorphisms), as was shown by *A. Banyaga* [16–18].

Now, we can try to characterise all the elements in $\Omega^k(M)$ with the above property. The answer to this problem was established by *Lee Hwa Chung* [243], who studied the uniqueness of the *integral invariant forms* by local transformations generated by the flux of the local Hamiltonian vector fields. Next, we state the geometrical version of this theorem and prove its partial result[3].

Theorem 2.5 (Lee Hwa Chung). *Let (M, Ω) be a symplectic manifold and $\alpha \in \Omega^k(M)$ an absolute invariant form for every $X \in \mathfrak{X}_{lh}(M)$. Then:*

(1) If k is odd, that is, $k = 2r - 1$ with $r \in \mathbb{N}$, then $\alpha = 0$.

(2) If k is even, that is, $k = 2r$ with $r \in \mathbb{N}$, then $\alpha = c \overbrace{\Omega \wedge \ldots \wedge \Omega}^{(r \ times)} \equiv c \bigwedge_r \Omega$, where $c \in \mathbb{R}$.

Proof. We prove the statement for the case $k \leq 2$, which is the only one we need later (for the proof of the general case, see [252]).

As α is invariant under the action of every $X \in \mathfrak{X}_{lh}(M)$, we have

$$0 = \mathrm{L}(X)\alpha = \mathrm{d}\, i(X)\alpha + i(X)\mathrm{d}\alpha \quad \Longleftrightarrow \quad \mathrm{d}\, i(X)\alpha = -\, i(X)\mathrm{d}\alpha. \qquad (2.6)$$

Consider now $X, X' \in \mathfrak{X}_{lh}(M)$ and $\mathrm{p} \in M$; there exist $U \subset M$, $\mathrm{p} \in U$, and $f, g \in C^\infty(U)$ such that $i(X)\Omega|_U = \mathrm{d}f$ and $i(X')\Omega|_U = \mathrm{d}g$ (from now on, we write $X|_U \equiv X_f$ y $X'|_U \equiv X_g$). The vector field X_h, given in U as $X_h|_U = fX_g + gX_f$, is locally Hamiltonian, and in U, its local Hamiltonian function is $h = fg \in C^\infty(U)$, since:

$$i(X_h)\Omega|_U = i(fX_g + gX_f)\Omega = f\, i(X_g)\Omega + g\, i(X_f)\Omega = f\mathrm{d}g + g\mathrm{d}f \equiv \mathrm{d}h.$$

Thus, we have

$$i(X_h)\alpha|_U = f\, i(X_g)\alpha + g\, i(X_f)\alpha,$$

and taking the exterior differential

$$\mathrm{d}\, i(X_h)\alpha|_U = \mathrm{d}f \wedge i(X_g)\alpha + f\mathrm{d}\, i(X_g)\alpha + \mathrm{d}g \wedge i(X_f)\alpha + g\mathrm{d}\, i(X_f)\alpha.$$

However, having in mind (2.6),

$$\mathrm{d}\, i(X_h)\alpha = -\, i(X_f h)\mathrm{d}\alpha|_U$$
$$= -f\, i(X_g)\mathrm{d}\alpha - g\, i(X_f)\mathrm{d}\alpha = f\mathrm{d}\, i(X_g)\alpha + g\mathrm{d}\, i(X_f)\alpha,$$

and comparing the last two equations, we conclude that

$$(\mathrm{d}f \wedge i(X_g)\alpha + \mathrm{d}g \wedge i(X_f)\alpha)|_U = 0. \qquad (2.7)$$

---

[3] The original proof of this theorem is local using Darboux coordinates.

Now, putting $X_f = X_g$ in this expression, that is, $f = g$, we obtain, for every $f \in C^\infty(U)$,

$$\mathrm{d}f \wedge i(X_f)\alpha|_U = 0.$$

Now, we have two options:

(1) If $k = 1$, then $i(X_f)\alpha \in C^\infty(M)$, and this last equality leads to $i(X_f)\alpha = 0$, for every $X_f \in \mathfrak{X}_{lh}(M)$. Considering that with Lemma 2.2, local Hamiltonian vector fields span locally TM, we obtain $i(X)\alpha|_U = 0$, for every $X \in \mathfrak{X}(U)$, and this implies necessarily that $\alpha|_U = 0$ (for every U) and then $\alpha = 0$.
(2) If $k = 2$, we have to conclude:

- either $i(X_f)\alpha = 0$,
- or $i(X_f)\alpha|_U = \eta_{X_f}\mathrm{d}f$, where $\eta_{X_f} \in C^\infty(U)$.

In the first case, reasoning as in the above item, we conclude that $\alpha = 0$. In the second case, going to the expression (2.7), we obtain

$$(\mathrm{d}f \wedge \mathrm{d}g\, \eta_{X_g} + \mathrm{d}g \wedge \mathrm{d}f\, \eta_{X_f})|_U = 0;$$

that is

$$(\mathrm{d}f \wedge \mathrm{d}g)(\eta_{X_g} - \eta_{X_f})|_U = 0,$$

for every $f, g \in C^\infty(U)$. Then it must be $\eta_{X_f} = \eta_{X_g} \equiv \eta$, that is, the function η does not depend on the local Hamiltonian vector field.

Therefore, for every $X_f \in \mathfrak{X}_{lh}(M)$, we have $i(X_f)\alpha|_U = \eta\, \mathrm{d}f$, with $\eta \in C^\infty(M)$, and then

$$i(X_f)\alpha|_U = \eta\, \mathrm{d}f = \eta\, i(X_f)\Omega = i(X_f)(\eta\Omega);$$

but considering Lemma 2.2, this equality leads to

$$i(X)(\alpha - \eta\Omega)|_U = 0,$$

for every $X \in \mathfrak{X}(M)$, and we conclude that $\alpha = \eta\Omega$. Finally, as α is invariant for every local Hamiltonian vector field, for every $Y \in \mathfrak{X}_{lh}(M)$, we have

$$0 = \mathrm{L}(Y)\alpha = \mathrm{L}(Y)(\eta\Omega) = (\mathrm{L}(Y)\eta)\Omega + \eta\, \mathrm{L}(Y)\Omega = (\mathrm{L}(Y)\eta)\Omega,$$

and then $\mathrm{L}(Y)\eta = 0$, and by Lemma 2.2, this result holds for every $Y \in \mathfrak{X}(M)$; therefore, $\eta = c$ (constant) and thus $\alpha = c\Omega$. $\qquad\square$

Remark 2.9.

(1) So, the absolute invariant forms for every local Hamiltonian vector field are in multiples of exterior products of the symplectic form and hence they must be of an even degree. The above proof is for 2-forms only.

(2) A similar result can be proven also for presymplectic manifolds [181] (see also [152] for another interesting generalisation of this theorem).

2.1.7   *Poisson brackets*

The symplectic form allows us to introduce certain well-known operations in analytical mechanics in a natural way.

Definition 2.9. Let (M, Ω) be a symplectic manifold. The **Lagrange bracket** of two vector fields $X, Y \in \mathfrak{X}(M)$ is the bilinear map

$$(,) : \mathfrak{X}(M) \times \mathfrak{X}(M) \longrightarrow C^\infty(M)$$
$$X, Y \qquad \mapsto \quad (X, Y) \ ,$$

defined by

$$(X, Y) := \Omega(X, Y) := i(Y)\, i(X)\Omega.$$

Remark 2.10.

(1) This bracket is not an internal operation in $\mathfrak{X}(M)$, as is the *Lie bracket*, whose result is another vector field.

(2) From the skew symmetry of Ω, we deduce immediately that the Lagrange bracket is skew symmetric, that is, $(X, Y) = -(Y, X)$.

Considering (2.1), from this concept, we obtain:

Definition 2.10. Let (M, Ω) be a symplectic manifold. The **Poisson bracket** of two functions $f, g \in C^\infty(M)$ is the Lagrange bracket of their associated Hamiltonian vector fields, that is, the bilinear map

$$\{,\} : C^\infty(M) \times C^\infty(M) \longrightarrow C^\infty(M)$$
$$f, g \qquad \mapsto \quad \{f, g\} \ ,$$

defined by

$$\{f, g\} := \Omega(X_f, X_g) := i(X_g)\, i(X_f)\Omega.$$

Local expressions: If $(U; x^i, y_i)$ is a symplectic chart and

$$X \mid_U = A^i \frac{\partial}{\partial x^i} + B_i \frac{\partial}{\partial y_i} \quad , \quad Y \mid_U = C^i \frac{\partial}{\partial x^i} + D_i \frac{\partial}{\partial y_i},$$

then

$$(X, Y) \mid_U = -B_i C^i + A^i D_i.$$

Furthermore,

$$\{f, g\} \mid_U = \frac{\partial f}{\partial x^i} \frac{\partial g}{\partial y_i} - \frac{\partial f}{\partial y_i} \frac{\partial g}{\partial x^i}.$$

In particular, for the canonical coordinates x^i, y_i we have

$$\{x^i, x^j\} = 0, \quad \{y_i, y_j\} = 0, \quad \{x^i, y_j\} = \delta^i_j.$$

The main properties of the Poisson bracket are collected in the following:

Proposition 2.6. *Let (M, Ω) be a symplectic manifold and $\{\, ,\, \}$ the associate Poisson bracket. Then*

(1) $\{f, g\} = -\{g, f\}$ (skew symmetry).
(2) $\{f, \{g, h\}\} + \{g, \{h, f\}\} + \{h, \{f, g\}\} = 0$ (Jacobi identity).
(3) $\{f, g\} = L(X_g)f = -L(X_f)g$.
(4) $X_{\{f,g\}} = [X_g, X_f]$.

Proof. (1) This follows logically from the definition.

(2) This is a consequence of Ω being closed.

(3) Considering the Cartan formula for the Lie derivative:

$$\{f, g\} = i(X_g)\, i(X_f)\Omega = i(X_g)\mathrm{d}f = L(X_g)f.$$

In an analogous way, we have $\{f, g\} = -L(X_f)g$.

(4) The statement is equivalent to $i([X_g, X_f])\Omega = \mathrm{d}\{f, g\}$, and remembering Proposition 2.5, we obtain

$$i([X_g, X_f])\Omega = L(X_g)\, i(X_f)\Omega = L(X_g)\mathrm{d}f$$
$$= \mathrm{d}\, L(X_g)f = \mathrm{d}\{f, g\}. \tag{2.8}$$

$\square$

Remark 2.11.

(1) The first two properties establish that $C^\infty(M)$ with the Poisson bracket is a real Lie algebra.

(2) The third property allows us to give a geometric interpretation of the Poisson bracket between two functions: it measures the variation of

one of them along the integral curves of the Hamiltonian vector field associated with the other.

(3) The fourth property tells us that the map $C^\infty(M) \longrightarrow \mathfrak{X}(M)$ given by $f \mapsto X_f$ is a Lie algebra (anti)-homomorphism between $(C^\infty(M), \{\,,\,\})$ and $(\mathfrak{X}(M), [\,,\,])$.

Using again the canonical isomorphism, and considering (2.8), we can establish the following generalisation:

Definition 2.11. Let (M, Ω) be a symplectic manifold. The **Poisson bracket** of two 1-form $\alpha, \beta \in \Omega^1(M)$ is the bilinear map

$$\{,\} : \Omega^1(M) \times \Omega^1(M) \longrightarrow \Omega^1(M)$$
$$\alpha, \beta \qquad \mapsto \{\alpha, \beta\} \,,$$

defined by

$$\{\alpha, \beta\} := i([X_\alpha, X_\beta])\Omega,$$

where $X_\alpha = \sharp_\Omega(\alpha)$ and $X_\beta = \sharp_\Omega(\beta)$.

The following is evident:

Proposition 2.7. *Let (M, Ω) be a symplectic manifold. Then, for every $f, g \in C^\infty(M)$,*

$$d\{f, g\} = -\{df, dg\}.$$

The properties of the Poisson bracket of 1-form type are clearly analogous to those of the Poisson bracket of functions.

2.1.8 *Canonical transformations and symplectomorphisms*

In the previous section, we saw how the properties of the symplectic structure allow us to introduce the concept of a local Hamiltonian vector field and how the integral curves of these fields are obtained as solutions to the Hamilton equations. We will see also that these kinds of vector fields are suitable to describe dynamical systems. This means that there is a deep connection between the dynamics of physical systems and the geometric properties of their phase spaces.

In this way, from a dynamical perspective, it is reasonable to suppose that the more relevant transformations among dynamical systems are those preserving the dynamical equations which, in our case, mean geometrically that they transform Hamiltonian vector fields into Hamiltonian vector fields. Consequently, we have:

Definition 2.12. Let (M_1, Ω_1) and (M_2, Ω_2) be symplectic manifolds and $\Phi : M_1 \longrightarrow M_2$ a diffeomorphism. We say that Φ is a **canonical transformation** if it maps local Hamiltonian vector fields into local Hamiltonian vector fields biunivocally, that is, $\Phi_*(\mathfrak{X}_{lh}(M_1)) = \mathfrak{X}_{lh}(M_2)$.

Concerning the geometrical aspects, the more interesting transformations between symplectic manifolds are the following:

Definition 2.13. Let (M_1, Ω_1) and (M_2, Ω_2) be symplectic manifolds and $\Phi : M_1 \longrightarrow M_2$ a diffeomorphism. We say that Φ is a **symplectomorphism** (or also a **symplectic transformation**) if it preserves their symplectic structures, that is, $\Phi^*\Omega_2 = \Omega_1$.

As Hamiltonian vector fields are defined using the symplectic form, we can expect that there is some connection between both kinds of transformations. In fact, Lee Hwa Chung's Theorem allows us to prove that both concepts are essentially the same:

Theorem 2.6. *Let (M_1, Ω_1) and (M_2, Ω_2) be symplectic manifolds and $\Phi \in \mathrm{Diff}\,(M_1, M_2)$. The necessary and sufficient condition for Φ to be a canonical transformation is $\Phi^*\Omega_2 = c\Omega_1$, with $c \in \mathbf{R}$.*

Proof. Suppose that Φ is a canonical transformation. For every $X_1 \in \mathfrak{X}_{lh}(M_1)$, we have $X_2 = \Phi_*X_1 \in \mathfrak{X}_{lh}(M_2)$, and as we know from Theorem 2.3, $\mathrm{L}(X_2)\Omega_2 = 0$, hence

$$0 = \Phi^*(\mathrm{L}(X_2)\Omega_2) = \mathrm{L}(\Phi_*^{-1}X_2)(\Phi^*\Omega_2) = \mathrm{L}(X_1)(\Phi^*\Omega_2).$$

However, this means that $\Phi^*\Omega_2$ is invariant by any element of $\mathfrak{X}_{lh}(M_1)$, and from Lee Hwa Chung's Theorem, we conclude that $\Phi^*\Omega_2 = c\Omega_1$, $c \in \mathbf{R}$ and Φ is a symplectomorphism.

Conversely, suppose that Φ is a symplectomorphism, $\Phi^*\Omega_2 = c\Omega_1$. Given $X_1 \in \mathfrak{X}_{lh}(M_1)$, we have $\mathrm{L}(X_1)\Omega_1 = 0$. Then

$$0 = \Phi^{*^{-1}}(\mathrm{L}(X_1)\Omega_1) = \mathrm{L}(\Phi_*X_1)(\Phi^{*^{-1}}\Omega_1) = \frac{1}{c}\mathrm{L}(\Phi_*X_1)\Omega_2;$$

hence, by Theorem 2.3, $\Phi_*X_1 \in \mathfrak{X}_{lh}(M_2)$, that is, Ω_2 is invariant by every element of $\mathfrak{X}_{lh}(M_2)$, and Φ is a canonical transformation. $\square$

Remark 2.12.

(1) The constant c that appears in this last theorem is called the **valence** of the canonical transformation. It is usual to consider transformations with $c = 1$ (i.e., symplectomorphisms), and they are called **univalent** or **restricted canonical transformations**. Another terminology is also used, calling canonical transformations those with valence $c = 1$, and then calling the rest **generalised canonical transformations** [313].

(2) Observe that a diffeomorphism is an *univalent canonical transformation* if, and only if, it is a *symplectomorphism*.

(3) All these definitions and properties are also valid for presymplectic manifolds. In this case, we talk about **presymplectomorphisms** (the study of this case is done in [28, 63, 64]).

Another fundamental result is:

Proposition 2.8. *Let (M, Ω) be a symplectic (respectively, presymplectic) manifold. A vector field $X \in \mathfrak{X}(M)$ is a local Hamiltonian vector field if, and only if, its flux is a group of local symplectomorphisms (respectively, local presymplectomorphisms).*

Proof. Let F_t be the flux of a vector field X. We know that $\mathrm{L}(X)\Omega = 0$ is equivalent to $F_t^*\Omega = \Omega$, hence the result follows directly. $\qquad\square$

Finally, it is easy to prove that:

Proposition 2.9. *The set of canonical transformations of a symplectic (respectively, presymplectic) manifold (M, Ω), with the operation of composition, is a group.*

The group of the symplectomorphisms of a symplectic manifold is denoted by $\mathbf{Sp}(M, \Omega)$, and it has crucial relevance in the study of *symmetries* of dynamical systems.

2.1.9 *Characterisation of canonical transformations*

The last theorem has some important corollaries, which give alternative characterisations for a transformation to be canonical (or a symplectomorphism). The most important of them uses the Poisson bracket of functions.

Previously, we have had to specify how the Hamiltonian functions associated with Hamiltonian vector fields are transformed under these kinds of transformations.

Proposition 2.10. *Let (M_1, Ω_1) and (M_2, Ω_2) be symplectic manifolds and $\Phi \in \mathrm{Diff}\,(M_1, M_2)$ a canonical transformation of valence c. If $X_1 \in \mathfrak{X}_{lh}(M_1)$, let $X_2 := \Phi_* X_1 \in \mathfrak{X}_{lh}(M_2)$, and $h_1 \in C^\infty(U_1)$ and $h_2 \in C^\infty(U_2)$ local Hamiltonian functions of X_1 and X_2 on $U_1 \subset M_1$ and $U_2 := \Phi(U_1) \subset M_2$, respectively. Then*

$$ch_1 = \Phi^* h_2 + k, \quad k \in \mathbf{R}.$$

Proof. Following the previous theorem, we have

$$d(\Phi^{*^{-1}} h_1) = \Phi^{*^{-1}} dh_1 = \Phi^{*^{-1}}(i(X_1)\Omega_1)\,|_{U_1}$$

$$= i(\Phi_* X_1)(\Phi^{*^{-1}}\Omega_1)\,|_{\Phi(U_1)} = \frac{1}{c}\,i(X_2)\Omega_2\,|_{U_2} = \frac{1}{c}dh_2 = d\left(\frac{1}{c}h_2\right),$$

that is, $d(ch_1) = d(\Phi^* h_2)$, and the result follows. $\qquad\square$

Bearing this in mind, we state:

Theorem 2.7. *Let (M_1, Ω_1) and (M_2, Ω_2) be symplectic manifolds. A diffeomorphism $\Phi : M_1 \longrightarrow M_2$ is a canonical transformation of valence c if, and only if,*

$$\Phi^*\{f_2, g_2\} = \frac{1}{c}\{\Phi^* f_2, \Phi^* g_2\}.$$

for every $f_2, g_2 \in C^\infty(M_2)$.

Proof. Let $f_2, g_2 \in C^\infty(M_2)$ and $X_{f_2}, X_{g_2} \in \mathfrak{X}_h(M_2)$ be the corresponding Hamiltonian vector fields.

If Φ is a canonical transformation, then $\Phi_*^{-1} X_{g_2} \in \mathfrak{X}_h(M_1)$, and by the above proposition, we know that $i(\Phi_*^{-1} X_{g_2})\Omega_1 = d(\frac{1}{c}\Phi^* g_2)$, that is, $\Phi_*^{-1} X_{g_2} = X_{\frac{1}{c}\Phi^* g_2}$, and then

$$\Phi^*\{f_2, g_2\} = \Phi^*(\mathrm{L}(X_{g_2})f_2) = \mathrm{L}(\Phi_*^{-1} X_{g_2})\Phi^* f_2 = \mathrm{L}(X_{\frac{1}{c}\Phi^* g_2})\Phi^* f_2$$

$$= \frac{1}{c}\{\Phi^* f_2, \Phi^* g_2\}.$$

Conversely, if the condition holds, given $f_2, g_2 \in C^\infty(M_2)$, we have

$$\Phi^*\{f_2, g_2\} = \mathrm{L}(\Phi_*^{-1} X_{g_2})\Phi^* f_2,$$

and furthermore,

$$\frac{1}{c}\{\Phi^*f_2, \Phi^*g_2\} = \mathrm{L}(X_{\frac{1}{c}\Phi^*g_2})\Phi^*f_2;$$

hence, combining these two last equalities,

$$\Phi_*^{-1}X_{g_2} = X_{\frac{1}{c}\Phi^*g_2} \in \mathfrak{X}_{lh}(M_1); \ \forall X_{g_2} \in \mathfrak{X}_{lh}(M_2).$$

Therefore, Φ is a canonical transformation with valence c as a consequence of Lee Hwa Chung's Theorem. $\qquad\square$

> **Remark 2.13.** This result states that a transformation is canonical if, and only if, the Poisson bracket is invariant, up to a multiplicative constant, by its action. This is an alternative way to say that the symplectic structure is invariant by the transformation.

From the local point of view, if $\Phi : M_1 \longrightarrow M_2$ is a canonical transformation with valence $c = 1$, and we have Darboux coordinates (x^i, y_i) in M_1 and $(\tilde{x}^i, \tilde{y}_i)$ in M_2, we have

$$\{\Phi^*\tilde{x}^i, \Phi^*\tilde{x}^j\} = \Phi^*\{\tilde{x}^i, \tilde{x}^j\},$$
$$\{\Phi^*\tilde{y}_i, \Phi^*\tilde{y}_j\} = \Phi^*\{\tilde{y}_i, \tilde{y}_j\},$$
$$\{\Phi^*\tilde{y}_i, \Phi^*\tilde{y}_j\} = \Phi^*\{\tilde{y}_i, \tilde{y}_j\},$$

that is, $(\Phi^*\tilde{x}^i, \Phi^*\tilde{y}_i)$ is a symplectic coordinate system in M_1 if, and only if, $(\tilde{x}^i, \tilde{y}_i)$ is a symplectic coordinate system in M_2.

2.1.10 *Generating functions of canonical transformations*

Another characterisation of canonical transformations is by means of the *generating functions*.

> **Proposition 2.11.** *Let (M_1, Ω_1) and (M_2, Ω_2) be symplectic manifolds and $\Phi \in \mathrm{Diff}\,(M_1, M_2)$. Let $U_1 \subset M_1$ and $U_2 := \Phi(U_1) \subset M_2$, and $\Theta_i \in \Omega^1(U_i)$ such that $\Omega_i\,|_{U_i} = \mathrm{d}\Theta_i$, (i = 1, 2). Then Φ is a canonical transformation, with valence c, if, and only if, there exists a function $F_1 \in \mathrm{C}^\infty(U_1)$ such that*
> $$(\Phi^*\Theta_2 - c\Theta_1 - \mathrm{d}F_1)\,|_{U_1} = 0,$$
> *or equivalently, there exists a function $F_2 \in \mathrm{C}^\infty(U_2)$ such that*
> $$(\Phi^{*^{-1}}\Theta_1 - \frac{1}{c}\Theta_2 - \mathrm{d}F_2)\,|_{U_2} = 0.$$

> *These functions F_1, F_2 are called (Poincaré) generating functions of the canonical transformation, and the relationship between them is $F_1 = c\Phi^* F_2 + k$, $k \in \mathbf{R}$.*

Proof. From Theorem 2.6, Φ is a canonical transformation if, and only if,

$$0 = (\Phi^*\Omega_2 - c\Omega_1)\,|_{U_1} = d(\Phi^*\Theta_2 - \Theta_1),$$

and then Poincaré's Lemma leads to the result.

The statement concerning F_2 is obtained in an analogous way, and comparing both results, we arrive at the relationship between these functions. $\qquad\square$

In classical texts of mechanics (see, for example, [165, 180, 237]), a more general concept of generating functions, which include the above ones, is studied. In the geometrical context, they are introduced as follows: [1] (without loss of generality, we will restrict ourselves to the case $c = 1$, that is, symplectomorphisms or univalent canonical transformations).

First, considering Proposition 2.4, we have:

> **Proposition 2.12.** *Let (M_1, Ω_1), (M_2, Ω_2) be symplectic manifolds with $\dim M_1 = \dim M_2$, $\Phi : M_1 \longrightarrow M_2$ a diffeomorphism, and $\jmath : graph\,\Phi \hookrightarrow M_1 \times M_2$ the natural embedding.*
> *The map Φ is a symplectomorphism if, and only if, $graph\,\Phi$ is a Lagrangian submanifold of the symplectic manifold $(M_1 \times M_2, \Omega_{12} := \pi_1^*\Omega_1 - \pi_2^*\Omega_2)$.*

Proof. Remember that *Lagrangian submanifolds* of a symplectic manifold $(\mathcal{M}, \Omega)$ are maximal isotropic submanifolds $\jmath \colon S \hookrightarrow \mathcal{M}$ or equivalently, such that S verifies that $\dim S = \frac{1}{2}\dim \mathcal{M}$ and $\jmath^*\Omega = 0$. Clearly, $\dim(graph\,\Phi) = \frac{1}{2}\dim(M_1 \times M_2)$, and as $\Phi^*\Omega_2 = \Omega_1$, the other condition holds trivially. $\qquad\square$

If $\Omega_j = -d\theta_j$, $j = 1, 2$, and Θ_j are local symplectic potentials for Ω_j, being $graph\,\Phi$ a Lagrangian submanifold, we have

$$0 = \jmath^*(\pi_1^*\Omega_1 - \pi_2^*\Omega_2) = d\jmath^*(\pi_2^*\Theta_2 - \pi_1^*\Theta_1), \qquad (2.9)$$

which, given a point in $graph\,\Phi$, is locally equivalent to

$$\jmath^*(\pi_2^*\Theta_2 - \pi_1^*\Theta_1)|_W = -d\mathcal{S}, \qquad (2.10)$$

where $\mathcal{S}$ is a function defined in an open neighbourhood of the given point, $W \subset graph\,\Phi$. This function depends on the choice of Θ_1 and Θ_2.

Definition 2.14. $\mathcal{S}$ is called a **Weinstein generating function** of the Lagrangian submanifold *graph* Φ and hence of the symplectomorphism Φ.

If $(U_1; x^i, y_i)$, $(U_2; \tilde{x}^i, \tilde{y}_i)$ are Darboux charts such that $W \subset U_1 \times U_2$, local coordinates in W can be chosen in several ways. This leads to six different possible choices for $\mathcal{S}$. Thus, for instance, if $(W; x^i, \tilde{x}^i)$ is a chart, then (2.10) gives the symplectomorphism explicitly as

$$\tilde{y}_i \mathrm{d}\tilde{x}^i - y_i \mathrm{d}x^i = -\mathrm{d}\mathcal{S}(x, \tilde{x}) \quad \Longleftrightarrow \quad \tilde{y}_i = -\frac{\partial \mathcal{S}}{\partial \tilde{x}^i}(x, \tilde{x}), \; y_i = \frac{\partial \mathcal{S}}{\partial x^i}(x, \tilde{x}).$$

Of course, the *Poincaré generating functions* are two particular choices of *Weinstein generating functions*.

2.2   Hamiltonian Dynamical Systems

Once the foundations of symplectic geometry are established, we are ready to use it for describing the autonomous Hamiltonian dynamical systems. As we will see in the next chapter, this geometric formulation includes, as particular cases, the Lagrangian and the canonical Hamiltonian formalisms of variational dynamical systems, that is, those which are described by Lagrangian functions.

2.2.1   *Hamiltonian systems*

Some of the authors who developed geometric mechanics chose an axiomatic manner in their exposition (see, for instance, [327]). Following their ideas, we first state the postulates for the geometric study of autonomous Hamiltonian dynamical systems.

The first postulate concerns the physical states:

Postulate 2.1 (First postulate of Hamiltonian mechanics). The **state space** or **phase space** of a dynamical system is a differentiable manifold M endowed with a closed form $\Omega \in Z^2(M)$ such that:

- If Ω is nondegenerate, that is *symplectic*, the system is *regular* and the dimension of M is twice the number of degrees of freedom of the system. In this case, every point of this manifold represents a physical state of the system.
- If Ω is degenerate, that is *presymplectic*, the system is *singular* (and the manifold M is not necessarily even-dimensional).

The second postulate refers to the *observables*, that is, the *physical magnitudes*:

Postulate 2.2 (Second postulate of Hamiltonian mechanics). The observables or physical magnitudes of a dynamical system are functions of $C^\infty(M)$.

The result of the measure of an observable is the value that the function that represents it takes a point in the phase space M (i.e., in a given state, according to the first postulate).

If, according to Postulate 2.1, a (symplectic or presymplectic) manifold (M, Ω) constitutes the phase space of a dynamical system, there is a very natural way of introducing the dynamics [4]. Thus, we state:

Postulate 2.3 (Third postulate of Hamiltonian mechanics). The dynamics of a dynamical system is given by a closed 1-form $\alpha \in Z^1(M)$, which is called the **Hamiltonian 1-form** of the system.

Finally, the dynamical equations are stated in:

Postulate 2.4 (Fourth postulate of Hamiltonian mechanics). The dynamical trajectories of the system are the integral curves of a vector field $X_\alpha \in \mathfrak{X}(M)$, if it exists, associated with the form α by map $\flat_\Omega$, that is, of the vector field solution to the equation

$$i(X_\alpha)\Omega = \alpha. \tag{2.11}$$

Then, the integral curves $c \colon I \subseteq \mathbb{R} \longrightarrow M$ of X_α are solutions to the equation

$$i(\widetilde{c})(\Omega \circ c) = \alpha \circ c. \tag{2.12}$$

The vector field X_α is called a (local or global) **Hamiltonian vector field** of the Hamiltonian system, and Eqs. (2.11) and (2.12) are the **Hamilton equation** for X_α and its integral curves.

These equations can be obtained from a variational principle called the *minimal action principle of Hamilton–Jacobi* [5].

---

[4] In some cases, the 2-form Ω can also contain dynamical information as it happens, for instance, in the Lagrangian formalism of the Lagrangian systems (see Chapter 3).

[5] See Section 3.6.4.

Then, we have:

Definition 2.15.

(1) A **regular** or **symplectic Hamiltonian dynamical system** is
    a triple (M, Ω, α), where (M, Ω) is a symplectic manifold and
    $\alpha \in Z^1(M)$ is the Hamiltonian 1-form of the system. If (M, Ω)
    is a presymplectic manifold, then (M, Ω, α) is said to be a **singular**
    or **presymplectic Hamiltonian dynamical system**.

(2) With Poincaré's Lemma, for every $p \in M$, there exists $U \subset M$, with
    $p \in U$, and $h \in C^\infty(U)$, such that $\alpha \mid_U = dh$, which is called a **local**
    Hamiltonian function of the system, and the abovementioned triple
    is said to be a **local Hamiltonian system**.

    If α is an exact form, then there exists $h \in C^\infty(M)$ such that
    $\alpha = dh$, which is called a **global Hamiltonian function** of the
    system, and the triple (M, Ω, h) is said to be a **global Hamiltonian**
    system.

2.2.2   *Hamilton equations*

Given a Hamiltonian dynamical system (M, Ω, α), the **Hamiltonian prob-**
lem posed by the system consists of finding a vector field $X_\alpha \in \mathfrak{X}(M)$
verifying Eq. (2.11).

Proposition 2.13. *If (M, Ω, α) is a regular Hamiltonian system, then*
there exists a unique vector field $X_\alpha \in \mathfrak{X}(M)$, which is a solution to Eq.
(2.11).

Proof. For regular systems, $\flat_\Omega$ is the canonical isomorphism of the symplectic manifold (M, Ω) and the existence and uniqueness of X_α are assured. $\square$

Remark 2.14. If (M, Ω, α) is a presymplectic Hamiltonian system, the
Hamilton equations are not necessarily compatible everywhere on M and
we need to look for the maximal subset of M where a solution exists. This
subset is obtained by a procedure called the *constraint algorithm* which, in
the most favourable cases, leads to finding a *final constraint submanifold*
$P_f \hookrightarrow M$, where there are Hamiltonian vector fields $X_\alpha \in \mathfrak{X}(M)$, tangent

to P_f, which are solutions to the Hamilton equations on P_f, although these vector field solutions are not necessarily unique. Functions vanishing on the final constraint submanifold are called *constraints*.

Singular or presymplectic Hamiltonian systems and the corresponding constraint algorithms have been widely studied in the literature. The interested reader can see, for instance, [60, 63, 149, 186, 196, 212, 219, 258, 280, 324] and other papers cited therein. Apart from some examples in the problems, in this book, we only work with regular Hamiltonian systems.

Local expressions: If (M, Ω, α) is a regular Hamiltonian system with $\alpha = dh$, in a symplectic chart $(U; x^i, y_i)$ of M, if $X|_U = f^i \dfrac{\partial}{\partial x^i} + g_i \dfrac{\partial}{\partial y_i} \in \mathfrak{X}(M)$ then, according to (2.4), we have

$$X_\alpha \,|_U = \frac{\partial h}{\partial y_i} \frac{\partial}{\partial x^i} - \frac{\partial h}{\partial x^i} \frac{\partial}{\partial y_i},$$

and the integral curves $c(t) = (x^i(t), y_i(t))$ of X_α are the solution to the Hamilton equations (2.2) whose local expression is

$$\frac{dx^i}{dt} = \frac{\partial h}{\partial y_i}(c(t)), \quad \frac{dy_i}{dt} = -\frac{\partial h}{\partial x^i}(c(t)).$$

Remark 2.15.

(1) In a Hamiltonian dynamical system, the physical observable represented by the Hamiltonian function is associated with the energy of the system.

(2) This presentation of dynamical systems is called *Hamiltonian formalism of mechanics* because the dynamical trajectories are given by the integral curves of a Hamiltonian vector field.

2.3   Symmetries of Regular Hamiltonian Systems

In this section, we introduce the ideas of *constants of motion*, or *conserved quantities*, and the *symmetries* of a dynamical system and the relationship between both notions. Noether's Theorem will be proven and actions of the Lie group on Hamiltonian systems together with reduction theory will also be introduced and developed.

2.3.1   *Preliminaries*

The appropriate method for conducting this study involves utilizing the theory of *actions of Lie groups* on symplectic and presymplectic manifolds (see, for instance, [1, 108, 156, 177, 207, 249, 260, 290]). Within this theory, the associated concept of the *momentum map* plays a crucial role in order to introduce conserved quantities and the subsequent *Marsden–Weinstein reduction Theorem*. Although this theorem was initially stated for regular autonomous Hamiltonian systems, the Marsden–Weinstein technique has been applied and generalised to many different situations, for instance, for time-dependent regular Hamiltonian systems (with regular values of the momentum map) in the framework of cosymplectic manifolds [4, 210], for autonomous and nonautonomous Hamiltonian systems with singular values of the momentum map [10, 140, 320], and also for singular systems in the autonomous (presymplectic) [51,156] and nonautonomous [136,210,220] cases. Further generalisations are reduction on Poisson and Jacobi manifolds [61, 193, 253, 264, 284, 285], Lie algebroids [76, 77], cotangent bundles of Lie groups [268], Lagrangian reduction [86, 269], Euler–Poincaré reduction [80,85], Routh reduction for regular and singular Lagrangians [82,239], reduction of nonholonomic systems [27,36,55,257], reduction of optimal control systems [35, 271, 334, 341], and in the context of Dirac structures and implicit Hamiltonian systems [31, 274]. Finally, although symmetries and conservation laws in field theories have been studied in several geometric frameworks (see, for instance, [133, 183, 205, 206, 261, 312] and the references quoted therein), the problem of reduction by symmetries of classical field theories has been solved only for particular situations such as Lagrangian and Poisson reduction [83], Euler–Poincaré reduction in principal fibre bundles [81, 84], reduction in multisymplectic and poly(co)symplectic manifolds [144,146,158,262], and for discrete field theories [343]. Of course, this list of references is far from being complete.

To develop our study, we followed the historical order, with a few exceptions, as far as possible: we begin by introducing the basic notions for conserved quantities and symmetries, paying particular attention to the case of *Noether symmetries* and *Noether's Theorem*. The primitive ideas were the constants of motion, the linear momentum and the angular momentum, and their relationship with symmetries as a global transformation of the space of positions of a system, and after that, the infinitesimal symmetries and the Noether Theorem. The modern approach comes from the action of Lie groups on the space of states of the system and the reduction theory, hence we review this theory, the notions of comomentum and momentum maps, and the subsequent *Marsden–Weinstein reduction theorem*. We consider only the case of regular dynamical systems (although some results can be generalised to the singular case).

Throughout this section, (M, Ω, α) will be a regular Hamiltonian dynamical system and $\alpha = dh$, where the Hamiltonian function h is locally or globally defined. Usually, we will write (M, Ω, h). Then, $X_h \in \mathfrak{X}_{lh}(M)$ denotes the dynamical vector field solution to the system.

2.3.2   *Conserved quantities (constants of motion)*

Let (M, Ω, h) be a regular Hamiltonian system and $X_h \in \mathfrak{X}_{lh}(M)$ the dynamical vector field.

The dynamical evolution of an observable, which is represented by a function $f \in C^\infty(M)$, is the variation of this function along the integral curves $c(t) = (q^i(t), y_i(t))$ of the vector field X_h, that is, it is given by

$$\frac{d(f \circ c)(t)}{dt} = ((L(X_h)f) \circ c)(t) = (X_h(f) \circ c)(t),$$

and considering the definition of the Poisson bracket, it can be written as

$$\frac{d(f \circ c)}{dt} = \{f, h\} \circ c,$$

or in a symplectic chart $(U; x^i, y_i)$ of M,

$$L(X_h)f) \mid_U = X_h(f) \mid_U = \frac{\partial h}{\partial y_i} \frac{\partial f}{\partial x^i} - \frac{\partial h}{\partial x^i} \frac{\partial f}{\partial y_i}.$$

Then, we define the following:

Definition 2.16. A function $f \in C^\infty(M)$ is a **conserved quantity** or a **constant of motion** if

$$L(X_h)f = 0;$$

that is, it is invariant by the dynamical vector field.

To say that $f \in C^\infty(M)$ is a conserved quantity of the system (M, Ω, h) means the following: if $p \in M$ and X_h is the Hamiltonian vector field of the system, let $c: (-\epsilon, \epsilon) \subset \mathbb{R} \longrightarrow M$ be the integral curve of X_h with initial condition $c(0) = p$. Then, if f is a conserved quantity, we have $f(c(t)) = f(p)$, for every $t \in (-\epsilon, \epsilon)$, that is, the image of c by f is contained in $S_p = \{q \in M \mid f(q) = f(p)\}$, the level surface of f passing through p.

One of the fundamental properties of the autonomous Hamiltonian dynamical systems is the *conservation of the Hamiltonian*. Remember that for physical systems, the Hamiltonian is the energy of the system. Now, we can state this result in a geometric way:

Proposition 2.14 (Conservation of energy). *Let (M, Ω, h) be a Hamiltonian system. The (local or global) Hamiltonian function h is a conserved quantity.*

Proof. It is straightforward since, as Ω is skew symmetric, we have

$$\mathrm{L}(X_h)h = i(X_h)\mathrm{d}h = \Omega(X_h, X_h) = 0. \qquad \square$$

Remark 2.16. We have seen how the fundamental properties of the symplectic form are essential to describe geometrically physical systems: the *nondegeneracy* allows us to ensure the existence (and uniqueness) of the dynamical Hamiltonian vector field, and hence, to determine the dynamical evolution of the system, and from skew symmetry and the fact to be closed, we obtain the conservation of the energy.

2.3.3 *Dynamical symmetries*

When talking about symmetries, it is usual to refer to the idea that "a symmetry of a dynamical system lets the solutions to the differential equations describing the dynamics of the system be invariant". In this way, we have:

Definition 2.17. A **dynamical symmetry** of the Hamiltonian dynamical system (M, Ω, h) is a diffeomorphism $\Phi \colon M \longrightarrow M$ satisfying

$$\Phi_* X_h = X_h,$$

that is, the dynamical vector field X_h is invariant by Φ.

Observe that if Φ is a dynamical symmetry, then so is Φ^{-1}.

Let $c \colon (-\epsilon, \epsilon) \subset \mathbb{R} \longrightarrow M$ be an integral curve of X_h, that is, $\dot{c} = X_h \circ c$, then $\dot{c} = X_h \circ c = \Phi_* X_h \circ c$. If X_h is invariant by Φ, we have:

$$X_h \circ (\Phi \circ c) = \Phi_* X_h \circ \Phi \circ c = \mathrm{T}\Phi \circ X_h \circ \Phi^{-1} \circ \Phi \circ c = \mathrm{T}\Phi \circ X_h \circ c = \mathrm{T}\Phi \circ \dot{c} = \overline{\Phi \circ c};$$

hence Φ transforms integral curves of X_h into integral curves of X_h. The converse is also true, that is, if Φ transforms integral curves of X_h into integral curves of X_h, then Φ lets the vector field X_h be invariant. This is another equivalent version of a dynamical symmetry.

If the diffeomorphism Φ is locally generated by a vector field, by means of the local group of diffeomorphisms generated by its flux, then the above definition leads to the infinitesimal version of symmetries:

Definition 2.18. An **infinitesimal dynamical symmetry** of a Hamiltonian system (M, Ω, h) is a vector field $Y \in \mathfrak{X}(M)$ such that the local diffeomorphisms generated by its flux are dynamical symmetries of the system, that is,

$$L(Y)X_h = [Y, X_h] = 0. \tag{2.13}$$

Recall that this definition is equivalent to stating that the flux of Y is made up of dynamical symmetries of X_h. Thus, if Φ_t is the flux of Y and c is an integral curve of X_h, the $\Phi_t \circ c$ is also an integral curve of X_h.

Remark 2.17. In the above definition, condition (2.13) is sometimes relaxed by setting

$$[Y, X_h] = gX_h, \ g \in C^\infty(M), \tag{2.14}$$

(and the original condition is recovered taking $g = 0$). This allows us to consider also as symmetries the reparametrisations of the integral curves of the dynamical vector field.

The infinitesimal dynamical symmetries of (M, Ω, h) have a natural structure of real Lie algebra. Clearly, they are closed by real linear combinations and we have the following proposition:

Proposition 2.15. *If $Y_1, Y_2 \in \mathfrak{X}(M)$ are infinitesimal dynamical symmetries, then $[Y_1, Y_2]$ is an infinitesimal dynamical symmetry.*

Proof. Using the Jacobi identity, we have

$$[[Y_1, Y_2], X_h] = [Y_2, [X_h, Y_1]] + [Y_1, [Y_2, X_h]] = 0. \qquad \square$$

A first result relating symmetries with conserved quantities is:

Proposition 2.16.

(1) Let $\Phi \colon M \longrightarrow M$ be a dynamical symmetry of (M, Ω, h). If $f \in C^\infty(M)$ is a constant of motion of the system, then $\Phi^* f$ is also a conserved quantity.

(2) Let $Y \in \mathfrak{X}(M)$ be an infinitesimal dynamical symmetry of (M, Ω, h). If $f \in C^\infty(M)$ is a constant of motion of the system, then $L(Y)f$ is also a conserved quantity.

Proof. (1) Directly, we have

$$L(X_h)(\Phi^* f) = \Phi^*(L(\Phi_* X_h) f) = 0.$$

(2) The same proof holds, taking the flux Φ_t of Y as a dynamical symmetry. Alternatively,

$$
\begin{aligned}
L(X_h) L(Y) f &= i(X_h) d L(Y) f = i(X_h) L(Y) d f \\
&= L(Y) i(X_h) d f - i([Y, X_h]) d f = L(Y) L(X_h) f = 0.
\end{aligned}
$$

$\square$

2.3.4 *Noether symmetries: Noether Theorem*

In the above study of symmetries of the Hamiltonian system $(M, \Omega, \alpha = dh)$, we have not used the specific structure of the system. In fact, the only element we have considered is the dynamical vector field $X_h \in \mathfrak{X}(M)$. No mention has been made of the symplectic structure of M, the Hamiltonian function h, or the Hamiltonian form α. We have studied symmetries of the dynamical vector field independently of its origin.

If we consider the other elements defining the system, we arrive at different types of dynamical symmetries depending on whether they leave invariant the geometric structure, that is, the symplectic form or the dynamical elements (i.e., the Hamiltonian function). For a more complete analysis of all these kinds of symmetries and how they generate conserved quantities, see, for instance, [87, 88, 310, 314, 315].

The following kinds of symmetries have a special relevance as generators of constants of motion:

Definition 2.19. A diffeomorphism $\Phi \in \mathrm{Diff}\,(M)$ is a **Noether symmetry** of the Hamiltonian system (M, Ω, h) if:

(1) Φ is a symplectomorphism in M, that is, $\Phi^* \Omega = \Omega$.
(2) Φ lets the dynamics be invariant, that is, $\Phi^* h = h$.

Remark 2.18. The second condition can also be expressed in the form $\Phi^* h = h + c$ (with $c \in \mathbb{R}$). This means that it transforms a Hamiltonian function into another Hamiltonian function of the same dynamical vector field X_h. More generically, if $\alpha = dh$ is the Hamiltonian 1-form, this is equivalent to stating that $\Phi^* \alpha = \alpha$.

Definition 2.20. A vector field $Y \in \mathfrak{X}(M)$ is an **infinitesimal Noether symmetry** of the system if the local diffeomorphisms generated by the flux of Y are dynamical symmetries of the system, that is,

(1) $\mathrm{L}(Y)\Omega = 0$, that is, $Y \in \mathfrak{X}_{lh}(M)$.
(2) $\mathrm{L}(Y)h = 0$.

Remark 2.19. In the infinitesimal case, as every infinitesimal Noether symmetry $Y \in \mathfrak{X}(M)$ is a locally Hamiltonian vector field, by the first item of the definition, for every $p \in M$ we have, there exists an open set $U_p \ni p$ and $f_Y \in C^\infty(U_p)$ such that $i(Y)\Omega = df_Y$, in U_p. The local Hamiltonian f is unique, up to the sum of constant functions.

A first relevant result is:

Proposition 2.17.

(1) Every Noether symmetry of (M, Ω, h) is a dynamical symmetry.
(2) Every infinitesimal Noether symmetry of (M, Ω, h) is an infinitesimal dynamical symmetry.

Proof. (1) Let $\Phi \in \mathrm{Diff}(M)$ be a Noether symmetry. If $X_h \in \mathfrak{X}(M)$ is a solution to the dynamical equations, then $0 = i(X_h)\Omega - dh$, and as $\Phi^*\Omega = \Omega$ and $\Phi^*h = h$, we have

$$0 = \Phi^*(i(X_h)\Omega - dh) = i(\Phi_*^{-1}X_h)\Phi^*\Omega - \Phi^*dh$$
$$= i(\Phi_*^{-1}X_h)\Omega - dh = i(\Phi_*^{-1}X_h)\Omega - dh;$$

but as the system is regular, the vector field solution is unique, and then $\Phi_*^{-1}X_h = X_h$, and the result holds.
(2) If $Y \in \mathfrak{X}(M)$ is an infinitesimal Noether symmetry, then $Y \in \mathfrak{X}_{lh}(M)$ and $\mathrm{L}(Y)h = 0$, hence

$$i([Y, X_h])\Omega = \mathrm{L}(Y)\,i(X_h)\Omega - i(X_h)\,\mathrm{L}(Y)\Omega = \mathrm{L}(Y)dh = d\,\mathrm{L}(Y)h = 0,$$

but with Ω being nondegenerate, this implies that $[Y, X_h] = 0$, and therefore Y is an infinitesimal dynamical symmetry. $\square$

Clearly, the set of infinitesimal Noether symmetries is a real vector space. Moreover, it is easy to prove that it is real Lie algebra:

Proposition 2.18. *If* $Y_1, Y_2 \in \mathfrak{X}(M)$ *are infinitesimal Noether symmetries, then* $[Y_1, Y_2]$ *is also an infinitesimal Noether symmetry.*

Proof. First, we have $L([Y_1, Y_2])\Omega = 0$, since $[Y_1, Y_2] \in \mathfrak{X}_{lh}(M)$, because $Y_1, Y_2 \in \mathfrak{X}_{lh}(M)$. Furthermore,

$$L([Y_1, Y_2])h = i([Y_1, Y_2])dh = L(Y_1)\, i(Y_2)dh - i(Y_2)\, L(Y_1)dh$$
$$= L(Y_1)\, L(Y_2)h - i(Y_2)d\, L(Y_1)h = 0.$$

$\square$

In addition, we get:

Proposition 2.19. *Let* $Y \in \mathfrak{X}(M)$ *be an infinitesimal Noether symmetry of* (M, Ω, h). *For every* $p \in M$, *there exists an open set* $U_p \ni p$ *such that if* $\vartheta \in \Omega^1(U_p)$ *is a symplectic potential of* Ω *in* U_p, *that is,* $d\vartheta = \Omega$, *then:*

(1) There exists $\zeta_Y \in C^\infty(U_p)$ *verifying that* $L(Y)\vartheta = d\zeta_Y$.

(2) If $f_Y \in C^\infty(U_p)$ *is a local Hamiltonian function of* Y, *then in* U_p, *we have that*

$$f_Y = \zeta_Y - i(Y)\vartheta \quad \text{(up to the sum of a constant function)}. \quad (2.15)$$

Proof. (1) Observe that $L(Y)\vartheta$ is a closed form in U_p since

$$d(L(Y)\vartheta) = L(Y)d\vartheta = L(Y)\Omega = 0.$$

Then, by Poincaré's Lemma, there exists $\zeta_Y \in C^\infty(U_p)$ such that $L(Y)\vartheta = d\zeta_Y$, in an open subset of U_p, and reducing the initial domain if necessary, we can suppose it is in U_p.

(2) If $i(Y)\Omega = df_Y$ in U_p, we obtain

$$d\zeta_Y = L(Y)\vartheta = d\, i(Y)\vartheta + i(Y)d\vartheta = d\, i(Y)\vartheta + i(Y)\Omega = d\{i(Y)\vartheta + f_Y\},$$

and the result holds. $\square$

Finally, the fundamental result related with infinitesimal Noether-type symmetries is the classical *Noether's Theorem*, whose geometric Hamiltonian version is as follows:

Theorem 2.8 (Noether). *If* $Y \in \mathfrak{X}(M)$ *is an infinitesimal Noether symmetry of a Hamiltonian system* (M, Ω, h), *then its Hamiltonian func-*

tion f_Y is a conserved quantity, that is,

$$L(X_h)f_Y = 0.$$

Proof. In fact, as $i(Y)\Omega = \mathrm{d}f_Y$, we have

$$L(X_h)f_Y = i(X_h)\mathrm{d}f_Y = i(X_h)\,i(Y)\Omega = -\,i(Y)\,i(X_h)\Omega$$
$$= -\,i(Y)\mathrm{d}h = -\,L(Y)h = 0.$$

$\square$

Noether's Theorem is very relevant, since it gives a way to associate a constant of motion with every Noether symmetry. The converse statement of Noether's Theorem also holds, and it allows associating a (Noether) symmetry with every conserved quantity:

Theorem 2.9 (Inverse Noether). *For every conserved quantity $f \in C^\infty(M)$, its Hamiltonian vector field $Y_f \in \mathfrak{X}_{lh}(M)$ is an infinitesimal Noether symmetry.*

Proof. As $Y_f \in \mathfrak{X}_{lh}(M)$, then $L(Y_f)\Omega = 0$. Furthermore, as f is a conserved quantity, then $L(X_h)f = 0$ and therefore

$$L(Y_f)h = i(Y_f)\mathrm{d}h = i(Y_f)\,i(X_h)\Omega = -\,i(X_h)\,i(Y_f)\Omega$$
$$= -\,i(X_h)\mathrm{d}f = -\,L(X_h)f = 0.$$

$\square$

In general, for symmetries that are not of Noether type, there is not such as direct way of obtaining constants of motion, except in some particular cases like in the following:

Theorem 2.10. *If $Y \in \mathfrak{X}(M)$ is an infinitesimal dynamical symmetry such that $L(Y)h \neq 0$, then the function $f = L(Y)h$ is a conserved quantity.*

Proof. We have

$$L(X_h)f = L(X_h)\,L(Y)h = L([X_h, Y])h + L(Y)\,L(X_h)h = 0.$$

$\square$

For other kinds of symmetries, the way to obtain conserved quantities is, in general, more complicated (see, for instance, [6,74,104,254], and [310] and the references quoted therein, where a complete classification of the symmetries of Hamiltonian systems is done).

2.3.5 *Actions of Lie groups on symplectic manifolds*

In this section, we analyse the transformations (i.e., diffeomorphisms) of a dynamical system, which are generated by Lie groups. In fact, a family of diffeomorphisms in a differentiable manifold endowed with the operation of composition has the structure of the a group and when the variations of these transformations are considered, the set of these variations has a smooth differentiable structure. In this case, the corresponding group of transformations is a Lie group. (For more information about actions of Lie groups, see, for instance, [1,11,91,108,177,207,249,260,290,327,357] [6]. See also Appendix A.2 for a review of Lie groups.)

Definition 2.21. Let G be a Lie group and M a differentiable manifold. A **left-action** of G on M is a map $\phi\colon G \times M \longrightarrow M$ verifying the following properties:

(i) $\phi(g_1 g_2, \mathrm{p}) = \phi(g_1, \phi(g_2, \mathrm{p}))$; for $g_1, g_2 \in G$ and $\mathrm{p} \in M$.
(ii) If $e \in G$ denotes the neutral element, then $\phi(e, \mathrm{p}) = \mathrm{p}$.

A **right action** is defined in an analogous way, changing the operation law in G by the opposite. In any case, M is said to be a **left G-manifold** (respectively, a **right G-manifold**).

Since, for every $g \in G$ and $p \in M$, the map $\Phi_g\colon \mathrm{p} \mapsto \phi(g, \mathrm{p})$ is a diffeomorphism in M, we have a left-action of G in M is also a homomorphism

$$\Phi : G \longrightarrow \mathrm{Diff}(M)$$
$$g \mapsto \Phi_g \qquad .$$

From now on, we only consider left-actions.

Definition 2.22. Let G be a Lie group, M a differentiable manifold, Φ an action of G on M, and $\mathrm{p} \in M$.

---

[6]We would also like to highlight the work of *J.A. Lázaro-Camí*, who compiled the main results presented in Sections 2.3.5–2.3.7 in an unpublished note.

(1) The **isotropy group** of p (with respect to Φ) is the subgroup of G

$$G_p := \{g \in G \mid \Phi_g(p) = p\}.$$

(2) The **orbit** of p (with respect to Φ) is the set

$$\mathcal{O}_p := \{p' \in M \mid p' = \Phi_g(p), \text{ for every } g \in G\}.$$

(3) The action is **effective** or **faithful** if $\bigcap\limits_{p \in M} G_p = \{e\}$ or equivalently, if $\Phi_g = \mathrm{Id}_M \iff g = e$ (i.e., Φ is injective).

(4) The action is **free** if the following map, defined for a fixed $p \in M$,

$$
\begin{aligned}
G &\longrightarrow M \\
g &\longmapsto \Phi_g(p)
\end{aligned}'
$$

is injective, for every $p \in M$ (this means that there are no invariant points in M under this action).

(5) The action is **transitive** if, for every $p_1, p_2 \in M$, there exists $g \in G$ such that $\Phi_g(p_1) = p_2$, or equivalently, if $\mathcal{O}_p = M$, for every $p \in M$ (i.e., Φ has only one orbit). In this case, M is said to be a **homogeneous G-space**.

(6) The action is **proper** if the anti-image of every compact set by the map $(g, p) \mapsto (\Phi_g(p), p)$, for every $g \in G$ and $p \in M$, is also a compact set.

Definition 2.23. Let M_1, M_2 be differentiable manifolds, G a Lie group and $\Phi_1 \colon G \times M_1 \longrightarrow M_1$, $\Phi_2 \colon G \times M_2 \longrightarrow M_2$ actions of G on M_1 and M_2. A map $F \colon M_1 \longrightarrow M_2$ is **equivariant** with respect to these actions if, for every $g \in G$, the following diagram commutes:

$$
\begin{array}{ccc}
M_1 & \xrightarrow{\Phi_{1_g}} & M_1 \\
F \downarrow & & \downarrow F \\
M_2 & \xrightarrow{\Phi_{2_g}} & M_2
\end{array}
$$

Remember that the *Lie algebra* of a Lie group G is $\mathbf{g} := T_e G$ or equivalently, the set of *left-invariant vector fields* of $\mathfrak{X}(G)$ (i.e., the set of vector fields that are invariant by the action induced on $\mathfrak{X}(G)$ by the left-action Φ of G on G). Then, every action of G on M induces a Lie algebra homomorphism

$$
\begin{aligned}
\xi : \mathbf{g} &\longrightarrow \mathfrak{X}(M) \\
X &\longmapsto \xi_X
\end{aligned}'
$$

defined as follows: the uniparametric subgroup of G generated by X determines another uniparametric subgroup of transformations in M

$$\sigma_t : M \longrightarrow M$$
$$\mathrm{p} \mapsto \Phi_{\alpha(-t)}(\mathrm{p}) := \Phi_{\exp(-tX_e)}(\mathrm{p}) \, ,$$

where $\exp \colon \mathrm{T}_e G \longrightarrow G$ denotes the *exponential map* of the group. Then, the map

$$\sigma : \mathbb{R} \times M \longrightarrow M$$
$$(t, \mathrm{p}) \mapsto \sigma_t(\mathrm{p})$$

is the flux of some vector field $\xi_X \in \mathfrak{X}(M)$, that is, for $\mathrm{p} \in M$ and $t \in \mathbb{R}$, $\sigma(t, \mathrm{p})$ is the integral curve of ξ_X passing through p. This vector field is given by

$$\xi_X(\mathrm{p}) = \frac{\mathrm{d}}{\mathrm{d}t}\sigma(t, \mathrm{p}) = \frac{\mathrm{d}}{\mathrm{d}t}\sigma_t(\mathrm{p}) = \frac{\mathrm{d}}{\mathrm{d}t}\Phi_{\exp(-tX_e)}(\mathrm{p})\Big|_{t=0}.$$

Hence, for every $f \in C^\infty(M)$, we have

$$(\xi_X(f))(\mathrm{p}) = \frac{\mathrm{d}}{\mathrm{d}t}f(\Phi_{\exp(-tX_e)}(\mathrm{p}))\Big|_{t=0}.$$

Definition 2.24. For $X \in \mathbf{g}$, the vector field $\xi_X \in \mathfrak{X}(M)$ is the **infinitesimal generator** or **fundamental vector field** of the action Φ associated with X.

Definition 2.25. Let Φ be an action of G on M.

(1) A k-form $\vartheta \in \Omega^k(M)$ is **G-invariant** by Φ if $\Phi_g^* \vartheta = \vartheta$, for every $g \in G$; or what is the equivalent, if $\mathrm{L}(\xi_Z)\vartheta = 0$, for every $Z \in \mathbf{g}$.
(2) A vector field $Z \in \mathfrak{X}(M)$ is **G-invariant** by Φ if $\Phi_{g*}Z = Z$, for every $g \in G$; or what is the equivalent, if $\mathrm{L}(\xi_X)Z = 0$, for every $X \in \mathbf{g}$.

Proposition 2.20. *Let $\psi \colon G_1 \longrightarrow G_2$ be a Lie group homomorphism. Then the following diagram commutes:*

$$
\begin{array}{ccc}
G_1 & \xrightarrow{\ \psi\ } & G_2 \\
{\scriptstyle \exp}\Big\uparrow & & \Big\uparrow{\scriptstyle \exp} \\
\mathrm{T}_{e_1}G_1 \simeq \mathbf{g}_1 & \xrightarrow{\ \psi_*\ } & \mathrm{T}_{e_2}G_2 \simeq \mathbf{g}_2
\end{array}
$$

that is, $\psi(\exp(tX_{e_1})) = \exp(t(\psi_* X)_{e_2})$.

Proof. It is immediate, observing that $t \mapsto \exp(t(\psi_* X)_{e_2})$ is the only uniparametric subgroup of G_2 whose tangent vector field at $t = 0$ is $(\psi_* X)_{e_2}$. $\qquad\square$

A very relevant type of group actions are the following:

Definition 2.26. Let G be a Lie group, (M, Ω) a symplectic manifold, and $\Phi: G \times M \longrightarrow M$ an action of G on M. We say that Φ is a **symplectic action** of G on M (also that G **acts symplectically** on M by Φ) if Φ_g is a symplectomorphism, for every $g \in G$, that is, $\Phi_g^* \Omega = \Omega$. Then M is said to be a **symplectic G-space**.

According to this definition, if G acts symplectically on M by Φ, then the fundamental vector field ξ_X associated with X by Φ is a locally Hamiltonian vector field, and conversely, if $\xi_X \in \mathfrak{X}_{lh}(M)$, for every $X \in \mathbf{g}$, then Φ is a symplectic action of G on M. Therefore:

Proposition 2.21. *Let $\xi: \mathbf{g} \longrightarrow \mathfrak{X}(M)$ be the map such that $\xi(X) := \xi_X$, for $X \in \mathbf{g}$. Then, Φ is a symplectic action of G on M if, and only if, $\operatorname{Im} \xi \subseteq \mathfrak{X}_{lh}(M)$.*

This means that $\mathrm{L}(\xi_X)\Omega = 0$, for every $X \in \mathbf{g}$ (i.e., $i(\xi_X)\Omega$) is a closed form.

Definition 2.27. Let G be a Lie group, (M, Ω) a symplectic manifold and $\Phi: G \longrightarrow M$ a symplectic action of G on M. We say that Φ is a **strongly symplectic action** of G on M if $\xi_X \in \mathfrak{X}_h(M)$, for every $X \in \mathbf{g}$; or what is the equivalent, $i(\xi_X)\Omega$ is an exact form.

In this case, M is said to be a **strongly symplectic G-space** and Φ is a **Hamiltonian action** of G on M. Otherwise, it is a **locally Hamiltonian action** of G on M.

To discuss the obstruction for a symplectic action to be strongly symplectic, let $\mathbf{g}_h := \{X \in \mathbf{g} \mid \xi(X) = \xi_X \in \mathfrak{X}_H(M)\}$ and consider the following sequences of Lie algebras:

$$
\begin{array}{ccccccccc}
0 & \longrightarrow & \mathbf{g}_h & \longrightarrow & \mathbf{g} & \longrightarrow & \mathbf{g}/\mathbf{g}_h & \longrightarrow & 0 \\
& & \downarrow \xi & & \downarrow \xi & & \downarrow \tilde{\xi} & & \\
0 & \longrightarrow & \mathcal{X}_h(M) & \longrightarrow & \mathcal{X}_{lh}(M) & \longrightarrow & H^1(M) & \longrightarrow & 0
\end{array}
\tag{2.16}
$$

where $H^1(M)$ denotes the first de Rham's cohomology group of M, and $\tilde{\xi}$ is a Lie algebra homomorphism, which makes the diagram commutative.

Then, the image of **g** by ξ is in $\mathcal{X}_h(M)$ (i.e., the action is strongly symplectic) if, and only if, $\tilde{\xi} = 0$.

There are two specially relevant cases for which every symplectic action is strongly symplectic: when M is simply connected (then every closed form is exact, therefore, $\mathcal{X}_{lh}(M) = \mathcal{X}_h(M)$) and if G is semisimple (then $\mathbf{g} = [\mathbf{g}, \mathbf{g}]$ and $[\mathcal{X}_{lh}(M), \mathcal{X}_{lh}(M)] \subset \mathcal{X}_h(M)$).

As a particular case, we have:

Definition 2.28. Let G be a Lie group, (M, Ω) an exact symplectic manifold (i.e., $\Omega = d\Theta$ for some $\Theta \in \Omega^1(M)$), and $\Phi \colon G \longrightarrow M$ an action of G to M. Φ is said to be an **exact action** of G on M if, $\Phi_g^* \Theta = \Theta$, for every $g \in G$.

Of course, every exact action is strongly symplectic.

2.3.6   *Comomentum and momentum maps*

Next, we introduce new geometric elements that are very relevant in the theory of symplectic group actions. First, following [327] we have:

Definition 2.29. Let G be a Lie group that acts symplectically on a symplectic manifold (M, Ω). A **comomentum map** associated with this action is every Lie algebra linear application (if it exists)

$$\begin{aligned} \mathrm{j}^* : \mathbf{g} &\longrightarrow C^\infty(M) \\ X &\mapsto f_X := \mathrm{j}^*(X) \end{aligned}$$

such that the following diagram commutes:

$$\begin{array}{ccccccccc} & & & & & & \mathbf{g} & & \\ & & & \mathrm{j}^* \nearrow & & & \downarrow \xi & & \\ 0 \to \mathbb{R} & \longrightarrow & C^\infty(M) & \xrightarrow[\sharp_\Omega \circ d]{} & \mathcal{X}_{lh}(M) & \longrightarrow & H^1(M) & \to 0 \end{array}$$

or equivalently,

$$i(\xi_X)\Omega = df_X.$$

The obstruction to the existence of comomentum maps is given in the following:

> **Proposition 2.22.** *Let G be a Lie group that acts symplectically on (M, Ω). A comomentum map associated with this action exists if, and only if, the map $\widetilde{\xi}: \mathbf{g}/\mathbf{g}_h \longrightarrow H^1(M)$ in (2.16) reduces to be $\widetilde{\xi} = 0$.*

Proof. By definition, if a comomentum map exists, then $\sharp_\Omega \circ d \circ j^* = \xi$. However, as $\text{Im } \xi \subset \text{Im } \sharp_\Omega$, this implies that $\widetilde{\xi} = 0$.

Conversely, if $\widetilde{\xi} = 0$, then $\text{Im } \xi := \xi(\mathbf{g}) \subset \mathcal{X}_h(M)$, and for all $X \in \mathbf{g}$, there exists $f_X \in C^\infty(M)$ such that $i(\xi_X)\Omega = df_X$. $\qquad\square$

Observe that this is also the obstruction for a symplectic action to be strongly symplectic. Hence:

> **Proposition 2.23.** *Let G be a Lie group that acts symplectically on (M, Ω). A comomentum map associated with this action exists if, and only if, the action is strongly symplectic.*

Remark 2.20. If a comomentum map exists, it is not unique. In fact, if $f_C: \mathbf{g} \longrightarrow \mathbb{R}$ is a continuous linear function, then $f_C \in \mathbf{g}^*$ and then $f_C(X) = ctn.$, for every $X \in \mathbf{g}$. Therefore, if j^* is a comomentum map, so is $j'^* = j^* + f_C$.

Furthermore, if a comomentum map exists, it is not a Lie algebra homomorphism necessarily, since for $X, Y \in \mathbf{g}$, we have

$$\sharp_\Omega \circ d\{f_X, f_Y\} = \sharp_\Omega \, i([\xi_X, \xi_Y])\Omega = -\sharp_\Omega \, i(\xi_{[X,Y]})\Omega = -\sharp_\Omega df_{[X,Y]};$$

therefore $d\{f_X, f_Y\} = -df_{[X,Y]}$ and then $\{f_X, f_Y\} = -f_{[X,Y]} + \sigma(X, Y)$, where $\sigma: \mathbf{g} \times \mathbf{g} \longrightarrow \mathbb{R}$ is a skew symmetric bilinear function whose existence measures the obstruction for the comomentum map to be a Lie algebra homomorphism. This leads us to state:

Definition 2.30. Let G be a Lie group that acts symplectically on (M, Ω). The action is said to be a **Poissonian action** or also a **strongly Hamiltonian action** if

(i)   There exists a comomentum map for this action.
(ii) The comomentum map is a Lie algebra (anti)homomorphism.

Then the triple (M, Ω, j^*) is called a **Hamiltonian G-space**.

As a particular case, we have:

Proposition 2.24. *If (M, Ω) is an exact symplectic manifold and the action of G on M is exact, then it is Poissonian.*

Proof. Let $\Omega = d\Theta$. Defining $f_X := -\Theta(\xi_X)$, then $j^*\colon X \mapsto f_X$ is a comomentum map and

$$f_{[X,Y]} = -\Theta(\xi_{[X,Y]}) = \Theta([\xi_X, \xi_Y]) = i([\xi_X, \xi_Y])\Theta$$
$$= L(\xi_X)\, i(\xi_Y)\Theta = -L(\xi_X)f_Y = -\{f_X, f_Y\}.$$

$\square$

Definition 2.31. [327]. Let G be a Lie group that acts in a strongly symplectic way on a symplectic manifold (M, Ω). A **momentum map** associated with this action is the dual map of a comomentum map, that is, a map

$$J : M \longrightarrow \mathbf{g}^*$$
$$\mathrm{p} \mapsto \mu \quad,$$

such that for every $X \in \mathbf{g}$ and for $\mathrm{p} \in M$,

$$\langle X, J(\mathrm{p}) \rangle := j^*(X)(\mathrm{p}) = f_X(\mathrm{p}).$$

(Observe that map $\mu := J(\mathrm{p})\colon \mathbf{g} \longrightarrow \mathbb{R}$ such that $X \mapsto f_X(\mathrm{p})$, for $X \in \mathbf{g}$, is, in fact, an element of $\mathbf{g}^*$.)

The obstruction to the existence of momentum maps is the same as that for the comomentum maps. In particular, as an immediate corollary of Proposition 2.24, we have:

Proposition 2.25. *If (M, Ω) is an exact symplectic manifold and the action of G on M is exact, then a momentum map exists. It is given by $\langle X, J(\mathrm{p}) \rangle := -\Theta(\xi_X)(\mathrm{p})$, for every $X \in \mathbf{g}$, for $\mathrm{p} \in M$.*

Remark 2.21. As a more general result, if G is a connected Lie group and Φ a strongly symplectic action of G on (M, Ω), then the action is Poissonian if, and only if, the momentum maps associated with this action are Ad^*-equivariant, that is, for every $g \in G$, we have $Ad^* \circ J = \Phi_g \circ J$; where $Ad^*\colon \mathbf{g}^* \longrightarrow \mathbf{g}^*$ is the *coadjoint action* of G (see [327] for the proof).

2.3.7   *Reduction by symmetries*

Finally, we introduce the concept of *group of symmetries* of a Hamiltonian system, its relationship with the momentum map, and give the main insights about the theory of reduction by symmetries.

> **Definition 2.32.** Let G be a Lie group, (M, Ω, h) a Hamiltonian system, and $\Phi \colon G \longrightarrow M$ an action of G on M. We say that G is a **symmetry group** of the Hamiltonian system if
>
> (i) Φ is a symplectic action.
> (ii) For every $g \in G$, we have $\Phi_g^* h = h$.
>
> The diffeomorphism Φ_g, for every $g \in G$, is called a **symmetry** of the Hamiltonian system.

From this definition, it is obvious that G is a symmetry group of a Hamiltonian system (M, Ω, h) if, and only if, $\mathrm{L}(\xi_X)h = 0$, for every $X \in \mathbf{g}$.

In this context, the Noether Theorem is stated as follows:

> **Theorem 2.11 (Noether's Theorem).** *Let G be a symmetry group that acts in a strongly symplectic way on the Hamiltonian system (M, Ω, h). Then, for every $X \in \mathbf{g}$, the functions f_X are constants of motion, that is, $\mathrm{L}(X_h) f_X = 0$.*

Proof. We have $i(X_h)\Omega = \mathrm{d}h$, then

$$\mathrm{L}(X_h) f_X = i(X_h)\mathrm{d}f_X = i(X_h)\, i(\xi_X)\Omega = -\, i(\xi_X)\, i(X_h)\Omega$$
$$= -\, i(\xi_X)\mathrm{d}h = -\, \mathrm{L}(\xi_X)h = 0.$$

$\square$

One of the most important features related to the study of Hamiltonian systems with symmetry is *reduction theory*. Now, we are going to state the main results on this subject (see, for instance, [107, 263, 265–267, 292–294, 358] for deeper explanations and details).

First, we have to introduce the following concept:

> **Definition 2.33.** Let G be a Lie group and a strongly symplectic action of G on (M, Ω). Let J be a momentum map associated with this action. Then, $\mu \in \mathbf{g}^*$ is a **weakly regular value** of J if:

(i) $J^{-1}(\mu)$ is a submanifold of M.

(ii) For every $p \in J^{-1}(\mu)$, $T_p(J^{-1}(\mu)) = \ker T_p J$.

If $T_p J$ is surjective, then μ is said to be a **regular value**. (Of course, every regular value is weakly regular.)

Then we have:

Theorem 2.12 (Marsden–Weinstein). *Let G be a Lie group acting on a symplectic manifold (M, Ω) such that the action is Poissonian, free and proper. Let μ be a weakly regular value of the momentum map associated with this action, and let G_μ be the isotropy group of $J^{-1}(\mu)$ for the action of G on M. Then:*

*(1) The submanifold $J^{-1}(\mu)$ is stable under the action of G_μ and so the quotient $J^{-1}(\mu)/G_\mu$ is well defined (it is called the **orbit space** of $J^{-1}(\mu)$).*

(2) Denote by $\pi \colon J^{-1}(\mu) \longrightarrow J^{-1}(\mu)/G_\mu$ the canonical projection and by $\iota \colon J^{-1}(\mu) \hookrightarrow M$ the embedding. Then, $J^{-1}(\mu)/G_\mu$ is a differentiable manifold endowed with a (unique) symplectic structure $\widehat{\Omega}$ such that $\iota^* \Omega = \pi^* \widehat{\Omega}$.

(3) If G is a symmetry group of the Hamiltonian system (M, Ω, h), then $(J^{-1}(\mu)/G_\mu, \widehat{\Omega}, \widehat{h})$ is a Hamiltonian system, where $\iota^* h = \pi^* \widehat{h}$.

Proof. (1) In fact, if $g \in G_\mu$, for all $p \in J^{-1}(\mu)$, we have

$$\Phi_g(p) = \Phi_g(J^{-1}(\mu)) = J^{-1}(g\mu)) = J^{-1}(\mu),$$

then $J^{-1}(\mu)$ is invariant under the action of G_μ and the quotient is well defined.

(2) We omit the proof of this statement because it is long and complex. It can be found in the abovementioned references.

(3) It is evident, since h is G-invariant (and hence it is G_μ-invariant), therefore it projects onto $J^{-1}(\mu)/G_\mu$. $\square$

2.4 Exercises

Exercise 2.1
Prove the first two statements of Proposition 2.1.

Exercise 2.2
Prove Proposition 2.4.

Exercise 2.3
Taking the condition (2.14) as the definition of infinitesimal dynamical symmetry, prove the properties stated in Propositions 2.15, 2.16(2), and Theorem 2.10.

Exercise 2.4
Let (M, Ω) be a symplectic manifold and X a locally Hamiltonian vector field. Let Λ be a 2-form on M with $\Lambda = \rho\Omega$, and ρ a function. Prove that Λ is invariant by X if, and only if, ρ is invariant by X.

Exercise 2.5
Consider the diffeomorphism ρ between the upper part of the cylinder $S^1 \times \mathbb{R}$ on $\mathbb{R}^2 - \{0\}$ given by $(r, \theta) \mapsto (r\cos\theta, r\sin\theta)$.
(***Comment:*** *Remember that although θ is not globally defined, $d\theta$ is.*)
Prove that $\Lambda = d(r^2/2) \wedge d\theta$ is a symplectic form on $S^1 \times \mathbb{R}$ and that ρ is a canonical transformation from $(S^1 \times \mathbb{R}, \Lambda)$ to $(\mathbb{R}^2, \Omega)$, where $\Omega = dx \wedge dy$. Compare this with the usual expression $dx \wedge dy = r dr \wedge d\theta$.

Exercise 2.6
The Hamiltonian function of a two-dimensional dynamical system is $h = q^1 p_1 - q^2 p_2 - a(q^1)^2 + b(q^2)^2$, where $a, b \in \mathbb{R}$.

(1) Prove that $C_1 = \dfrac{1}{q^1}(p_2 - bq^2)$ and $C_2 = q^1 q^2$ are constants of motion.
(2) Discuss if there are other constants of motion independent of the above ones.
(3) Using the definition, prove explicitly that $\{C_i, C_j\}$ is a constant of motion.

Exercise 2.7
In the phase plane $(\mathbb{R}^2, \Omega)$, with coordinates (q, p) and $\Omega = dq \wedge dp$, determine for which values of the parameters $a, b \in \mathbb{R}$ the following transformations $(q, p) \mapsto (\tilde{q}, \tilde{p})$ are canonical:

(1) $\tilde{q} = q + ap$, $\tilde{p} = p + bq$.
(2) $\tilde{q} = q^a \cos(bp)$, $\tilde{p} = q^a \sin(bp)$.

Study if a rotation on the phase plane,

$$\tilde{q} = q\cos\theta - p\sin\theta, \quad \tilde{p} = q\sin\theta + p\cos\theta,$$

is a canonical transformation.

Exercise 2.8 (Symplectic connections).
(*Comment: Given a Riemannian manifold $(M, \mathbf{g})$, we know that there exists a unique connection ∇ that is symmetric (i.e., torsionless) and Riemannian (i.e., $\nabla\mathbf{g} = 0$): the Levi–Civita connection. The aim of this problem is to analyse the same situation for a symplectic manifold, changing the metric $\mathbf{g}$ by the symplectic form Ω.*

Observe that taking $M = \mathbb{R}^2$ and $\Omega = dx \wedge dy$, then any connection ∇, with Christoffel symbols $\Gamma^1_{12}, \Gamma^1_{21}$ arbitrary and all the others $\Gamma^k_{ij} = 0$, is torsionless and $\nabla(\Omega) = 0$. Hence, we cannot expect unicity (but maybe we can obtain some other results).)

Let M be a differentiable manifold and $\Omega \in \Omega^2(M)$ nondegenerate. Denote by $Con(M)$ the set made of all the connections on M, and by $Con(M, \Omega)$ the subset given by those for which Ω is parallel, that is, $\nabla(\Omega) = 0$ (the elements of $Con(M, \Omega)$ are called Ω-connections). Given $\nabla \in Con(M)$, we write T^{∇} for the torsion of ∇.

(1) Let $\nabla \in Con(M, \Omega)$ and $\nabla' \in Con(M)$. Prove that $\nabla' \in Con(M, \Omega)$ if, and only if, the tensor field $S = \nabla - \nabla' \in \mathfrak{T}^1_2(M)$ satisfies that

$$\Omega(S(X, Y), Z) = \Omega(S(X, Z), Y), \quad X, Y, Z \in \mathfrak{X}(M).$$

(2) Let $\nabla \in Con(M, \Omega)$ and $X, Y, Z \in \mathfrak{X}(M)$. Prove that

$$d\Omega(X, Y, Z) = \Omega(T^{\nabla}(X, Y), Z) + \Omega(T^{\nabla}(X, Z), Y) + \Omega(T^{\nabla}(Z, X), Y).$$

(Observe that from this equation, a necessary condition for $Con(M, \Omega)$ to contain torsionless elements is that the 2-form Ω must be closed, that is, symplectic.)

(3) Let $\nabla' \in Con(M)$ with $T^{\nabla'} = 0$.

(a) If $N \in \mathfrak{T}^1_2(M)$ is defined by

$$\Omega(N(X, Y), Z) = (\nabla'_X \Omega)(Y, Z); \quad X, Y, Z \in \mathfrak{X}(M),$$

prove that N is well defined and

$$\Omega(N(X, Y), Z) = -\Omega(N(X, Z), Y).$$

(b) Prove that, for $X, Y, Z \in \mathfrak{X}(M)$, we have

$$\Omega(N(X, Y), Z) + \Omega(N(Z, X), Y) + \Omega(N(Y, Z), X) = d\Omega(X, Y, Z).$$

(c) Let $S \in \mathcal{T}_2^1(M)$ be defined by

$$S(X,Y) = \frac{1}{3}\left(N(X,Y) + N(Y,X)\right),$$

and $\nabla = \nabla' + S$. Prove that $T^{\nabla} = 0$ and Ω is parallel with respect to ∇.

(*Comment: In this exercise, we have proven the existence of symplectic connections in any paracompact manifold. Observe that we have assumed the existence of a torsionless connection, for example, the Levi–Civita connection of a Riemannian metric. Furthermore, we have studied the structure of the set $Con(M,\Omega)$.*)

Exercise 2.9 (Non-Noether symmetries).

Let (M,Ω,h) be a regular Hamiltonian system, $X_h \in \mathfrak{X}(M)$ the Hamiltonian vector field, and $Y \in \mathfrak{X}(M)$ an infinitesimal symmetry of the system.

(1) Suppose that $L(Y)\Omega \neq c\Omega$ and that $L(Y)h \neq k$; with $c, k \in \mathbb{R}$.

   (a) Prove that there exists a closed 2-form $\mu \in \Omega^2(M)$ and a function $f \in C^\infty(M)$ such that $i(X_h)\mu = \mathrm{d}f$.
   (*Comment: We say that X_h is a bi-Hamiltonian vector field.*)
   (b) Obtain a conserved quantity associated with the symmetry Y (proving that it is so).
   (c) Considering the above results, what happens if $L(Y)\Omega = \Omega$?, and if $L(Y)h = h$? What is the relationship between these two cases?

(2) Suppose that $L(Y)h = 0$ and that $L(Y)\Omega \neq 0$, but $L^2(Y)\Omega = 0$.

   (a) Prove that $\beta = i(Y)\Omega$ is not a closed form, but using β, we can obtain an associated closed 1-form in M.
   (b) Using this last item, obtain a conserved quantity associated with the infinitesimal symmetry Y (proving that it is so).

   (*Comment: We say that Y is a second-order Noether symmetry.*)

(*Solution: See [310].*)

Exercise 2.10 (Isotropic two-dimensional harmonic oscillator).

Consider the manifold $Q = \mathbb{R}^2$, with coordinates (q^1, q^2), and the Hamiltonian function

$$h = \frac{1}{2}\left[(p_1)^2 + (p_2)^2 + K\left((q^1)^2 + (q_2)^2\right)\right] \in C^\infty(T^*Q), \quad K \in \mathbb{R}^+.$$

Let $Y = q^2\frac{\partial}{\partial q^1} + q^1\frac{\partial}{\partial q^2} + p_2\frac{\partial}{\partial p_1} + p_1\frac{\partial}{\partial p_2} \in C^\infty(T^*Q).$

(1) Obtain the Hamiltonian vector field and the Hamilton equations that determine the dynamical trajectories of the system.
(2) Prove that Y is an infinitesimal symmetry of the system.
(3) Study if Y is a Noether symmetry.
(4) Give a conserved quantity associated with Y (proving that it is so).
(5) Obtain a Noether symmetry that generates the same conserved quantity as Y.

(*Solution:* See [74, 254, 310].)

Chapter 3

Symplectic Mechanics (II): Autonomous Lagrangian Dynamical Systems

The aim of this chapter is to state a geometrical formulation for a particular kind of dynamical systems, variational systems, that are described by Lagrangian functions. They can also be described as Hamiltonian dynamical systems, but have their own relevant and particular characteristics. This formulation gives rise to two formalisms: the *Lagrangian formalism* and its associated *canonical Hamiltonian formalism*. These include a wide range of mechanical systems that are very relevant in physics, including the *conservative Newtonian systems*, which are those whose configuration space is endowed with a metric (they are studied in Chapter 5).

There are several alternative ways to state the geometric description for these kinds of systems. The geometric framework of the theory is the tangent bundle of a manifold, TQ, which represents the phase space of the system. A selected function on it, the *Lagrangian* $\mathcal{L}$, is a depository of the physical information of the system. Then, a first formulation consists of using $\mathcal{L}$ to define the so-called *Legendre transformation*, which connects TQ and T^*Q, and if this map is a (local) diffeomorphism, it is used to translate the Hamiltonian description previously stated in T^*Q to TQ (see, for instance, [1]).

The second and more general approach consists of using the canonical structures of the tangent bundle to construct, starting from $\mathcal{L}$, the geometrical objects needed to establish the dynamics of the system. This is the so-called *Klein* or *Cartan formulation* of the Lagrangian formalism, which was stated in the classical references [164, 178, 200, 201, 227] and developed later by many other authors [58, 103, 105, 108, 138, 316]. It is more direct and does not need any previously established Hamiltonian formalism (in fact, the existence of this Hamiltonian formalism is not assured for some singular Lagrangian systems). This is the procedure that we follow in this exposition.

Another elegant alternative was developed in [338, 339], using the concept of *special symplectic manifold* and *Lagrangian submanifold*, and obtaining both the Lagrangian and Hamiltonian formalisms for Lagrangian systems and their equivalence. Finally, another unified Lagrangian–Hamiltonian formalism for these systems is given in [322]. These two approaches have subsequently been generalised and applied in many physical contexts.

The structure of the chapter is as follows: first, we introduce the mathematical framework that is needed to state the Lagrangian formalism and its associated canonical Hamiltonian formalism, that is, the canonical geometric structures of the tangent and the cotangent bundles of a manifold. Next, we develop the Lagrangian formalism for *dynamical Lagrangian systems* and its associated *canonical Hamiltonian formalism*, their equivalence, and the *Hamilton–Jacobi theory* for the Hamiltonian formalism. Furthermore, we describe the *Skinner–Rusk formalism*, which is a nice formulation that unifies both the Lagrangian and the Hamiltonian formalisms. We also discuss the symmetries and conserved quantities in both formalisms, paying special attention to those that are canonical lifts of diffeomorphisms and vector fields (which are called *natural*), and studying the equivalence of Lagrangians in this context. Finally, we study the variational formulation introducing the *Hamilton* and the *Hamilton–Jacobi variational principles* for the Lagrangian and Hamiltonian formalisms, respectively. Some relevant physical systems, the harmonic oscillator and the Kepler problem, are analysed using this geometric treatment.

3.1 Geometric Structures of the Tangent and Cotangent Bundles

(See [103, 105, 108, 138, 164, 178, 200, 201, 227, 316].)

The tangent bundle of a manifold Q, denoted as TQ (whose construction and main characteristics are reviewed in Appendix A.1), is endowed with three canonical geometric structures: the *vertical subbundle*, the *vertical endomorphism*, and the *Liouville vector field*. Furthermore, there are some characteristic vector fields in TQ whose integral curves are obtained as solutions to *second-order differential equations* in Q. Similarly, the *cotangent bundle* of Q, denoted as T^*Q, is endowed with some canonical differential forms. Next, we present and discuss all these topics in detail.

In this section, Q is a differentiable manifold with $\dim Q = n$.

3.1.1 *The vertical subbundle: Vertical lift*

Let us consider the canonical projection $\tau_Q \colon TQ \longrightarrow Q$ and its tangent map $T\tau_Q \colon TTQ \longrightarrow TQ$.

> **Definition 3.1.** Let $(q, u) \in TQ$ and $V_{(q,u)}(\tau_Q) := \ker T_{(q,u)}\tau_Q$. The **vertical subbundle** of TTQ is the vector bundle (of rank n) $V(\tau_Q) \longrightarrow TQ$, where
>
> $$V(\tau_Q) := \bigcup_{(q,u) \in TQ} V_{(q,u)}(\tau_Q).$$
>
> The sections of this bundle $V(\tau_Q) \longrightarrow TQ$ are called **vertical vector fields**, and the set of all these vector fields is denoted by $\mathfrak{X}^{V(\tau_Q)}(TQ)$.

It is straightforward to prove that, in natural coordinates of TQ, the expression of these vector fields is $f^i \dfrac{\partial}{\partial v^i}$, that is, $\mathfrak{X}^{V(\tau_Q)}(TQ)$ is locally generated by the set $\left\{ \dfrac{\partial}{\partial v^i} \right\}$.

Another interpretation of the fibres of $V(\tau_Q)$ is the following: for every $q \in Q$, consider the n-dimensional vector space T_qQ as a differentiable manifold and the natural immersion

$$
\begin{aligned}
j_q \colon T_qQ &\longrightarrow TQ \\
u &\longmapsto (q, u)
\end{aligned}.
$$

Observe that $\tau_Q \circ j_x$ is the constant map equal to x. If $u \in T_qQ$, we have

$$T_u j_q \colon T_u T_q Q \longrightarrow T_{(q,u)} TQ;$$

but as $\tau_Q \circ j_q$ is a constant map, then $T_u(\tau_Q \circ j_q) = T_{(q,u)}\tau_Q \circ T_u j_q = 0$. This is equivalent to stating that $\operatorname{Im} T_u j_q \subseteq \ker T_{(q,u)}\tau_Q = V_{(q,u)}(\tau_Q)$, and as j_q is an immersion,

$$\dim \operatorname{Im} T_u j_q = \dim \left(T_u T_q Q = n = \dim V_{(q,u)}(\tau_Q) \right);$$

hence, $\operatorname{Im} T_u j_q = V_{(q,u)}(\tau_Q)$, since both of them have the same dimension and the first one is a subset of the second one.

As a consequence of the above discussion, $V_{(q,u)}(\tau_Q)$ is identified naturally with $T_u T_q Q$ through the isomorphism induced by $T_u j_q$ onto its image. Furthermore, as $T_q Q$ is a vector space, if $u \in T_q Q$, we have $T_q Q$, which is canonically identified with $T_u(T_qQ)$ by means of the directional derivative [1]. Thus, $V_{(q,u)}(\tau_Q)$ can also be identified with T_qQ. In other words, $V(\tau_Q)$ is the pull-back to TQ of the bundle TQ over Q by means of the map τ_Q, that is:

$$V(\tau_Q) \simeq \tau_Q^*(TQ) \longrightarrow TQ$$
$$\tau_{TQ} \downarrow \qquad\qquad \downarrow \tau_Q$$
$$TQ \xrightarrow{\ \tau_Q\ } Q$$

In this way, we have constructed the isomorphism of vector spaces

$$T_qQ \longrightarrow T_uT_qQ \simeq V_{(q,u)}(\tau_Q) \subset T_{(q,u)}TQ$$
$$v \mapsto D_v(q,u)$$

where, if $f \in C^\infty(TQ)$, then

$$(D_v(q,u))f \equiv D_vf(q,u) = \lim_{t \to 0} \frac{f(q, u + tv) - f(q,u)}{t}.$$

Definition 3.2. The vector $D_v(q,u)$ is called the **vertical lift** of v to the point (q,u), and the map

$$\lambda_q^{(q,u)} : T_qQ \longrightarrow T_{(q,u)}TQ$$
$$v \mapsto D_v(q,u)$$

which implements the above isomorphism is called the **vertical lift**.

In natural coordinates, if $v = \lambda^i \frac{\partial}{\partial q^i}\Big|_{(q,u)}$, then $\lambda_q^{(q,u)}(v) = \lambda^i \frac{\partial}{\partial v^i}\Big|_{(q,u)}$.

The vertical lift is naturally extended to the vector fields of $\mathfrak{X}(Q)$, and the mapping $\lambda^V : \mathfrak{X}(Q) \longrightarrow \mathfrak{X}^{V(\tau_Q)}(TQ)$ is $C^\infty(Q)$-lineal.

In natural coordinates, if $X = f^i \frac{\partial}{\partial q^i}$, then its vertical lift is

$$\lambda^V(X) \equiv X^V = \tau_Q^* f^i \frac{\partial}{\partial v^i}.$$

---

[1]Remember that if F is an n-dimensional real vector space and $u \in F$, the natural identification between F and T_uF is given as follows:

$$F \longrightarrow T_uF$$
$$v \mapsto D_v(u)$$

where $D_v(u)$ denotes the directional derivative with respect to the vector v at the point u, that is, if $f \colon \mathbb{R}^n \simeq F \longrightarrow \mathbb{R}$ is any differentiable function, then

$$(D_v(u))f \equiv D_vf(u) := \lim_{t \to 0} \frac{f(u + tv) - f(u)}{t}.$$

If $x^1, \ldots, x^n$ are coordinates in F and $v = (\lambda^1, \ldots, \lambda^n)$, then $(D_v(u))f = \lambda^i \frac{\partial f}{\partial x^i}\Big|_u$; hence, $D_v(u) = \lambda^i \frac{\partial}{\partial x^i}\Big|_u$ and the identification is straightforward.

3.1.2 The canonical (or vertical) endomorphism

Definition 3.3. Let $(q, u) \in TQ$. The map

$$J_{(q,u)} : \mathrm{T}_{(q,u)}TQ \longrightarrow \mathrm{T}_{(q,u)}TQ$$
$$Y \longmapsto \mathrm{D}_{(\mathrm{T}_{(q,u)}\tau_Q)Y}(q, u) = \lambda_q^{(q,u)}(\mathrm{T}_{(q,u)}\tau_Q(Y))$$

is called the **canonical** or **vertical endomorphism** (at (q, u)).

Observe that $J_{(q,u)}Y$ is the vertical lift of $(\mathrm{T}_{(q,u)}\tau_Q)Y$ to the point (q, u) (i.e., $J_{(q,u)}$ consists of projecting to $\mathrm{T}_q Q$ and to lift vertically).

It is clear that the image of $J_{(q,u)}$ is in $\mathrm{V}_{(q,u)}(\tau_Q)$, and as $\mathrm{T}_{(q,u)}\tau_Q$ is a surjective map, it coincides with $\mathrm{V}_{(q,u)}(\tau_Q) = \ker J_{(q,u)}$. Therefore:

Proposition 3.1. *The canonical endomorphism has the following properties for $(q, u) \in \mathrm{T}_q Q$:*

(1) $\operatorname{Im} J_{(q,u)} = \mathrm{V}_{(q,u)}(\tau_Q) = \ker J_{(q,u)}$.
(2) $(J_{(q,u)})^2 = 0$.

The action of J is extended in a natural way to vector fields

$$J : \mathfrak{X}(TQ) \longrightarrow \mathfrak{X}^{V(\tau_Q)}(TQ) \subset \mathfrak{X}(TQ)$$

and to differential forms:

$$J^* : \Omega^k(TQ) \longrightarrow \Omega^k(TQ)$$
$$\alpha \longmapsto i(J)\alpha \quad,$$

where $J^*\alpha = i(J)\alpha$ is defined as

$$[i(J)\alpha](X_1, \ldots, X_k) := \alpha(J(X_1), \ldots, J(X_i), \ldots, J(X_k)).$$

In particular, if $k = 1$, then we have $i(J)\alpha = \alpha \circ J$, which is a 1-form verifying that $i(X)(i(J)\alpha) = 0$, for every $X \in \mathfrak{X}^{V(\tau_Q)}(TQ)$. These kinds of differential forms are called τ_Q-**semibasic forms**.

Local expressions: If (q^i, v^i) are natural coordinates in TQ, we have

$$J_{(q,u)} \frac{\partial}{\partial q^i}\bigg|_{(q,u)} = \mathrm{D}_{\frac{\partial}{\partial q^i}\big|_q}(q, u) = \frac{\partial}{\partial v^i}\bigg|_{(q,u)}, \quad J_{(q,u)} \frac{\partial}{\partial v^i}\bigg|_{(q,u)} = \mathrm{D}_0(q, u) = 0,$$

then the local expression of $J_{(q,u)}$ is

$$J_{(q,u)} = \mathrm{d}q^i\bigg|_{(q,u)} \otimes \frac{\partial}{\partial v^i}\bigg|_{(q,u)},$$

and by extension,

$$J = \mathrm{d}q^i \otimes \frac{\partial}{\partial v^i}.$$

Observe that J is a tensor field of type $(1, 1)$ in TQ.

3.1.3   *The Liouville vector field*

Definition 3.4. Let $q \in Q$ and $(q, v) \in TQ$. Consider the vertical lift of the vector $v \in T_q Q$ to the point (q, v), that is, $D_v(q, v)$. This operation allows us to construct a vertical vector field $\Delta \in \mathfrak{X}(TQ)$, which is called the **Liouville vector field,** as follows:

$$\Delta \colon TQ \longrightarrow TTQ$$
$$(q, v) \mapsto ((q, v), D_v(q, v))\,.$$

Local expression: In local coordinates (q^i, v^i) in TQ, let $f \colon TQ \longrightarrow \mathbb{R}$ be a function and $(q, v) \in TQ$; then we have

$$\Delta_{(q,v)} f = D_v(q, v) f = \lim_{t \to 0} \frac{f(q, v + tv) - f(q, v)}{t} \; ;$$

therefore,

$$\Delta_{(q,v)} q^i = \lim_{t \to 0} \frac{q^i(q, v + tv) - q^i(q, v)}{t} = \lim_{t \to 0} \frac{q^i(q) - q^i(q)}{t} = 0,$$

$$\Delta_{(q,v)} v^i = \lim_{t \to 0} \frac{v^i(q, v + tv) - v^i(q, v)}{t} = \lim_{t \to 0} \frac{(v + tv)(q^i) - v(q^i)}{t}$$
$$= v(q^i) = v^i(q, v),$$

and hence, the local expression of Δ is

$$\Delta = v^i \frac{\partial}{\partial v^i}.$$

Considering that the fibres $T_q Q$ of the tangent bundle are vector spaces, and that $\Delta_{(q,v)} = v \in T_q Q$ for every $(q, v) \in TQ$, typical then Δ is the vector field that generates the dilatations along the fibres of TQ. Another way to see this interpretation is considering the local expression of Δ and its associated system of differential equations

$$\frac{dq^i}{dt} = 0, \quad \frac{dv^i}{dt} = v^i$$

whose general solution is

$$q^i(t) = A^i, \quad v^i(t) = B^i e^t \quad ; \quad (A^i, B^i \text{ ctns.}).$$

Thus, the flux of Δ is

$$F^\Delta \colon \mathbb{R} \times TQ \longrightarrow TQ$$
$$(t, q^i, v^i) \mapsto (q^i, v^i e^t)\,.$$

Hence, the elements of the uniparametric local group of diffeomorphisms generated by Δ are

$$F_t^\Delta \colon TQ \longrightarrow TQ$$
$$(q^i, v^i) \mapsto (q^i, v^i e^t)\,;$$

that is, they are homothetics on the fibres with positive reason.

3.1.4 Holonomic curves: Second-order differential equations

A curve in TQ is not necessarily the canonical lift of a curve in the base manifold Q [2]. Then we have:

> **Definition 3.5.** A curve $\sigma\colon (a,b) \subseteq \mathbb{R} \longrightarrow TQ$ is **holonomic** if there exists $\gamma\colon (a,b) \subseteq \mathbb{R} \longrightarrow Q$ such that $\sigma = \widetilde{\gamma}$.

In natural coordinates, $\sigma(t) = (q^i(t), v^i(t))$ is a canonical lift if, and only if, $v^i(t) = \dot{q}^i(t)$.

> **Definition 3.6.** Let $\sigma\colon (a,b) \subseteq \mathbb{R} \longrightarrow TQ$ be a curve in TQ. Then $\tau_Q \circ \sigma \colon (a,b) \subseteq \mathbb{R} \longrightarrow Q$ is a curve in Q, which is called the **curve in the base manifold** associated with σ.

Using this, we can characterise holonomic curves as follows:

> **Proposition 3.2.** *A curve* $\sigma\colon (a,b) \subseteq \mathbb{R} \longrightarrow TQ$ *is holonomic if, and only if,* $\widetilde{(\tau_Q \circ \sigma)} = \sigma$.

Proof. ($\Longrightarrow$) If σ is holonomic, then there exists $\gamma\colon (a,b) \subseteq \mathbb{R} \longrightarrow Q$ such that $\sigma = \widetilde{\gamma}$; therefore,

$$\tau_Q \circ \sigma = \tau_Q \circ \widetilde{\gamma} = \gamma \quad \Longrightarrow \quad \widetilde{(\tau_Q \circ \sigma)} = \widetilde{\gamma} = \sigma.$$

($\Longleftarrow$) The converse is apparent. $\qquad\square$

Finally, we have the following fundamental result:

> **Proposition 3.3.**
>
> *(1) Let* $\varphi\colon Q \longrightarrow Q$ *be a diffeomorphism and* $T\varphi\colon TQ \longrightarrow TQ$ *be its canonical lift to* TQ. *Then,*
>
> $$(T\varphi)^* J = J, \quad (T\varphi)_* \Delta = \Delta.$$
>
> *(2) Let* $Z \in \mathfrak{X}(Q)$ *and* $Z^C \in \mathfrak{X}(TQ)$ *its canonical lift to* TQ. *Then the canonical endomorphism* J *and the Liouville vector field* Δ *are*

---

[2]See Appendix A.1.1.

> *invariant by the uniparametric group of local diffeomorphisms generated by Z^C.*

Proof. (1) The result for J is a straightforward consequence of the local expressions of J and $T\varphi$.

The result for Δ is a straightforward consequence of the property $T\varphi \circ F_t^{\Delta} = F_t^{\Delta} \circ T\varphi$, where F_t^{Δ} is an element uniparametric group of local diffeomorphisms generated by Δ.

(2) It is apparent from the above result, taking the uniparametric group of local diffeomorphisms generated by Z and Z^C. □

This means that the canonical lifts of diffeomorphisms and vector fields to the tangent bundle preserve the canonical structures of this bundle.

> **Definition 3.7.** A vector field $X \in \mathfrak{X}(TQ)$ is a **second-order differential equation** (SODE) and a **holonomic vector field** if its integral curves are holonomic. (It is also said that X satisfies the **second-order condition**.)

The local interpretation of this definition is as follows:

> **Proposition 3.4.** *The necessary and sufficient condition for a vector field $X \in \mathfrak{X}(TQ)$ to be a SODE is that its local expression in any natural system of coordinates in TQ is*
>
> $$X = v^i \frac{\partial}{\partial q^i} + g^i \frac{\partial}{\partial v^i}, \tag{3.1}$$
>
> *that is, $X_{(q,v)} = (v^i, g^i(q, v))$, for every $(q, v) \in TQ$.*

Proof. ($\Longrightarrow$) In natural coordinates, the general expression of a vector field $X \in \mathfrak{X}(TQ)$ is $X = f^i \frac{\partial}{\partial q^i} + g^i \frac{\partial}{\partial v^i}$. Let $\sigma \colon (a, b) \subset \mathbb{R} \longrightarrow TQ$ be an integral curve of X, with $\sigma(t) = (q^i(t), v^i(t))$; then

$$\frac{dq^i}{dt} = (f^i \circ \sigma)(t) = f^i(q^j(t), v^j(t)), \quad \frac{dv^i}{dt} = (g^i \circ \sigma)(t) = g^i(q^j(t), v^j(t)). \tag{3.2}$$

Furthermore, if X is a SODE, by definition, there exists $\gamma \colon (a, b) \subset \mathbb{R} \longrightarrow Q$, with $\gamma(t) = (q^i(t))$, such that $\sigma = \widetilde{\gamma}$, that is, $\sigma(t) = (q^i(t), \dot{q}^i(t))$. Therefore, we have $\dot{q}^i(t) = v^i(t)$, for $i = 1, \ldots, n$; hence, going to the first group of equations (3.2), we obtain

$$v^i(t) = \frac{dq^i}{dt} = f^i(q^j(t), v^j(t)), \quad t \in (a, b),$$

and the result follows.

($\Longleftarrow$) If $X = v^i \dfrac{\partial}{\partial q^i} + g^i \dfrac{\partial}{\partial v^i}$, then its integral curves $\sigma\colon (a, b) \subset \mathbb{R} \longrightarrow TQ$, with $\sigma(t) = (q^i(t), v^i(t))$, are determined by the system

$$\frac{dq^i}{dt} = (v^i \circ \sigma)(t) = v^i(t), \quad \frac{dv^i}{dt} = (g^i \circ \sigma)(t) = g^i(q^j(t), v^j(t)),$$

that is, $\sigma(t) = (q^i(t), \dot{q}^i(t))$. Therefore, they are holonomic curves, and X is a SODE. $\qquad\square$

> **Remark 3.1.** Observe that with the above conditions, the second group of equations (3.2) is written as
>
> $$\frac{d^2 q^i}{dt^2} = f^i(q^j, \dot{q}^j),$$
>
> which is a system of second-order ordinary differential equations, whose solution completely determines the integral curves of X. This fact justifies the name SODE for this kind of vector fields.

From this result, we obtain the following intrinsic characterisation:

> **Proposition 3.5.** *The necessary and sufficient condition for a vector field $X \in \mathfrak{X}(TQ)$ to be a SODE is that*
>
> $$J(X) = \Delta.$$

Proof. In natural coordinates, the general expression of $X \in \mathfrak{X}(TQ)$ is $X = f^i \dfrac{\partial}{\partial q^i} + g^i \dfrac{\partial}{\partial v^i}$, so

$$J(X) = f^i \frac{\partial}{\partial v^i} = v^i \frac{\partial}{\partial v^i} \equiv \Delta \quad \Longleftrightarrow \quad f^i = v^i,$$

and the result follows from the above proposition. $\qquad\square$

There is another intrinsic characterisation as follows: the bundle TTQ has two natural projections

$$TTQ \overset{T\tau_Q}{\longrightarrow} TQ$$
$$\tau_{TQ} \downarrow \qquad \downarrow \tau_Q$$
$$TQ \overset{\tau_Q}{\longrightarrow} Q$$

Taking natural coordinates in the bundles TQ and TTQ and recalling the expression (A.1), then for every $((q, v), Y_{(q,v)}) \equiv (q^i, v^i; u^i, w^i) \in TTQ$, we have

$$\tau_{TQ}((q, v), Y_{(q,v)}) = \tau_{TQ}(q^i, v^i; u^i, w^i) = (q^i, v^i),$$
$$T\tau_Q((q, v), Y_{(q,v)}) = \tau_{TQ}(q^i, v^i; u^i, w^i) = (\tau_Q(q^i, v^i), T_{(q^i, v^i)}\tau_Q(u^i, w^i))$$
$$= (q^i, u^i).$$

By definition, $X \in \mathfrak{X}(TQ)$ is a section of the projection τ_{TQ}, that is, a map $X \colon TQ \longrightarrow TTQ$ such that $\tau_{TQ} \circ X = \mathrm{Id}_{TQ}$. Then:

Proposition 3.6. *The necessary and sufficient condition for a vector field $X \in \mathfrak{X}(TQ)$ to be a SODE is that X is a section of the projection $T\tau_Q$, that is, a map $X \colon TQ \longrightarrow TTQ$ such that*

$$T\tau_Q \circ X = \mathrm{Id}_{TQ}.$$

Proof. In natural coordinates, the general expression of $X \in \mathfrak{X}(TQ)$ is $X = f^i \dfrac{\partial}{\partial q^i} + g^i \dfrac{\partial}{\partial v^i}$, that is, for every $(q, v) \equiv (q^i, v^i) \in TQ$, it is a map

$$X(q, v) = X(q^i, v^i) = (q^i, v^i; f^i(q, v), g^i(q, v)).$$

Therefore,

$$(T\tau_Q \circ X)(q, v) = (T\tau_Q \circ X)(q^i, v^i)$$
$$= T\tau_Q(q^i, v^i; f^i(q, v), g^i(q, v)) = (q^i, f^i(q, v)),$$

and as $\mathrm{Id}_{TQ}(q, v) = (q^i, v^i)$, we have

$$(T\tau_Q \circ X)(q, v) = \mathrm{Id}_{TQ}(q, v) \quad \Longleftrightarrow \quad f^i = v^i,$$

and the result follows. $\qquad\qquad\qquad\qquad\qquad\qquad\qquad\qquad\qquad\quad\square$

Remark 3.2. In summary, we have proven that the following assertions are equivalent:

(1) A vector field $X \in \mathfrak{X}(TQ)$ is a SODE.
(2) The integral curves of X are canonical lifts of curves in Q.
(3) The expression of X in a system of natural coordinates in TQ is (3.1).
(4) $J(X) = \Delta$.
(5) $T\tau_Q \circ X = \mathrm{Id}_{TQ}$.

3.1.5 *Canonical forms in the cotangent bundle*

An essential characteristic of the cotangent bundle is the following:

Theorem 3.1. *The cotangent bundle* T^*Q *is endowed with a canonical 2-form, which is closed and nondegenerate, and hence, it is a symplectic manifold.*

Proof. As we know, $T^*Q := \{p = (q, \xi) \mid q \in Q, \xi \in T_q^*Q\}$. As T^*Q is a differentiable manifold, we can consider its tangent bundle TT^*Q and the tangent structure induced by $\pi_Q : T^*Q \longrightarrow Q$, which must be understood as a fibre map $T_{(q,\xi)}\pi_Q : T_{(q,\xi)}T^*Q \longrightarrow T_qQ$. Now, we prove the theorem constructively using the fibre projection.

(1) We construct the differential 1-form $\Theta \in \Omega^1(T^*Q)$ as follows: for every $p = (q, \xi) \in T^*Q$ and $X_p \in T_pT^*Q$,

$$\Theta_p(X_p) := \xi(T_p\pi_Q(X_p)).$$

Its expression in a natural chart of coordinates of T^*Q is obtained in the following way: as we have seen in Proposition A.4 of Appendix A.1.2, $\xi = p_i(p)\, dq^i \mid_q$; furthermore, every $X_p \in T_p(T^*Q)$ has the local expression $X_p = A^j \dfrac{\partial}{\partial q^j}\Big|_p + B_j \dfrac{\partial}{\partial p_j}\Big|_p$, and the general expression of Θ_p is $\Theta_p = a_i\, dq^i \mid_p + b^i\, dp_i \mid_p$, then

$$\Theta_p(X_p) = a_iA^i + b^iB_i \,,$$

but, from the definition,

$$\Theta_p(X_p) := \xi(T_p\pi_Q(X_p)) = (p_i(m)dq^i \mid_q)\left[T_p\pi_Q\left(A^j \frac{\partial}{\partial q^j}\Big|_p + B_j \frac{\partial}{\partial p_j}\Big|_p\right)\right]$$

$$= (p_i(p)dq^i \mid_x)\left(A^j \frac{\partial}{\partial q^j}\Big|_q\right) = p_i(p)A^j\delta_j^i = p_i(p)A^i,$$

and as this holds for every X_p (i.e., for every A^i, B_i), from the above expressions, we obtain $a_i = p_i(m)$ and $b^i = 0$, that is, the final local expression of Θ is [3]

$$\Theta = p_i\, dq^i.$$

---

[3]The 1-form Θ is also called the *tautological form* of T^*Q because at every point $p = (q, \xi) \in T^*Q$, it assigns the form ξ.

(2) We define the differential 2-form

$$\Omega := -d\Theta,$$

whose local expression in a natural chart is

$$\Omega = dq^i \wedge dp_i.$$

Clearly, Ω is closed (because it is exact) and nondegenerate (as we can see from its local expression in coordinates). $\qquad\square$

Then, cotangent bundles of manifolds are the canonical models of symplectic manifolds, since they carry a canonical symplectic form. Furthermore, their natural coordinates are also the Darboux coordinates for this canonical symplectic form.

Definition 3.8. The forms $\Theta \in \Omega^1(T^*Q)$ and $\Omega \in \Omega^2(T^*Q)$ are the 1 and 2 **canonical forms** of the cotangent bundle.

From Proposition 2.3, we have:

Proposition 3.7. *The cotangent bundle of a differentiable manifold is an oriented manifold.*

The canonical lift of a vector field in Q to the cotangent bundle T^*Q [4] can be characterised using the canonical forms. First, we have:

Definition 3.9. Every vector field $Z \in \mathfrak{X}(Q)$ induces a function $F_Z \in C^\infty(T^*Q)$ defined as

$$F_Z : \quad T^*Q \quad \longrightarrow \quad \mathbb{R}$$
$$p \equiv (q, \xi) \quad \mapsto \quad \xi(Z_q).$$

Considering the definition of the canonical 1-form Θ of T^*Q and Proposition A.6, we have, for every $p \equiv (q, \xi) \in T^*Q$,

$$F_Z(p) \equiv F_Z(q, \xi) = \xi(Z_q) = \xi(T\pi_Q(Z_p^*)) = \Theta_p(Z_p^*),$$

and thus, we have proven that:

Proposition 3.8. *If $Z \in \mathfrak{X}(Q)$, then $F_Z = \Theta(Z^*)$.*

---

[4]See Appendix A.1.2.

Bearing this in mind, we obtain:

Proposition 3.9. *If $Z \in \mathfrak{X}(Q)$, then the canonical lift of Z to T^*Q is the only vector field $Z^* \in \mathfrak{X}(T^*Q)$ such that $i(Z^*)\Omega = dF_Z$.*

Proof. In fact, using the Cartan formula and considering Proposition 3.10,

$$i(Z^*)\Omega = -i(Z^*)d\Theta = d\,i(Z^*)\Theta - L(Z^*)\Theta = d(\Theta(Z^*)) = dF_Z,$$

since, according to Proposition 3.10, as the local uniparametric groups of diffeomorphisms of Z^* are canonical lifts, they let the canonical forms of T^*Q become invariant, and then $L(Z^*)\Theta = 0$ and $L(Z^*)\Omega = 0$ [5]. $\square$

Local expression: Considering this characterisation, it is simple to obtain the local expression of Z^* in a chart of canonical coordinates $(U; q^i, p_i)$ of T^*Q. So, if $Z|_{\pi_Q(U)} = f^i(q)\dfrac{\partial}{\partial q^i}$, we have $F_Z(q^i, p_i) = p_i f^i(q^j)$, and then

$$Z^*|_U = f^i \frac{\partial}{\partial q^i} - p_j \frac{\partial f^j}{\partial q^i} \frac{\partial}{\partial p_i}. \tag{3.3}$$

Another relevant property is:

Proposition 3.10. *If $\varphi: Q \longrightarrow Q$ is a diffeomorphism, then $(T^*\varphi)^*\Theta = \Theta$, and as a consequence, $(T^*\varphi)^*\Omega = \Omega$. (With the terminology introduced in Section 2.1.8, we say that $T^*\varphi$ is a simplectomorphism.)*

Proof. For every $(q, \xi) \in T^*Q$ and $V \in T_{(\varphi(q),\xi)}(T^*Q)$, using the definition of Θ and the first property of the above proposition, we obtain

$$\big((T^*\varphi)^*\Theta\big)_{(\varphi(q),\xi)}(V) = \Theta_{(q, T^*\varphi(\xi))}\big(T_{(\varphi(q),\xi)}(T^*\varphi)(V)\big)$$

$$= (\varphi_q^*\xi)\Big(T_{(q,T^*\varphi(\xi))}\pi_Q\big(T_{(\varphi(q),\xi)}(T^*\varphi)(V)\big)\Big)$$

$$= (\varphi_q^*\xi)\big(T_{(\varphi(q),\xi)}(\pi_Q \circ T^*\varphi)(V)\big)$$

$$= \xi\Big(T_{\varphi(q)}\varphi\big(T(\pi_Q \circ T^*\varphi)(V)\big)\Big)$$

$$= \xi\Big(T_{(\varphi(q),\xi)}\big(\varphi \circ \pi_Q \circ T^*\varphi)(V)\big)\Big)$$

$$= \xi\big(T_{(\varphi(q),\xi)}\pi_Q(V)\big) = \Theta_{(\varphi(q),\xi)}(V).$$

The result for Ω is straightforward from here. $\square$

---

[5] With the terminology of Section 2.1.5, Z^* is the (global) Hamiltonian vector field associated with the function F_Z.

Finally, from Proposition 3.10, we obtain:

Proposition 3.11. *If $Z \in \mathfrak{X}(Q)$, then*

$$\mathrm{L}(Z^*)\Theta = 0, \quad \mathrm{L}(Z^*)\Omega = 0.$$

Proof. To prove this, it is sufficient to take the local uniparametric groups of diffeomorphisms of Z and their canonical lifts. $\qquad\square$

3.1.6 *Fibre derivative of a function*

Let $F \in C^\infty(TQ)$. Given $q \in Q$, we consider the function $F_q : T_qQ \longrightarrow \mathbb{R}$, which is the restriction of F to the fibre T_qQ. If $(q,u) \in TQ$, then the differential of F_q at this point is an element of T_q^*Q, so $\mathrm{D}_{(q,u)}F_q \in T_q^*Q$. Then:

Definition 3.10. The **fibre derivative** of F is the map

$$
\begin{aligned}
\mathcal{F}F : TQ &\longrightarrow \quad T^*Q \\
(q,u) &\longmapsto (q, \mathrm{D}F_q(u))
\end{aligned}.
$$

Observe that $\pi_Q \circ \mathcal{F}F = \tau_Q$, that is, $\mathcal{F}F$ preserves the fibres.

Local expressions: Consider a natural chart $(U; q^i, v^i)$ of TQ and the corresponding canonical chart $(\mathcal{F}F(U); q^i, p_i)$ in T^*Q. Observe that a basis of T_qQ is $\left\{ \left.\dfrac{\partial}{\partial q^i}\right|_q \right\}$, and the coordinates in T_qQ are $(v^1, \dots, v^n)$. Then, the Jacobian matrix of $\mathrm{D}_{(q,u)}F_q$ is

$$
\left(\left.\frac{\partial F}{\partial v^1}\right|_{(q,u)} \cdots \left.\frac{\partial F}{\partial v^n}\right|_{(q,u)} \right),
$$

and if $v = \lambda^i \left.\dfrac{\partial}{\partial q^i}\right|_q \in T_qQ$, then

$$
(\mathrm{D}_{(q,u)}F_q)(v) = \left(\left.\frac{\partial F}{\partial v^1}\right|_{(q,u)} \cdots \left.\frac{\partial F}{\partial v^n}\right|_{(q,u)} \right) \begin{pmatrix} \lambda^1 \\ \vdots \\ \lambda^n \end{pmatrix},
$$

and therefore

$$
\mathcal{F}F(x,u) = \left(q, \left.\frac{\partial F}{\partial v^i}\right|_{(q,u)} \left.dq^i\right|_q \right),
$$

that is, the expression in coordinates of $\mathcal{F}F$ is

$$
q^i \circ \mathcal{F}F = q^i, \quad p_i \circ \mathcal{F}F = \frac{\partial F}{\partial v^i}.
$$

3.2 Lagrangian Formalism for Lagrangian Dynamical Systems

Next, we introduce the so-called *Lagrangian formalism* of the Lagrangian dynamical systems (see [58, 103, 105, 108, 138, 164, 178, 200, 201, 227, 316] as general references).

First, we state what these kinds of systems are.

3.2.1 *Lagrangian dynamical systems*

The physical and geometrical foundations of this formulation are the following:

> **Postulate 3.1 (First postulate of Lagrangian mechanics).** In a physical Lagrangian system, the n possible degrees of freedom of the system are described by the domain of variation of a set of n *generalised coordinates*, which determine locally the **configuration space** of the system. The *states* of the system are locally described using generalised coordinates and their corresponding generalised velocities.
>
> Geometrically, this means that the configuration space Q of a dynamical system with n degrees of freedom is an n-dimensional differentiable manifold, and the phase space is the tangent bundle TQ of the configuration manifold Q, which is called the **state space** or **phase space** of **positions-velocities** of the system.

If we have a mechanical system, the generalised coordinates correspond to "kinematic" degrees of freedom (position coordinates). However, if we have a more generic physical system, such as an electric or a thermodynamical system, then the generalised coordinates correspond to other kinds of physical magnitudes (for instance, electric charge and temperature).

> **Postulate 3.2 (Second postulate of Lagrangian mechanics).** The observables or physical magnitudes of a dynamical system are functions of $C^\infty(TQ)$.
>
> The result of a measure of an observable is the value that the function which represents the observable takes at a point of the phase space TQ (i.e., in some states, following the first postulate).

Postulate 3.3 (Third postulate of Lagrangian mechanics). There is a function $\mathcal{L} \in C^\infty(TQ)$, called the **Lagrangian function**, which carries the dynamical information of the system.

As we will see, from this function and using the geometrical structures of the tangent bundle, we can construct a differential form $\Omega_{\mathcal{L}} \in \Omega^2(TQ)$ and the *Lagrangian energy function* $E_{\mathcal{L}} \in C^\infty(TQ)$. We can also set the dynamical equations of the system (Postulate 3.4).

Taking all of this into account, we have:

Definition 3.11. A **Lagrangian dynamical system** is a pair $(TQ, \mathcal{L})$, where Q is a manifold representing the configuration space of a physical system and $\mathcal{L} \in C^\infty(TQ)$ is the Lagrangian function of the system.

Remark 3.3. It is important to point out that the dynamical systems described in this way are *autonomous*, that is, *independent of time* (a geometrical description of *nonautonomous dynamical systems* is presented in Chapter 4). They are also *first-order* systems, which are those whose Lagrangian function depends locally on the coordinates of position and velocity, in contrast with those for which the dependence is also on the *generalised accelerations* or, in general, higher order time derivatives of the generalised positions (for the geometrical description of *higher order* systems, see [42, 53, 71, 139, 167, 197, 198, 231, 296–298]).

In the next section, we explain how, using the canonical structures of the tangent bundle introduced in Section 3.1 and starting from the Lagrangian function, we can construct the dynamical-geometric structures that allow us to set the dynamical equations intrinsically. In this way, we will complete the set of postulates of the Lagrangian formalism.

3.2.2 *Geometric structures induced by the dynamics*

Given a Lagrangian function $\mathcal{L} \in C^\infty(TQ)$ and using the canonical structures of the tangent bundle (the vertical endomorphism $J \colon \mathfrak{X}(TQ) \longrightarrow \mathfrak{X}^{V(\tau_Q)}(TQ)$ and the Liouville vector field $\Delta \in \mathfrak{X}^{V(\tau_Q)}(TQ)$), we can define the following elements:

Definition 3.12. The **Cartan** or **Lagrangian** 1 and **2-forms** associated with $\mathcal{L}$ are

$$\Theta_{\mathcal{L}} := i(J)\mathrm{d}\mathcal{L} = \mathrm{d}\mathcal{L} \circ J \in \Omega^1(TQ),$$
$$\Omega_{\mathcal{L}} := -\mathrm{d}\Theta_{\mathcal{L}} \in \Omega^2(TQ).$$

The **Lagrangian energy** associated with $\mathcal{L}$ is the function

$$E_{\mathcal{L}} := \Delta(\mathcal{L}) - \mathcal{L} \in C^\infty(TQ).$$

The function $A_{\mathcal{L}} := \Delta(\mathcal{L}) \in C^\infty(TQ)$ is sometimes referred to as the **Lagrangian action function** associated with $\mathcal{L}$.

Remark 3.4.

(1) The physical observable represented by the Lagrangian energy is the total energy of the system, and this justifies this terminology.

(2) There are Lagrangian functions that, being different, give the same Lagrangian form $\Omega_{\mathcal{L}}$ and the same Lagrangian energy $E_{\mathcal{L}}$. They are called *gauge-equivalent Lagrangians* (and they are studied in Section 3.5.3).

(3) It is important to point out that for an arbitrary function $\mathcal{L} \in C^\infty(TQ)$, the form $\Omega_{\mathcal{L}}$ does not have a constant rank in TQ necessarily. The Lagrangian $\mathcal{L}$ is said to be **geometrically admissible** when this rank is constant. The theory that we are developing here concerns Lagrangian systems of this kind. When this is not the case, there exists a more general framework of *Poisson manifolds* as the phase states for the formalism (see, for instance, [249]).

Local expressions: Consider a natural chart $(U; q^i, v^i)$ in TQ. Remember that the local expression of the vertical endomorphism and the Liouville vector field on this chart are $J = \mathrm{d}q^i \otimes \dfrac{\partial}{\partial v^i}$ and $\Delta = v^i \dfrac{\partial}{\partial v^i}$. Then, given a Lagrangian function $\mathcal{L} = \mathcal{L}(q^i, v^i)$, the local expressions of the action and the Lagrangian energy associated with the Lagrangian $\mathcal{L}$ are

$$A_{\mathcal{L}} = v^i \frac{\partial \mathcal{L}}{\partial v^i}, \quad E_{\mathcal{L}} = v^i \frac{\partial \mathcal{L}}{\partial v^i} - \mathcal{L}.$$

The more general expression for a 1-form in TQ is

$$\Theta_{\mathcal{L}} = a_i(q^j, v^j)\mathrm{d}q^i + b_i(q^j, v^j)\mathrm{d}v^i,$$

then, for an arbitrary $Y \in \mathfrak{X}(TQ)$,

$$Y = A^i(q^j, v^j)\frac{\partial}{\partial q^i} + B^i(q^j, v^j)\frac{\partial}{\partial v^i},$$

so we have

$$\Theta_{\mathcal{L}}(Y) = a_i A^i + b_i B^i,$$

and furthermore, by definition,

$$\Theta_{\mathcal{L}}(Y) := (d\mathcal{L} \circ J)(Y) = \left(\frac{\partial \mathcal{L}}{\partial q^i} dq^i + \frac{\partial \mathcal{L}}{\partial v^i} dv^i \right) \left(A^i \frac{\partial}{\partial v^i} \right) = A^i \frac{\partial \mathcal{L}}{\partial v^i}.$$

Therefore, from both expressions, we obtain $a_i = \dfrac{\partial \mathcal{L}}{\partial v^i}$ and $b_i = 0$, that is,

$$\Theta_{\mathcal{L}} = \frac{\partial \mathcal{L}}{\partial v^i} dq^i,$$

and as a consequence,

$$\Omega_{\mathcal{L}} = \frac{\partial^2 \mathcal{L}}{\partial q^j \partial v^i} dq^i \wedge dq^j + \frac{\partial^2 \mathcal{L}}{\partial v^j \partial v^i} dq^i \wedge dv^j.$$

It is important to note that for an arbitrary function $\mathcal{L} \in C^\infty(TQ)$, although the Lagrangian 2-form $\Omega_{\mathcal{L}}$ is always closed, the function is not necessarily nondegenerate (since, as it is seen from the above local expression, its rank is determined by the second derivatives with respect to the velocities of the function $\mathcal{L}$). Then:

Definition 3.13. $\mathcal{L} \in C^\infty(TQ)$ is a **regular Lagrangian function** (and $(TQ, \mathcal{L})$ is a **regular Lagrangian system**) if the Lagrangian 2-form $\Omega_{\mathcal{L}}$ is nondegenerate. Otherwise, $\mathcal{L}$ is a **singular Lagrangian function** (and $(TQ, \mathcal{L})$ is said to be a **singular Lagrangian system**).

In this exposition, we are mainly interested in regular Lagrangian systems.

The nondegeneracy of the Lagrangian 2-form can be characterised locally as follows:

Proposition 3.12. *Let $(TQ, \mathcal{L})$ be a Lagrangian system. Then $\mathcal{L}$ is a regular Lagrangian if, and only if, in a natural chart $(U; q^i, v^i)$ of TQ, the partial Hessian matrix $W = \left(\dfrac{\partial^2 \mathcal{L}}{\partial v^j \partial v^i} \right)$ is regular everywhere on U.*

Proof. We have to prove that $\Omega_{\mathcal{L}}$ is nondegenerate if, and only if, this condition holds. Then, is it sufficient to consider that $\Omega_{\mathcal{L}} \in \Omega^2(TQ)$ is nondegenerate if, and only if, $(\Omega_{\mathcal{L}})^n$ is a volume form in TQ. Therefore,

$$\Omega_{\mathcal{L}}^n = n! \det \left(\frac{\partial^2 \mathcal{L}}{\partial v^j \partial v^i} \right)^n dq^1 \wedge \ldots dq^n \wedge dv^1 \wedge \ldots dv^n,$$

which is a non-vanishing form at every point when $\det\left(\dfrac{\partial^2 \mathcal{L}}{\partial v^j \partial v^i}\right) \neq 0$ everywhere. $\qquad\qquad\qquad\qquad\qquad\qquad\qquad\qquad\qquad\qquad\qquad\quad\square$

3.2.3 Lagrangian dynamical equations: Euler–Lagrange equations

The dynamical equation in the Lagrangian formalism is stated in the following:

> **Postulate 3.4 (Fourth postulate of Lagrangian mechanics).** The dynamical trajectories of a Lagrangian system $(TQ, \mathcal{L})$ are the integral curves of a vector field $X_\mathcal{L} \in \mathfrak{X}(TQ)$ such that:
>
> (1) $X_\mathcal{L}$ is a solution to the equation
>
> $$i(X_\mathcal{L})\Omega_\mathcal{L} = \mathrm{d}E_\mathcal{L}. \tag{3.4}$$
>
> (2) $X_\mathcal{L}$ is a SODE, that is,
>
> $$J(X_\mathcal{L}) = \Delta. \tag{3.5}$$

Equation (3.4) is called the **Lagrangian equation for vector fields**, and a vector field $X_\mathcal{L}$ solution to (3.4) (if it exists) is a **Lagrangian dynamical vector field**. If, in addition, the condition (3.5) holds, then $X_\mathcal{L}$ is called a **Euler–Lagrange vector field** of the system.

Considering the definition of integral curve and Eq. (2.11), it is straightforward to prove that:

> **Theorem 3.2.** *The vector field $X_\mathcal{L} \in \mathfrak{X}(TQ)$ is a Lagrangian vector field for a Lagrangian system $(TQ, \mathcal{L})$ if, and only if, the integral curves $c\colon I \subset \mathbb{R} \longrightarrow M$ of $X_\mathcal{L}$ are the solutions to the equations*
>
> $$i(\tilde{c})(\Omega_\mathcal{L} \circ c) = \mathrm{d}E_\mathcal{L} \circ c. \tag{3.6}$$
>
> *Equation (3.6) is the **Lagrangian equation** for the integral curves of $X_\mathcal{L}$. If, in addition, $X_\mathcal{L}$ is a SODE, then c is a holonomic curve and (3.6) is the **Euler–Lagrange equation** for the integral curves of $X_\mathcal{L}$.*

Remark 3.5. It is typical to require that the Lagrangian dynamical equations are obtained from a variational principle (as discussed in Section 3.6). A necessary condition for this is that the integral curves of the dynamical vector field $X_{\mathcal{L}}$ must be canonical lifts of curves in the base Q of the bundle TQ, which represents the Lagrangian phase space. This is why $X_{\mathcal{L}}$ is required to be a SODE.

Definition 3.14. Given a Lagrangian dynamical system $(TQ, \mathcal{L})$, the **Lagrangian problem** posed by the system consists of finding a vector field $X_{\mathcal{L}} \in \mathfrak{X}(TQ)$ verifying the conditions (3.4) and (3.5).

Local expressions: Consider a natural chart $(U; q^i, v^i)$ of TQ. Considering the local expressions of the several geometric elements appearing in the dynamical equations, we obtain $X_{\mathcal{L}} = A^i(q^j, v^j)\dfrac{\partial}{\partial q^i} + B^i(q^j, v^j)\dfrac{\partial}{\partial v^i}$, Eq. (3.4), which, written in coordinates, is:

$$\frac{\partial^2 \mathcal{L}}{\partial v^j \partial v^i} B^i - \left(\frac{\partial^2 \mathcal{L}}{\partial q^j \partial v^i} + \frac{\partial^2 \mathcal{L}}{\partial v^j \partial q^i} \right) A^i + \frac{\partial^2 \mathcal{L}}{\partial v^i \partial q^j} v^i - \frac{\partial \mathcal{L}}{\partial q^j} = 0, \quad (3.7)$$

$$\frac{\partial^2 \mathcal{L}}{\partial v^j \partial v^i}(A^i - v^i) = 0. \tag{3.8}$$

As we know, condition (3.5) is locally equivalent to the demand that $A^i = v^i$. Furthermore, the integral curves $\sigma: I \subseteq \mathbb{R} \longrightarrow TQ$ of $X_{\mathcal{L}}$ are canonical lifts of curves $\gamma: I \subseteq \mathbb{R} \longrightarrow Q$; therefore, if $\gamma(t) = (q^i(t))$, then $\sigma(t) = (q^i(t), \dot{q}^i(t))$, and

$$A^i = v^i = \frac{dq^i}{dt}, \quad B^i = \frac{d^2 q^i}{dt^2},$$

and the combination of these expressions with Eqs. (3.7) and (3.8) leads to the equation of the integral curves which is

$$\left(\frac{\partial^2 \mathcal{L}}{\partial v^j \partial v^i} \circ \sigma \right) \frac{d^2 q^i}{dt^2} = -\left(\frac{\partial^2 \mathcal{L}}{\partial v^j \partial q^i} \circ \sigma \right) \frac{dq^i}{dt} + \frac{\partial \mathcal{L}}{\partial q^j} \circ \sigma.$$

This can be written as

$$W_{ji}(q^k(t), \dot{q}^k(t)) \frac{d^2 q^i}{dt^2} = F_j(q^k(t), \dot{q}^k(t)),$$

or in an equivalent form as

$$\frac{d}{dt}\left(\frac{\partial \mathcal{L}}{\partial v^j} \circ \sigma \right) - \frac{\partial \mathcal{L}}{\partial q^j} \circ \sigma = 0. \tag{3.9}$$

This is the classical coordinate expression of the Euler–Lagrange equation (as we will see in Section 3.6).

Therefore, for the case of regular Lagrangians, we have the following result:

Theorem 3.3. *Let* $(TQ, \mathcal{L})$ *be a regular Lagrangian system. Then, there exists a unique vector field* $X_\mathcal{L} \in \mathfrak{X}(TQ)$ *which is the solution to the Lagrangian equation (3.4) and is a SODE.*

Proof. Existence and uniqueness are straightforward consequences of the fact that $\Omega_\mathcal{L}$ is nondegenerate.

Furthermore, if the Lagrangian is regular, the Hessian matrix $H_\mathcal{L}$ is regular, and Eq. (3.8) leads to $A^i = v^i$, and thus, the vector field $X_\mathcal{L}$ is a SODE. (Observe that in this case, all the coefficients B^i are determined by Eq. (3.7).) $\qquad\square$

Assuming that the Lagrangian $\mathcal{L}$ is regular, from (3.7) and (3.8), we can obtain the local expression of the Lagrangian dynamical vector field:

$$X_\mathcal{L} = v^i \frac{\partial}{\partial q^i} + W^{ik} \left(\frac{\partial \mathcal{L}}{\partial q^i} - v^j \frac{\partial^2 \mathcal{L}}{\partial q^j \partial v^i} \right) \frac{\partial}{\partial q^k}, \tag{3.10}$$

where W^{ik} is the inverse matrix of the partial Hessian matrix of $\mathcal{L}$, that is, $W_{ji}W^{ik} = \delta^k_j$.

Remark 3.6.

(1) For Lagrangian systems, the triple $(TQ, \Omega_\mathcal{L}, dE_\mathcal{L})$ is a Hamiltonian system, which is regular or singular depending on the regularity of the Lagrangian $\mathcal{L}$. Thus, the Lagrangian formalism of these kinds of systems is a Hamiltonian formalism with certain additional characteristics.

(2) If the Lagrangian system is singular, then $(TQ, \Omega_\mathcal{L}, dE_\mathcal{L})$ is a presymplectic Hamiltonian system and then Eq. (3.4) is not necessarily compatible everywhere on TQ. Even in the case that it has a solution, it is not unique and not necessarily a SODE. In fact, if $X_\mathcal{L}^0$ is a solution, then $X_\mathcal{L}^0 + Z$, with $Z \in \ker \Omega_\mathcal{L}$, is also a solution. Then, to obtain the Euler–Lagrange equations (3.6), the condition $J(X_\mathcal{L}) = \Delta$ must be added to the above Lagrangian equations. In general, solutions $X_\mathcal{L}$ could exist only in some submanifold $S_f \hookrightarrow TQ$, and a suitable *constraint algorithm* must be implemented to find this *final constraint submanifold* S_f (if it exists) where there are SODE vector fields $X_\mathcal{L} \in \mathfrak{X}(TQ)$, tangent to S_f, which are (not necessarily unique) solutions to the Lagrangian equations on S_f. All these problems have also been widely studied (see, for instance, [28, 29, 72, 73, 184, 185, 223, 281]).

3.3 Canonical Hamiltonian Formalism for Lagrangian Systems

The canonical Hamiltonian formalism of mechanics was initially introduced by *Hamilton, Lagrange, Poisson, Ostrogadsky,* and *Donkin,* and it constituted the dual formulation of the Lagrangian formalism.

3.3.1 Legendre map

The construction of the canonical Hamiltonian formalism is based on the introduction of a map: the *fibre derivative* of the Lagrangian function:

Definition 3.15. Let $(TQ, \mathcal{L})$ be a Lagrangian dynamical system. The **Legendre map** associated with this system is the fibre derivative of $\mathcal{L}$; that is the map

$$\mathcal{FL} : TQ \longrightarrow T^*Q$$
$$(q, u) \mapsto (q, \mathrm{D}\mathcal{L}_q(u))^{\cdot}$$

Local expression: Given a natural chart $(U; q^i, v^i)$ in TQ and the corresponding canonical coordinates $(\mathcal{FL}(U); q^i, p_i)$ in T^*Q, the local expression of $\mathcal{FL}$ is

$$q^i \circ \mathcal{FL} = q^i, \quad p_i \circ \mathcal{FL} = \frac{\partial \mathcal{L}}{\partial v^i},$$

and the coordinates p_i are called the **generalised momenta** associated with the generalised coordinates q^i.

A relevant characteristic of the Legendre map is given by the following:

Theorem 3.4. *If Θ and Ω are the canonical forms in T^*Q, then $\mathcal{FL}^*\Theta = \Theta_{\mathcal{L}}$ and $\mathcal{FL}^*\Omega = \Omega_{\mathcal{L}}$.*

Proof. The local expression of Θ is $\Theta = p_i \mathrm{d}q^i$; therefore,

$$\mathcal{FL}^*\Theta = (\mathcal{FL}^* p_i)\mathrm{d}(q^i \circ \mathcal{FL}) = \frac{\partial \mathcal{L}}{\partial v^i}\mathrm{d}q^i = \Theta_{\mathcal{L}},$$

and considering that $\mathcal{FL}^*$ commutes with the exterior differential, we obtain the result for Ω. □

Working also with local coordinates, it is straightforward to prove that:

Proposition 3.13. $\mathcal{L}$ *is a regular Lagrangian if, and only if,* $\mathcal{FL}$ *is a local diffeomorphism.*

Proof. As $\mathcal{L}$ is C^∞, the necessary and sufficient condition for $\mathcal{FL}$ to be a local diffeomorphism is that, for all $p \in TQ$, the differential of $\mathcal{L}$ at p, $D_p\mathcal{FL}$, is an isomorphism. Then it is sufficient to analyse the local expression of the Jacobian matrix of $\mathcal{FL}$ at p, which is

$$\mathcal{H}_{\mathcal{FL}}(p) = \begin{pmatrix} (\text{Id})_{n \times n} & (0)_{n \times n} \\ \left(\dfrac{\partial^2 \mathcal{L}}{\partial q^i_j \partial v^i}(p) \right) & \left(\dfrac{\partial^2 \mathcal{L}}{\partial v^j \partial v^i}(p) \right) \end{pmatrix} ;$$

therefore, $\det \mathcal{H}_{\mathcal{FL}}(p) \neq 0$ if, and only if, $\det \left(\dfrac{\partial^2 \mathcal{L}}{\partial v^i \partial v^j}(p) \right) \neq 0$. $\square$

This leads to:

Definition 3.16. $\mathcal{L}$ is a **hyperregular Lagrangian** if $\mathcal{FL}$ is a (global) diffeomorphism.

Regarding the singular Lagrangians, the most interesting ones are:

Definition 3.17. $\mathcal{L}$ is an **almost-regular Lagrangian** if:

(1) $\mathcal{FL}(TQ) \equiv P$ is a closed submanifold of T^*Q.
(2) $\mathcal{FL}$ is a submersion onto its image.
(3) For every $p \in TQ$, the fibres $\mathcal{FL}^{-1}(\mathcal{FL}(p))$ are connected submanifolds of TQ.

3.3.2 Canonical Hamiltonian formalism: Equivalence with the Lagrangian formalism

We essentially studied the case of hyperregular systems. Nevertheless, all the results hold also for the case of regular systems in general, changing T^*Q by $\mathcal{FL}(TQ) \subset T^*Q$ (or locally, at least). First, as $\mathcal{FL}$ is a diffeomorphism, we have:

Proposition 3.14. *Let* $(TQ, \mathcal{L})$ *be a hyperregular Lagrangian system. Then, there is a unique function* $h \in C^\infty(T^*Q)$ *such that* $\mathcal{FL}^*h = E_{\mathcal{L}}$, *which is called the* **Hamiltonian function** *associated with the system*

$(TQ, \mathcal{L})$, and the triple (T^*Q, Ω, h) is the **canonical Hamiltonian system** associated with $(TQ, \mathcal{L})$.

As $E_\mathcal{L}$ represents the energy of the system in the Lagrangian formalism, the function h represents the same observable in the canonical Hamiltonian formalism.

Therefore, we have the regular Hamiltonian system (T^*Q, Ω, dh), where Ω is the canonical 2-form in T^*Q fulfilling the postulates and results established in Section 2.2. In particular, the Hamiltonian equations for vector fields and their integral curves read

$$i(X_h)\Omega = dh, \qquad X_h \in \mathfrak{X}(T^*Q). \qquad (3.11)$$

$$i(\widetilde{c})(\Omega \circ c)) = dh \circ c, \qquad c: I \subseteq \mathbb{R} \longrightarrow T^*Q. \qquad (3.12)$$

Their local expressions in a natural chart of coordinates $(U; q^i, p_i)$ in T^*Q are the following: if $X_h|_U = f^i \dfrac{\partial}{\partial q^i} + g_i \dfrac{\partial}{\partial p_i} \in \mathfrak{X}(T^*Q)$, then,

$$f^i = \frac{\partial h}{\partial p_i}; \quad g_i = -\frac{\partial h}{\partial q^i},$$

and its integral curves $c(t) = (q^i(t), p_i(t))$ solution to (3.12) are the solutions to the system of first-order differential equations

$$\frac{dq^i}{dt} = \frac{\partial h}{\partial p_i}(c(t)), \quad \frac{dp_i}{dt} = -\frac{\partial h}{\partial q^i}(c(t)). \qquad (3.13)$$

The relationship between the Lagrangian and the canonical Hamiltonian formalisms of a (hyper)regular Lagrangian system is stated in the following:

Theorem 3.5 (Equivalence Theorem). *Let* $(TQ, \mathcal{L})$ *be a (hyper)regular Lagrangian system.*

(1) If $X_\mathcal{L}$ is the Lagrangian vector field solution to Eqs. (3.4) and (3.5), then there exists a unique vector field $\mathcal{F}\mathcal{L}_* X_\mathcal{L} \equiv X_h \in \mathfrak{X}(T^*Q)$, which is the solution to Eq. (3.11).

Conversely, if X_h is the Hamiltonian vector field solution to Eq. (3.11), then there is a unique vector field $\mathcal{F}\mathcal{L}_*^{-1} X_h \equiv X_\mathcal{L} \in \mathfrak{X}(TQ)$, which is the solution to Eqs. (3.4) and (3.5).

(2) Equivalently, if $\gamma: I \subset \mathbb{R} \longrightarrow Q$ is a curve and its canonical lift $\widetilde{\gamma}: I \subset \mathbb{R} \longrightarrow TQ$ is a solution to Eq. (3.6), then $\zeta = \mathcal{F}\mathcal{L} \circ \widetilde{\gamma}$ is a curve solution to Eq. (3.12).

> *Conversely, if* $\zeta \colon I \subset \mathbb{R} \longrightarrow T^*Q$ *is a curve solution to Eq. (3.12),*
> *then* $\widetilde{\gamma} = \widetilde{\pi_Q \circ \zeta} \colon I \subset \mathbb{R} \longrightarrow TQ$ *is a curve solution to Eq. (3.6).*

Proof. (1) For the first item, we have

$$0 = i(X_{\mathcal{L}})\Omega_{\mathcal{L}} - dE_{\mathcal{L}} = i(X_{\mathcal{L}})(\mathcal{FL}^*\Omega) - d(\mathcal{FL}^*h) = \mathcal{FL}^*[i(X_h)\Omega - dh],$$

and as $\mathcal{FL}$ is a diffeomorphism, this is equivalent to Eq. (3.11). The proof that $X_{\mathcal{L}}$ is a SODE is a consequence of the regularity.

(2) For the second item, if $\widetilde{\gamma}$ is a solution to Eq. (3.6) then, by Theorem 3.2, it is an integral curve of the Euler–Lagrange vector field $X_{\mathcal{L}}$, and this means that $\dot{\widetilde{\gamma}} = X_{\mathcal{L}} \circ \widetilde{\gamma}$. Then, as $\zeta = \mathcal{FL} \circ \widetilde{\gamma}$ and $\mathcal{FL}_* X_{\mathcal{L}} = X_h$, we have

$$\dot{\zeta} = T\mathcal{FL} \circ \dot{\widetilde{\gamma}} = T\mathcal{FL} \circ X_{\mathcal{L}} \circ \widetilde{\gamma} = X_h \circ \mathcal{FL} \circ \widetilde{\gamma} = X_h \circ \zeta,$$

so ζ is an integral curve of the Hamiltonian vector field X_h, and by Theorem 2.2, it is a solution to Eq. (3.12).

Conversely, from a solution ζ to Eq. (3.12), we construct the curve $\gamma = \pi_Q \circ \zeta$ and its canonical lift $\widetilde{\gamma}$. Then, as mentioned above, we conclude that $\widetilde{\gamma}$ is a solution to (3.6).

The following (commutative) diagram summarises the situation:

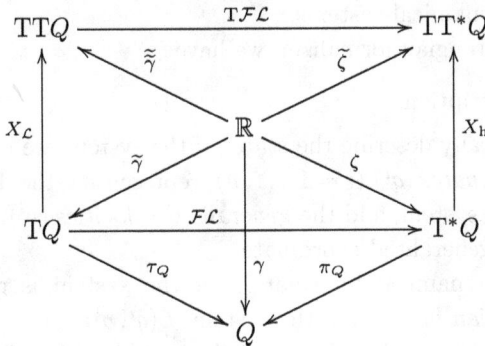

$\square$

Remark 3.7. If $(TQ, \mathcal{L})$ is an almost-regular Lagrangian system and $j_0 \colon P \hookrightarrow T^*Q$ is the natural embedding of $P \equiv \mathcal{FL}(TQ)$ in T^*Q, then it can be proven that there exists $h_0 \in C^\infty(P)$ such that $\mathcal{FL}_0^* h_0 = E_{\mathcal{L}}$, where $\mathcal{FL}_0 \colon TQ \longrightarrow P$ is defined by $\mathcal{FL} = j_0 \circ \mathcal{FL}_0$. This function is the **canonical Hamiltonian function** of the system and has the same physical interpretation as h. Now, taking $\Omega_0 = j_0^* \Omega$, the triple (P, Ω_0, dh_0) is in this case the **canonical Hamiltonian system** associated with $(TQ, \mathcal{L})$, which is the equivalent to $(T^*Q, \Omega, \mathrm{dh})$ in the regular case and is a presymplectic Hamiltonian system, in general. In particular, the equation equivalent to (3.11) is

$$i(X_{h_0})\Omega_0 = \mathrm{dh}_0; \quad X_{h_0} \in \mathfrak{X}(P),$$

which, if Ω_0 is a presymplectic form, is incompatible in general, and in the most interesting cases, X_{h_0} exists only in some submanifold $P_f \hookrightarrow P$, and is tangent to it. Moreover, the solution is not unique, since if X_{h_0} is a solution, then $X_{h_0} + Z$, for every $Z \in \ker \Omega_0$ tangent to P_f, is also a solution.

Details on the construction of the canonical Hamiltonian formalism for almost-regular Lagrangians and a deeper study on the equivalence of both the Lagrangian and Hamiltonian formalisms for this case can be found, for instance, in [28, 70, 72, 73, 184, 195, 223, 322].

3.3.3 *Discussion and comparison*

As a summary of the results in this section, we state the essential characteristics of the Lagrangian and canonical Hamiltonian formalisms of the Lagrangian dynamical systems.

In the Lagrangian formalism, we have

(1) Local description:

   (a) To locally describe the *states* of the system, we use the *generalised coordinates* (q^i) $(i = 1, \ldots, n)$, representing the degrees of freedom of the system, and the *generalised velocities* (v^i), corresponding to each generalised coordinate.

   (b) The dynamical information of the system is given by the *Lagrangian function* of the system, $\mathcal{L}(q^i, v^i)$.

   (c) The dynamical evolution of the system is described by the *Euler–Lagrange equations* (3.9), which are a second-order system of n differential equations for the functions $q^i(t)$.

   (d) The dynamical system is *regular* if $\det\left(\dfrac{\partial^2 \mathcal{L}}{\partial v^i \partial v^j}(q, v)\right) \neq 0$ for all the points (q, v). In this case, given a set of initial conditions (i.e., a state of the system), the local solution to the equations is unique.

(2) In the geometrical description, the above characteristics are translated as:

(a) The phase space of the system is the tangent bundle TQ of the manifold Q, which constitutes the configuration space of the system.

(b) The Lagrangian function $\mathcal{L} \in C^\infty(TQ)$ contains the dynamical information of the system.

(c) Starting from this function and using the geometric structures of the tangent bundle (the *canonical endomorphism* and the *Liouville vector field*), one can construct the *Lagrange 2-form* $\Omega_{\mathcal{L}}$, and the *Lagrangian energy* of the system, $E_{\mathcal{L}}$, and write the Lagrangian dynamical equations that are $i(X_{\mathcal{L}})\Omega_{\mathcal{L}} = dE_{\mathcal{L}}$, together with the second-order condition $J(X_{\mathcal{L}}) = \Delta$. The integral curves of $X_{\mathcal{L}} \in \mathfrak{X}(TQ)$ are the dynamical trajectories of the system.

(d) The dynamical system is *regular* if $\Omega_{\mathcal{L}}$ is a nondegenerate form. In this case, the vector field $X_{\mathcal{L}}$ is unique and necessarily a SODE.

Concerning to the canonical Hamiltonian formalism, we have

(1) Local description:

(a) The states of the system are now described by generalised coordinates of position (q^i) and their corresponding generalised momenta (p_i).

(b) The dynamical information of the system is given by the *Hamiltonian function* of the system, $h(q^i, p_i)$.

(c) The dynamical evolution of the system is obtained from the *Hamilton equations* (3.13), which are a first-order system of $2n$ differential equations.

(d) The dynamical system is *regular* if $\mathcal{F}\mathcal{L}$ is a (local) diffeomorphism. In this case, given a set of initial conditions (i.e., a state of the system), the solution to the equations is unique.

(2) In the geometrical description, the above characteristics are translated as:

(a) The phase space of the system is $\mathcal{F}\mathcal{L}(TQ)$, and it is, whether the cotangent bundle T^*Q of the manifold Q, which constitutes the configuration space of the system, an open set $U \subset T^*Q$, or a submanifold $P \hookrightarrow T^*Q$ (depending on the regularity of $\mathcal{L}$).

(b) The dynamical information is given by $h \in C^\infty(T^*Q)$, (respectively, $h_0 \in C^\infty(P)$): the *Hamiltonian function*.

(c) The Hamiltonian dynamical equations are obtained using this function and the canónical 2-form of the cotangent bundle Ω, and they are $i(X_h)\Omega = dh$ (respectively, $i(X_{h_0})\Omega_0 = dh_0$). The integral curves of the vector field solution are the dynamical trajectories of the system.

(d) The dynamical system is *regular* if $\mathcal{FL}$ is a (local) diffeomorphism.

The canonical Hamiltonian formalism is especially interesting because of the following facts:

- Locally, it manifests the asymmetry between the sets of variables q^i and v^i in the Lagrangian dynamical equations. This does not happen in the dynamical equations of the Hamiltonian formalism, with the coordinates q^i and p_i.

- Geometrically, in the Lagrangian formalism, the Lagrange 2-form and the Lagrangian energy, which appear in the dynamical equations, are obtained from the Lagrangian function and the canonical structures of the tangent bundle. This means that the dynamical information is present in both the Lagrangian 2-form and the Lagrangian energy.

  On the contrary, in the canonical Hamiltonian formalism, the corresponding geometric elements appearing in the dynamical equations are the canonical 2-form of the cotangent bundle and the Hamiltonian function; the first of them contains only the geometric information, and the second contains only the dynamical information.

- Finally, the characteristics of the canonical Hamiltonian formalism are suitable in order to make the quantisation of the physical system, and in particular, to implement the *geometric quantisation procedure* (see, for instance, [32, 157, 166, 182, 208, 226, 319, 325, 327, 340, 363]).

In conclusion, a *Lagrangian dynamical system* can be thought as a triple $(TQ, \Omega_\mathcal{L}, E_\mathcal{L})$ in the Lagrangian formalism, or as a triple (T^*Q, Ω, h) (or (P, Ω_0, h_0)), in the canonical Hamiltonian formalism. In both cases, the phase spaces are differential manifolds endowed with closed differential 2-forms (symplectic or presymplectic, depending on the regularity of the system).

3.3.4   *Geometric Hamilton–Jacobi theory*

One of the most relevant features of the study of Hamiltonian systems is the *Hamilton–Jacobi theory*, which provides a way to integrate Hamilton equations and systems of first-order ordinary differential equations in general. The classical theory is based on using canonical transformations [11, 165, 222, 237, 313] and also extends to singular systems [151] and higher order dynamics [98]. The geometric description of the standard theory and other geometric descriptions (for autonomous Hamiltonian systems) is well established in several texts, as [1, 11, 30, 222, 249, 259]. Following the ideas stated in [1, 232], the Hamilton–Jacobi theory has been formulated recently in a new, more general geometric perspective both for the Lagrangian and Hamiltonian formalisms and for autonomous and nonautonomous mechanical systems [65]. This framework has been extended to many other situations, namely, to singular Lagrangian and singular Hamiltonian systems [134, 135, 244], holonomic and nonholonomic mechanics [26, 66, 129, 221, 244, 287, 289], higher order dynamical systems [95, 96], control theory [22, 355], systems described using Poisson manifolds [135, 203], Lie algebroids [14, 245], dissipative systems described using contact manifolds [119, 120, 204], discretisation of dynamical systems [23, 288], first-order classical field theories [48, 128, 132, 137, 142], higher order field theories [349, 351], partial differential equations in general [350, 352], and other geometric applications and generalisations [21, 67, 202, 356] (a review on this approach and some of its applications is given in [311]).

Next, we review the foundations of this geometric framework of the Hamilton–Jacobi theory for the Hamiltonian formalisms of autonomous dynamical systems.

Let (T^*Q, ω, h) be a Hamiltonian system and $X_h \in \mathfrak{X}(T^*Q)$ be the Hamiltonian vector field.

> **Definition 3.18.** The **generalised Hamiltonian Hamilton–Jacobi problem** consists of finding a vector field $X \in \mathfrak{X}(Q)$ and a 1-form $\alpha \in \Omega^1(Q)$ such that if $\gamma \colon \mathbb{R} \longrightarrow Q$ is an integral curve of X, then $\alpha \circ \gamma \colon \mathbb{R} \longrightarrow T^*Q$ is an integral curve of X_h, that is, if $X \circ \gamma = \dot{\gamma}$, then $\overline{\alpha \circ \gamma} = X_h \circ (\alpha \circ \gamma)$. The pair (X, α) is said to be a **solution to the generalised Hamiltonian Hamilton–Jacobi problem.**

We have the following diagram:

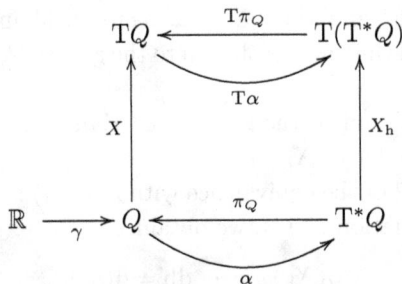

$$
\begin{array}{ccc}
TQ & \xleftarrow{\ \ T\pi_Q\ \ } & T(T^*Q) \\[2mm]
{\scriptstyle X}\Big\uparrow & {\scriptstyle T\alpha} & \Big\uparrow{\scriptstyle X_h} \\[2mm]
\mathbb{R} \xrightarrow{\ \gamma\ } Q & \xleftarrow{\ \ \pi_Q\ \ } & T^*Q \\
& {\scriptstyle \alpha} &
\end{array}
$$

Solutions to the generalised Hamiltonian Hamilton–Jacobi problem are characterised as follows:

Theorem 3.6. *The following statements are equivalent:*

(1) The pair (X, α) is a solution to the generalised Hamiltonian Hamilton–Jacobi problem.

(2) The vector fields X and X_h are α related, that is, $X_h \circ \alpha = T\alpha \circ X$. Therefore, $X = T\pi_Q \circ X_h \circ \alpha$, and as a consequence, the integral curves of X_h with initial conditions in $\operatorname{Im}\alpha$ project onto the integral curves of X.

   *X is called the **vector field associated with the form** α.*

(3) The submanifold $\operatorname{Im}\alpha$ of T^*Q is invariant by X_h (i.e., X_h is tangent to $\operatorname{Im}\alpha$).

(4) $i(X)d\alpha = -d(\alpha^*H)$.

Proof. The equivalence between items (1) and (2) is as follows: If (X, α) satisfies the condition in Definition 3.18, then for every integral curve of X, $\gamma \colon \mathbb{R} \longrightarrow Q$; by definition $X \circ \gamma = \dot{\gamma}$, then

$$X_h \circ \alpha \circ \gamma = \overline{\alpha \circ \gamma} = T\alpha \circ \dot{\gamma} = T\alpha \circ X \circ \gamma,$$

and hence, $X_h \circ \alpha = T\alpha \circ X$. The proof of the converse is straightforward.

From here, composing both members of this equality with $T\pi_Q$ and considering $\pi_Q \circ \alpha = \operatorname{Id}_Q$, we obtain $X = T\pi_Q \circ X_h \circ \alpha$. As a consequence, we conclude that the π_Q-projection of the integral curves of X_h on $\operatorname{Im}\alpha$ is the integral curves of X.

For the equivalence between items (2) and (3), if $X_h \circ \alpha = T\alpha \circ X$, then $X_h(\alpha(q)) = T\alpha(X(q))$, for every $q \in Q$ and therefore X_h is tangent to $\operatorname{Im}\alpha$. Conversely, if $\operatorname{Im}\alpha$ is invariant by X_h, then $X_h(\alpha(q)) \in T_{\alpha(q)}\operatorname{Im}\alpha$, which implies that there exists $u \in T_qQ$ such that $X_h(\alpha(q)) = T_q\alpha(u)$. Defining X by $T_q\alpha(X_q) = X_h(\alpha(q))$, we have X is differentiable, since $X = T\pi_Q \circ X_h \circ \alpha$, and thus X is a vector field in Q satisfying that $X_h \circ \alpha = T\alpha \circ X$, and then α is a solution to the generalised Hamilton–Jacobi problem.

If (X, α) is a solution to the generalised Hamilton–Jacobi problem, by item 2 we have $X = T\pi_Q \circ X_h \circ \alpha$.

Finally, the proof of the equivalence with item (4) is the following: from the Hamiltonian equation (3.11), we obtain

$$\alpha^* i(X_h)\omega = \alpha^*dh = d(\alpha^*h).$$

As Θ is the canonical form of T^*Q, then $\alpha^*\Theta = \alpha$, and

$$\alpha^*\Omega = -\alpha^*d\Theta = -d(\alpha^*\Theta) = -d\alpha, \qquad (3.14)$$

and as X and X_h are α related, we have

$$\alpha^* i(X_h)\omega = i(X)\alpha^*\Omega = -i(X)d\alpha,$$

which yields condition 4. For the converse, first define $Z_h = X_h \circ \alpha - T\alpha \circ X: Q \longrightarrow T(T^*Q)$, which is a vector field along α; then, we have to prove that $Z_h = 0$. First, we have Z_h is π_Q-vertical since, as $\pi_Q \circ \alpha = \mathrm{Id}_Q$,

$$\begin{aligned}
T\pi_Q \circ Z_h &= T\pi_Q \circ (X_h \circ \alpha - T\alpha \circ X) \\
&= T\pi_Q \circ (X_h \circ \alpha - T\alpha \circ T\pi_Q \circ X_h \circ \alpha) \\
&= T\pi_Q \circ X_h \circ \alpha - T\pi_Q \circ X_h \circ \alpha = 0.
\end{aligned}$$

From Eq. (3.11) and the hypothesis, as $\alpha^*\Omega = -d\alpha$, we obtain

$$\alpha^* i(X_h)\Omega = \alpha^*dh = d(\alpha^*h),$$
$$i(X)\alpha^*\Omega = -i(X)d\alpha = d(\alpha^*h);$$

hence, $\alpha^* i(X_h)\Omega - i(X)\alpha^*\Omega = 0$. Therefore, for every $q \in Q$ and $Y_q \in T_qQ$, we have

$$\begin{aligned}
0 &= (\alpha^* i(X_h)\Omega - i(X)\alpha^*\Omega)_q(Y_q) \\
&= \Omega_{\alpha(q)}(X_h(\alpha(q)), T_q\alpha(Y_q)) - \Omega_{\alpha(q)}(T_q\alpha(X_q), T_q\alpha(Y_q)) \\
&= \Omega_{\alpha(q)}(Z_h(q), T_q\alpha(Y_q)).
\end{aligned}$$

Furthermore, as $V(\pi_Q)$ is a Lagrangian distribution in (T^*Q, Ω), for every π_Q-vertical vector field $V \in \mathfrak{X}(T^*Q)$, we have $\Omega_{\alpha(q)}(Z_h(q), V(\alpha(q))) = 0$, but $T_{\alpha(q)}T^*Q = T_{\alpha(q)}(\mathrm{Im}\,\alpha) \oplus V_{\alpha(q)}(\pi_Q)$, and thus we have proven that

$$\Omega_{\alpha(q)}(Z_h(q), Z(\alpha(q))) = 0,$$

for every $Z \in \mathfrak{X}(T^*Q)$. Then, we conclude that $Z_h = 0$, since Ω is nondegenerate, or the equivalent, and X and X_h are α related. Thus, (X, α) is a solution to the generalised Hamilton–Jacobi problem. $\qquad\square$

In general, to solve the generalised Hamilton–Jacobi problem is a very difficult task, so we usually state a less general version of the problem, which leads to the standard version of the Hamilton–Jacobi problem:

Definition 3.19. The **Hamiltonian Hamilton–Jacobi problem** consists of finding a closed 1-form $\alpha \in \Omega^1(T^*Q)$, which is a solution to the generalised Hamiltonian Hamilton–Jacobi problem. This form α is said to be a **solution to the Hamiltonian Hamilton–Jacobi problem**.

As $d\alpha = 0$, for every point in Q, there exists a function S in a neighbourhood $U \subset Q$ such that $\alpha = dS$. Then, S is a **local generating function** of the solution α.

Theorem 3.7. *The following statements are equivalent:*

(1) The form $\alpha \in \Omega^1(Q)$ is a solution to the Hamiltonian Hamilton–Jacobi problem.

(2) $\operatorname{Im}\alpha$ is a Lagrangian submanifold of T^*Q, which is invariant by X_h, and S is a local generating function of this Lagrangian submanifold.

(3) The condition $d(\alpha^*h) = 0$ holds, or what is the equivalent, the function $h \circ dS \colon Q \longrightarrow \mathbb{R}$ is locally constant.

Proof. These statements are consequences of Theorem 3.6 and Definition 3.19. In fact, if α is a solution to the Hamilton–Jacobi problem, as a consequence of (3.14), $d\alpha = 0$ is equivalent to $\alpha^*\Omega = 0$. Then $\operatorname{Im}\alpha$ is a Lagrangian submanifold of (T^*Q, Ω), which is contained in a level set of h, because the condition $i(X)d\alpha = -d(\alpha^*h)$ implies that $d(\alpha^*h) = 0$. Notice that $\dim \operatorname{Im}\alpha = n$, and if $j_\alpha \colon \operatorname{Im}\alpha \hookrightarrow T^*Q$ is the natural embedding, we have $j_\alpha^*\Omega = 0$. $\qquad\square$

In natural coordinates of T^*Q, condition (3) is the classical form of the Hamiltonian Hamilton–Jacobi equation:

$$H\left(q^i, \frac{\partial S}{\partial q^i}\right) = E \ (ctn.). \tag{3.15}$$

Until now, we have only considered particular solutions α to the (generalised) Hamilton–Jacobi problem, which are given by particular solutions to the partial differential equation (3.15). Nevertheless, we are also interested in the general solution. Then:

Definition 3.20. Let $\Lambda \subseteq \mathbb{R}^n$. A family of solutions $\{\alpha_\lambda; \lambda \in \Lambda\}$, which depends on n parameters $\lambda \equiv (\lambda_1, \ldots, \lambda_n) \in \Lambda$, is a **complete solution** to the Hamilton–Jacobi problem if the map

$$\begin{aligned}\Phi : Q \times \Lambda &\longrightarrow T^*Q \\ (q, \lambda) &\mapsto \alpha_\lambda(q)\end{aligned}$$

is a local diffeomorphism.

Remark 3.8.

(1) Given a complete solution $\{\alpha_\lambda; \lambda \in \Lambda\}$, since $d\alpha_\lambda = 0$, for every $\lambda \in \Lambda$, there is a family of functions $\{S_\lambda\}$ defined in open sets $U_\lambda \subset Q$ such that $\alpha_\lambda = dS_\lambda$. Therefore, we have a function

$$\mathcal{S} : \bigcap U_\lambda \times \Lambda \subset Q \times \Lambda \longrightarrow \mathbb{R}$$
$$(q, \lambda) \longmapsto S_\lambda(q),$$

that is locally defined and is called a **local generating function** of the complete solution $\{\alpha_\lambda; \lambda \in \Lambda\}$.

(2) Every complete solution defines a Lagrangian foliation in T^*Q, which is transverse to the fibres, and such that X_h is tangent to the leaves. This foliation is locally defined by a family of functions, which are the components of a map $F: T^*Q \xrightarrow{\Phi^{-1}} Q \times \Lambda \longrightarrow \Lambda \subset \mathbb{R}^n$. Furthermore, these functions are a set of constants of motion of X_h.

Conversely, from a set n first integrals $f_1, \ldots, f_n$ of X_h in involution, such that $df_1 \wedge \ldots \wedge df_n \neq 0$, we can define a π-transversal Lagrangian foliation of T^*Q taking $f_i = \lambda_i$, with $\lambda_i \in \mathbb{R}$, and in this way, we obtain a local complete solution $\{\alpha_\lambda, \lambda \in \Lambda\}$. Then, from equations $f_i = \lambda_i$, we can locally isolate $p_i = p_i(q, \lambda)$, replace them in the expression of X_h, and finally project to the basis, then obtaining the family of vector fields $\{X_\lambda\}$ associated with the local complete solution. From a complete solution $\{\alpha_\lambda; \lambda \in \Lambda\}$, all the integral curves of X_h are obtained starting from the integral curves of the vector fields $\{X_\lambda\}$ associated with this complete solution.

(3) This geometric framework for the Hamilton–Jacobi theory can also be stated in a very natural way for the Lagrangian formalism (see [65] and Exercise 3.33).

The "classical" Hamilton–Jacobi problem for a (regular) Hamiltonian system (T^*Q, Ω, h) consists of obtaining a canonical transformation, which leads the system to equilibrium [11, 222, 237, 313]. This transformation is given by a generating function, which is the solution to the *Hamilton–Jacobi equation*. From a geometric point of view, this canonical transformation is associated with a foliation in the phase space of the system, T^*Q, which has the following properties: it is invariant by the dynamics, transverse to the canonical projection of the cotangent bundle, and is Lagrangian with respect to the canonical symplectic structure of T^*Q. Then, the restriction of the Hamiltonian vector field $X_h \in \mathfrak{X}(T^*Q)$ to each leaf S_λ of this foliation projects onto a vector field $X_\lambda \in \mathfrak{X}(Q)$, and the integral curves of X_h and X_λ are one-to-one related. In this way, all the dynamical trajectories are recovered from the integral curves of all these vector fields $\{X_\lambda\}$. Thus, the geometric Hamilton–Jacobi problem consists of finding this foliation and the vector fields $\{X_\lambda\}$.

Bearing this in mind, the relationship between the classical and the geometric Hamilton–Jacobi theories is established through the equivalence of complete solutions and canonical transformations (see [311, 352]).

Theorem 3.8. *A complete solution $\{\alpha_\lambda; \lambda \in \Lambda\}$ to the Hamilton–Jacobi problem provides a canonical transformation $\Phi \colon \mathrm{T}^*Q \longrightarrow \mathrm{T}^*Q$ leading the system to equilibrium, and the converse is also true.*

Proof. Let $\{\alpha_\lambda; \lambda \in \Lambda\}$ be a complete solution, and let $\mathcal{S}$ be its generating function in a neighbourhood of every point of T^*Q. As $\mathcal{S} = \mathcal{S}(q^i, \lambda^i)$, the set λ^i can be identified with a subset of coordinates $\lambda^i \equiv \tilde{q}^i$ in $\mathrm{T}^*Q \times \mathrm{T}^*Q$, and therefore, $\mathcal{S} = \mathcal{S}(q^i, \tilde{q}^i)$ can be thought as the local expression of a generating function of a local canonical transformation Φ, and hence of an open set W of the Lagrangian submanifold $graph\,\Phi \hookrightarrow \mathrm{T}^*Q \times \mathrm{T}^*Q$. When this construction is done in every chart, we obtain the transformation Φ and the submanifold $graph\,\Phi$. Finally, since (3.15) holds for every particular solution S_λ, we have

$$E = \mathrm{h}\left(q^i(\tilde{q}, \tilde{p}), \frac{\partial \mathcal{S}}{\partial q^i}(q(\tilde{q}, \tilde{p}), \tilde{q}) \right) = \mathrm{h}(\tilde{q}^i, \tilde{p}_i).$$

Conversely, let us start from a canonical transformation Φ and a generating function $\mathcal{S} = \mathcal{S}(q^i, \tilde{q}^i)$; if we take $\tilde{q} \equiv (\tilde{q}^i) = (\lambda^i) \equiv \lambda$, we obtain a family of functions $\{S_\lambda\}$, and then we get a local complete solution $\{\alpha_\lambda = dS_\lambda; \lambda \in \Lambda\}$ to the Hamiltonian Hamilton–Jacobi problem. Doing this construction in every chart, we have the complete solution. This means that on each local chart of T^*Q, fixing the coordinates $\tilde{q}^i = \lambda^i$ of a point, we obtain a local submanifold of T^*Q whose image by Φ^{-1} is the image of a local section $\alpha_\lambda \colon Q \longrightarrow \mathrm{T}^*Q$, which constitutes a particular solution to the Hamiltonian Hamilton–Jacobi problem. $\qquad\square$

3.4 Skinner–Rusk Unified Lagrangian–Hamiltonian Formalism

In their seminal articles of 1983 [322, 323], R. *Skinner* and R. *Rusk* proposed a new geometric framework to unify the Lagrangian and the Hamiltonian formalisms of first-order autonomous mechanical systems into a single one formulation. This is a simple and elegant framework, which is particularly suitable for the treatment of singular systems. Later, this nice formalism was generalised to many other types of physical systems, such as nonautonomous dynamical systems [25,52,194], vakonomic and nonholonomic mechanics [100], control systems [24,94], higher order mechanics [296,297], dissipative systems (first-order and higher order contact mechanics) [113,114], and first-order [56,126,153,305,306] and higher order [49,299,348] classical field theories.

In this section, we describe the main features of this *Skinner–Rusk* or *unified Lagrangian–Hamiltonian formalism* for autonomous dynamical systems.

3.4.1 *Unified bundle: Unified formalism*

This formalism is developed in the following bundle:

> **Definition 3.21.** The **unified bundle** or **Pontryagin bundle** is $\mathcal{W} = TQ \times_Q T^*Q$ and has natural projections
>
> $$\varrho_1 : \mathcal{W} \longrightarrow TQ, \quad \varrho_2 : \mathcal{W} \longrightarrow T^*Q, \quad \varrho_0 : \mathcal{W} \longrightarrow Q.$$
>
> Natural coordinates in $\mathcal{W}$ are (q^i, v^i, p_i).

> **Definition 3.22.** A curve $c \colon \mathbb{R} \to \mathcal{W}$ is **holonomic** in $\mathcal{W}$ if $\varrho_1 \circ c \colon \mathbb{R} \longrightarrow TQ$ is holonomic.
>
> A vector field $\Gamma \in \mathfrak{X}(\mathcal{W})$ is a **holonomic vector field** in $\mathcal{W}$ if its integral curves are holonomic in $\mathcal{W}$.

The coordinate expressions of holonomic curves and vector fields in $\mathcal{W}$ are the following:

$$c(t) = \left(q^i(t), \tfrac{dq^i}{dt}(t), p_i(t), \right),$$
$$\Gamma = v^i \frac{\partial}{\partial q^i} + F^i \frac{\partial}{\partial v^i} + G_i \frac{\partial}{\partial p_i}.$$

> **Definition 3.23.** The unified bundle $\mathcal{W}$ is endowed with the following canonical structures:

(1) The **coupling function** is the map $\mathcal{C} \colon \mathcal{W} \longrightarrow \mathbb{R}$ defined by

$$\mathcal{C} \colon TQ \times_Q T^*Q \longrightarrow \mathbb{R}$$
$$(q, v_q, \xi_q) = (q^i, v^i, p_i) \longmapsto \langle v_q \mid \xi_q \rangle = v^i p_i .$$

(2) If $\Theta \in \Omega^1(T^*Q)$ and $\Omega = -\mathrm{d}\Theta \in \Omega^2(T^*Q)$ are the canonical forms in T^*Q, the **canonical forms** in $\mathcal{W}$ are

$$\Theta_{\mathcal{W}} := \varrho_2^* \Theta \in \Omega^1(\mathcal{W}), \quad \Omega_{\mathcal{W}} := -\mathrm{d}\Theta_{\mathcal{W}} = \varrho_2^* \Omega \in \Omega^2(\mathcal{W}).$$

Using the coupling function, we introduce:

Definition 3.24. Given a Lagrangian function $\mathcal{L} \in C^\infty(TQ)$, if $\mathfrak{L} = \varrho_1^* \mathcal{L} \in C^\infty(\mathcal{W})$, the **Hamiltonian function** is defined as

$$\mathcal{H} := \mathcal{C} - \mathfrak{L} \in C^\infty(\mathcal{W}).$$

The coordinate expressions of these elements are given as

$$\Theta_{\mathcal{W}} = p_i \mathrm{d}q^i, \quad \Omega_{\mathcal{W}} = \mathrm{d}q^i \wedge \mathrm{d}p_i, \quad \mathcal{H} = v^i p_i - \mathfrak{L}(q^i, v^i).$$

The triple $(\mathcal{W}, \Omega_{\mathcal{W}}, \mathcal{H})$ is a presymplectic Hamiltonian system since $\ker \Omega_{\mathcal{W}} = \left\langle \dfrac{\partial}{\partial v^i} \right\rangle$. Then, the **dynamical problem** for this system consists of finding $X_{\mathcal{H}} \in \mathfrak{X}(\mathcal{W})$, which is a solution to the Hamiltonian equations

$$i(X_{\mathcal{H}})\Omega_{\mathcal{W}} = \mathrm{d}\mathcal{H}, \tag{3.16}$$

and then the integral curves $c \colon \mathbb{R} \longrightarrow \mathcal{W}$ of $X_{\mathcal{H}}$ are solutions to the equations

$$i(\widetilde{c})(\Omega_{\mathcal{W}} \circ c) = \mathrm{d}\mathcal{H} \circ c. \tag{3.17}$$

As $(\mathcal{W}, \Omega_{\mathcal{W}}, \mathcal{H})$ is a presymplectic Hamiltonian system, these equations are not compatible in $\mathcal{W}$. In fact, for an arbitrary vector field in $\mathfrak{X}(\mathcal{W})$,

$$X_{\mathcal{H}} = f^i \frac{\partial}{\partial q^i} + F^i \frac{\partial}{\partial v^i} + G_i \frac{\partial}{\partial p_i}.$$

Eq. (3.16) gives

$$f^i = v^i, \quad G_i = \frac{\partial \mathfrak{L}}{\partial q^i}, \quad p_i = \frac{\partial \mathfrak{L}}{\partial v^i}. \tag{3.18}$$

- The first equation ensures that $X_{\mathcal{H}}$ is a holonomic vector field in $\mathcal{W}$ (regardless of the regularity of the Lagrangian function).
- The second equation allows us to determine the component functions G_i.

- The third equation represents compatibility conditions, that is, they are *compatibility constraints* defining a submanifold $\mathcal{W}_0 \hookrightarrow \mathcal{W}$ where vector fields $X_{\mathcal{H}}$ solution to (3.16) are defined. Observe that these constraints give the Legendre map and hence $\mathcal{W}_0 = \text{graph}(\mathcal{FL})$.

Thus, we have

$$X_{\mathcal{H}}|_{\mathcal{W}_0} = v^i \frac{\partial}{\partial q^i} + F^i \frac{\partial}{\partial v^i} + \frac{\partial \mathcal{L}}{\partial q^i} \frac{\partial}{\partial p_i},$$

where the functions F^i are still undetermined. Nevertheless, the constraint algorithm for presymplectic systems continues by demanding that $X_{\mathcal{H}}$ is tangent to $\mathcal{W}_0$, that is, we have $X_{\mathcal{H}} \left(p_i - \dfrac{\partial \mathcal{L}}{\partial v^i} \right) \Big|_{\mathcal{W}_0} = 0$, which gives the equations for the remaining coefficients F^i,

$$\frac{\partial^2 L}{\partial v^i \partial v^j} F^j + \frac{\partial^2 L}{\partial q^j \partial v^i} v^j - \frac{\partial \mathcal{L}}{\partial q^i} = 0 \quad (\text{on } \mathcal{W}_0). \tag{3.19}$$

If $\mathcal{L}$ is regular, these equations are compatible and define a unique vector field $X_{\mathcal{H}}$ solution to (3.16) on $\mathcal{W}_0$, and the last system of equations gives the dynamical trajectories. If $\mathcal{L}$ is singular, Eq. (3.19) can be compatible or not, and eventually, new compatibility constraints can appear, defining a new submanifold $\mathcal{W}_1 \hookrightarrow \mathcal{W}_0$. In that case, the constraint algorithm continues by demanding the tangency of solutions to the new constraint submanifold $\mathcal{W}_1$ and so on. In the most favourable cases, there is a submanifold $\mathcal{W}_f \hookrightarrow \mathcal{W}_0$ (it could be $\mathcal{W}_f = \mathcal{W}_0$) such that there exist holonomic vector fields $X_{\mathcal{H}} \in \mathfrak{X}(\mathcal{W})$ defined on $\mathcal{W}_0$ and tangent to $\mathcal{W}_f$, which are solutions to Eq. (3.16) at support on $\mathcal{W}_f$.

3.4.2 *Recovering the Lagrangian and Hamiltonian formalisms*

Next, we study the equivalence of the unified formalism with the Lagrangian and the Hamiltonian formalisms. We restrict our analysis to the hyperregular case (the regular case is the same, at least locally).

Denoted by $\jmath_0 \colon \mathcal{W}_0 \hookrightarrow \mathcal{W}$, the natural embedding, we have

$$(\varrho_1 \circ \jmath_0)(\mathcal{W}_0) = \mathrm{T}Q, \quad (\varrho_2 \circ \jmath_0)(\mathcal{W}_0) = \mathrm{T}^*Q,$$

and being $\mathcal{W}_0$ the graph of the Legendre map, the projection $\varrho_1 \circ \jmath_0$ is a diffeomorphism between $\mathcal{W}_0$ and $\mathrm{T}Q$. The diagram below summarises the situation:

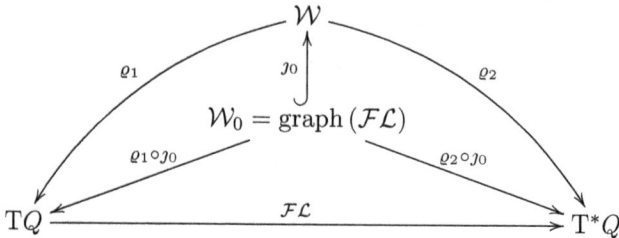

Therefore, functions, differential forms, and vector fields in $\mathcal{W}$ tangent to $\mathcal{W}_0$ can be restricted to $\mathcal{W}_0$, and then they can be translated to the Lagrangian side by using this diffeomorphism and to the Hamiltonian side using the Legendre map and the projection ϱ_2.

In particular, if $c(t) = (q^i(t), v^i(t), p_i(t))$ is a solution to Eq. (3.16) (or what is the equivalent, $c(t)$ is an integral curve of the vector field $X_{\mathcal{H}}$ solution to the dynamical Eq. (3.16)), then Eq. (3.19) leads to

$$\frac{d}{dt}\left(\frac{\partial \mathcal{L}}{\partial v^i} \circ c\right) = \frac{\partial \mathcal{L}}{\partial q^i} \circ c, \qquad (3.20)$$

and from Eq. (3.18), we obtain

$$\frac{dq^i}{dt} = v^i, \quad \frac{dp_i}{dt} = \frac{\partial \mathcal{L}}{\partial q^i} \circ c = -\frac{\partial \mathcal{H}}{\partial q^i} \circ c, \quad p_i = \frac{\partial \mathcal{L}}{\partial v^i} \circ c. \qquad (3.21)$$

From the first group of Eq. (3.21), together with (3.20), we recover the Euler–Lagrange equations for the curves $c_{\mathcal{L}}(t) = (q^i(t), v^i(t))$. Furthermore, considering the local expression of $\mathcal{H}$, we have $\dfrac{\partial \mathcal{L}}{\partial q^i} = -\dfrac{\partial \mathcal{H}}{\partial q^i}$, and hence, the second group of Eq. (3.21) reads

$$\frac{dp_i}{dt} = -\frac{\partial \mathcal{H}}{\partial q^i} \circ c,$$

and using the local expression of $\mathcal{H}$ again and the first group of equations (3.21), we get

$$\frac{\partial \mathcal{H}}{\partial q^i} \circ c = v^i = \frac{dq^i}{dt}.$$

Finally, using the third group of equations (3.18) (i.e., the Legendre map), we have $\mathcal{H} = \mathcal{FL}^*h$, and these last equations become the *Hamilton equations* for the curves $c_{\rm h}(t) = (q^i(t), p_i(t))$.

In this way, for the dynamical trajectories, we can state.

Theorem 3.9. *Every curve* $c\colon \mathbb{R} \longrightarrow \mathcal{W}$ *taking values in* $\mathcal{W}_0$ *can be viewed as* $c = (c_{\mathcal{L}}, c_{\mathrm{h}})$, *where* $c_{\mathcal{L}} = \varrho_1 \circ c \colon \mathbb{R} \longrightarrow \mathrm{T}Q$ *and* $c_{\mathrm{h}} = \mathcal{F}\mathcal{L} \circ c_{\mathcal{L}} \colon \mathbb{R} \longrightarrow \mathrm{T}^*Q$.

If $c \colon \mathbb{R} \longrightarrow \mathcal{W}$, *with* $\operatorname{Im} c \subset \mathcal{W}_0$, *is a curve fulfilling Eq.* (3.17), *then* $c_{\mathcal{L}}$ *is the lift to* $\mathrm{T}Q$ *of the projected curve* $c_o = \varrho_0 \circ c \colon \mathbb{R} \longrightarrow Q$ (*i.e.,* $c_{\mathcal{L}}$ *is a holonomic curve), and it is a solution to Eq.* (3.6), *where* $E_{\mathcal{L}} \in \mathrm{C}^\infty(\mathrm{T}Q)$ *is such that* $\mathcal{H} = \varrho_1^* E_{\mathcal{L}}$. *Moreover, the curve* $c_{\mathrm{h}} = \varrho_2 \circ c = \mathcal{F}\mathcal{L} \circ c_{\mathcal{L}}$ *is a solution to Eq.* (3.12), *where* $\mathrm{h} \in \mathrm{C}^\infty(\mathrm{T}^*Q)$ *is such that* $\mathcal{H} = \varrho_1^* \mathrm{h}$.

Conversely, if $c_o \colon \mathbb{R} \longrightarrow Q$ *is a curve, such that* $\widetilde{c}_o \equiv c_{\mathcal{L}}$ *is a solution to Eq.* (3.6), *then the curve* $c = (c_{\mathcal{L}}, \mathcal{F}\mathcal{L} \circ c_{\mathcal{L}})$ *is a solution to Eq.* (3.17) *and* $\mathcal{F}\mathcal{L} \circ c_{\mathcal{L}}$ *is a solution to Eq.* (3.12) .

Now, curves $c \colon \mathbb{R} \longrightarrow \mathcal{W}$ that serve as the solution to Eq. (3.17) are the integral curves of a holonomic vector field $X_{\mathcal{H}} \in \mathfrak{X}(\mathcal{W})$, which is the solution to (3.16); curves $c_{\mathcal{L}} \colon \mathbb{R} \longrightarrow \mathrm{T}Q$ are the integral curves of the holonomic vector field $X_{\mathcal{L}} \in \mathfrak{X}(\mathrm{T}Q)$, which is the solution to (3.4); and curves $c_{\mathrm{h}} \colon \mathbb{R} \longrightarrow \mathrm{T}^*Q$ are the integral curves of the vector field $X_{\mathrm{h}} \in \mathfrak{X}(\mathrm{T}^*Q)$, which is the solution to (3.11). Then, as a corollary of the above theorem, for the dynamical vector fields, we have:

Theorem 3.10. *Let* $X_{\mathcal{H}} \in \mathfrak{X}(\mathcal{W})$ *be the solution to Eq.* (3.16) (*on* $\mathcal{W}_0$), *which is tangent to* $\mathcal{W}_0$.

Then, the vector field $X_{\mathcal{L}} \in \mathfrak{X}(\mathrm{T}Q)$, *defined by* $X_{\mathcal{L}} \circ \varrho_1 = \mathrm{T}\varrho_1 \circ X_{\mathcal{H}}$, *is the solution to Eqs.* (3.4) *and* (3.5), *where* $E_{\mathcal{L}} \in \mathrm{C}^\infty(\mathrm{T}Q)$ *is such that* $\mathcal{H} = \varrho_1^* E_{\mathcal{L}}$.

The vector field $X_{\mathrm{h}} \in \mathfrak{X}(\mathrm{T}^*Q)$, *defined by* $X_{\mathrm{h}} \circ \varrho_2 = \mathrm{T}\varrho_2 \circ X_{\mathcal{H}}$, *is the solution to Eq.* (3.11), *where* $\mathrm{h} \in \mathrm{C}^\infty(\mathrm{T}^*Q)$ *is such that* $\mathcal{H} = \varrho_1^* \mathrm{h}$. *Furthermore,* $\mathcal{F}\mathcal{L}_* X_{\mathcal{L}} = X_{\mathrm{h}}$.

In summary, the main features of the unified formalism are as follows:

- It assures holonomy (even in the non-regular case).
- It provides the Legendre map.
- It gives the Euler–Lagrange and the Hamilton equations.

3.5 Symmetries of Regular Lagrangian Systems

For Lagrangian dynamical systems, the phase space of the system is either $M = \mathrm{T}Q$ in the Lagrangian formalism or $M = \mathrm{T}^*Q$ in the canonical Hamiltonian formalism. In these cases, there exist distinguished symplectic potentials: the Lagrangian 1-form $\Theta_{\mathcal{L}} \in \Omega^1(TQ)$ and the canonical 1-form $\Theta \in \Omega^1(T^*Q)$. Furthermore, the symmetries of the dynamical systems were canonical lifts of diffeomorphisms or vector fields on the base manifold Q. All of these lead to the introduction of new kinds of symmetries in such cases, whose properties are studied next.

3.5.1 *Symmetries in the canonical Hamiltonian formalism*

Consider a (regular) canonical Hamiltonian system (T^*Q, Ω, h), and let $X_h \in \mathfrak{X}_{lh}(T^*Q)$ be the Hamiltonian vector field of the system.

All we have stated for Hamiltonian dynamical systems in Section 2.3 holds for this particular case. Nevertheless, new kinds of symmetry can be introduced for this situation. Thus, in addition to the previous symmetries already defined, we can consider the following particular cases:

Definition 3.25. A dynamical symmetry $\Phi \in \mathrm{Diff}\,(T^*Q)$ of the canonical Hamiltonian system is a **natural dynamical symmetry** if there exists $\varphi \in \mathrm{Diff}\,(Q)$ such that $\Phi = T^*\varphi$ (i.e., Φ is the canonical lift of a diffeomorphism in Q).

Definition 3.26. An infinitesimal dynamical symmetry $Y \in \mathfrak{X}(Q)$ of the canonical Hamiltonian system is a **natural infinitesimal dynamical symmetry** if there exists $Z \in \mathfrak{X}(Q)$ such that $Y = Z^*$, that is, Y is the canonical lift of a vector field in Q. (The terminology "natural symmetries" refers both to the diffeomorphism φ and the vector field Z).

Remember that (T^*Q, Ω) is an exact symplectic manifold and a symplectic potential of Ω is the canonical 1-form $\Theta \in \Omega^1(T^*Q)$. Furthermore, for every $\varphi \in \mathrm{Diff}\,(Q)$, we have $(T^*\varphi)^*\Theta = \Theta$, and hence $(T^*\varphi)^*\Omega = \Omega$. In the same way, for every $Z \in \mathfrak{X}(Q)$, we have $\mathrm{L}(Z^*)\Theta = 0$, and then $\mathrm{L}(Z^*)\Omega = 0$; therefore, if Z^* is an infinitesimal natural dynamical symmetry, then $Z^* \in \mathfrak{X}_H(T^*Q)$, and as we saw in Proposition 3.9, the global Hamiltonian function of Z^* is $f_Z = i(Z^*)\Theta$ (up to constants). This leads to the introduction of the following particular type of Noether symmetries for the canonical Hamiltonian system (T^*Q, Ω, h):

Definition 3.27. A diffeomorphism $\Phi \in \text{Diff}(T^*Q)$ is a *natural Noether symmetry* if:

(1) There exists a diffeomorphism $\varphi \in \text{Diff}(Q)$ such that $\Phi = T^*\varphi$.
(2) $\Phi^*h = (T^*\varphi)^*h = h$.

Definition 3.28. A vector field $Y \in \mathfrak{X}(T^*Q)$ is an *infinitesimal natural Noether symmetry* if:

(1) There exists a vector field $Z \in \mathfrak{X}(Q)$ such that $Y = Z^*$.
(2) $L(Y)h = L(Z^*)h = 0$.

Other particular cases of Noether symmetries in this formalism are

Definition 3.29. A Noether symmetry is *exact* if $\Phi^*\Theta = \Theta$.

Definition 3.30. An infinitesimal Noether symmetry is *exact* if $L(Y)\Theta = 0$.

For infinitesimal exact Noether symmetries, local Hamiltonian functions are expressed as $f_Y = i(Y)\Theta$ (see Proposition 2.19).

Every (infinitesimal) natural Noether symmetry is a (infinitesimal) natural dynamical symmetry. Moreover, as every canonical lift preserves the canonical forms $\Theta \in \Omega^1(T^*Q)$ and $\Omega \in \Omega^2(T^*Q)$ (Propositions 3.10 and 3.11), we have:

Proposition 3.15. *Every (infinitesimal) natural Noether symmetry is an (infinitesimal) exact Noether symmetry.*

At this point, it is possible to state the Noether Theorem as in Theorem 2.8. In particular, for infinitesimal exact Noether symmetries, the associated conserved quantities are $f_Y = i(Y)\Theta$ (up to constants).

Summarising the following table recovers the relationships among the several types of symmetries of the canonical Hamiltonian systems:

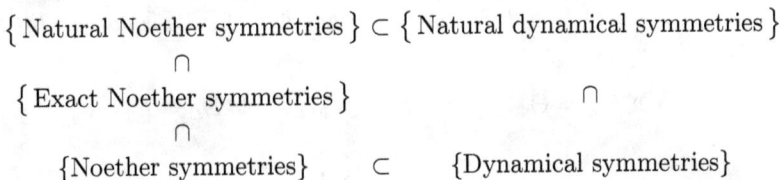

$\{$ Natural Noether symmetries $\} \subset \{$ Natural dynamical symmetries $\}$

$\cap$

$\{$ Exact Noether symmetries $\}$ $\cap$

$\{$Noether symmetries$\}$ $\subset$ $\{$Dynamical symmetries$\}$

3.5.2 Lagrangian formalism: Lagrangian symmetries and Noether's Theorem

Let $(TQ, \Omega_{\mathcal{L}}, E_{\mathcal{L}})$ be a regular Lagrangian system and $X_{\mathcal{L}} \in \mathfrak{X}(TQ)$ be the Euler–Lagrange vector field solution of the system.

Also, in this situation, all the concepts and results about symmetries established in Section 2.3 are true, and in this way, we can also introduce the concepts of *(infinitesimal) Lagrangian dynamical symmetry*, *(infinitesimal) Lagrangian Noether symmetry*, and *(infinitesimal) exact Lagrangian Noether symmetry*, and their properties and relationships, including Noether's Theorem. However, in addition, the study of symmetries and the Lagrangian formalism presents some nuances that should be highlighted.

First, we have:

Definition 3.31. A diffeomorphism $\Phi\colon TQ \longrightarrow TQ$ is a **natural Lagrangian dynamical symmetry** if:

(1) There exists a diffeomorphism $\varphi\colon Q \longrightarrow Q$ such that $\Phi = T\varphi$.
(2) $\Phi_* X_{\mathcal{L}} = X_{\mathcal{L}}$.

Definition 3.32. A vector field $Y \in \mathfrak{X}(TQ)$ is an **infinitesimal natural Lagrangian dynamical symmetry** if

(1) There exists $Z \in \mathfrak{X}(Q)$ such that $Y = Z^C$.
(2) $[Y, X_{\mathcal{L}}] = [Z^C, X_{\mathcal{L}}] = 0$.

As in Remark 2.17, this condition can be generalised by demanding that

$$[Y, X_{\mathcal{L}}] = [Z^C, X_{\mathcal{L}}] = g X_{\mathcal{L}}, \quad (\text{ for } g \in C^\infty(TQ)).$$

As in the canonical Hamiltonian formalism, lifts of diffeomorphisms and vector fields preserve the canonical structures of TQ (Proposition 3.3). A consequence of this is the following:

Proposition 3.16.

(1) Let $\varphi\colon Q \longrightarrow Q$ be a diffeomorphism and $\Phi = T\varphi$ its canonical lift to TQ. Then

$$\Phi^* \Theta_{\mathcal{L}} = \Theta_{\Phi_* \mathcal{L}}, \quad \Phi^* \Omega_{\mathcal{L}} = \Omega_{\Phi_* \mathcal{L}}, \quad \Phi^* E_{\mathcal{L}} = E_{\Phi_* \mathcal{L}}.$$

(2) *Let $Z \in \mathfrak{X}(Q)$ and its canonical lift Z^C to TQ. Then*
$$L(Z^C)\Theta_{\mathcal{L}} = 0, \quad L(Z^C)\Omega_{\mathcal{L}} = 0, \quad L(Z^C)E_{\mathcal{L}} = 0.$$

Proof. It is a straightforward consequence of Proposition 3.3 and the definitions of $\Theta_{\mathcal{L}}$, $\Omega_{\mathcal{L}}$, and $E_{\mathcal{L}}$. In fact:

(1) For $\Phi = T\varphi$, we obtain
$$\Phi^*\Theta_{\mathcal{L}} = \Phi^*(d\mathcal{L} \circ J) = d(\Phi^*\mathcal{L}) \circ J = \Theta_{\Phi^*\mathcal{L}},$$
$$\Phi^*\Omega_{\mathcal{L}} = \Phi^*(-d\Theta_{\mathcal{L}}) = -d\Phi^*\Theta_{\mathcal{L}} = \Omega_{\Phi^*\mathcal{L}},$$
$$\Phi^*E_{\mathcal{L}} = \Phi^*(\Delta(\mathcal{L}) - \mathcal{L}) = \Delta(\Phi^*\mathcal{L}) - \Phi^*\mathcal{L} = E_{\Phi^*\mathcal{L}}.$$

(2) They are using the uniparametric groups of diffeomorphisms generated by the fluxes of Z and Z^C, and the above item. $\qquad\square$

Nevertheless, the Lagrangian forms $\Theta_{\mathcal{L}}$ and $\Omega_{\mathcal{L}}$ are not canonical structures of TQ, since they depend on the choice of a Lagrangian function $\mathcal{L}$ and hence are not invariant by these lifts, necessarily. Thus, for Lagrangian Noether symmetries, we can state the following definitions:

Definition 3.33. A diffeomorphism $\Phi\colon TQ \longrightarrow TQ$ is a **natural Lagrangian Noether symmetry** if there exists a diffeomorphism $\varphi\colon Q \longrightarrow Q$ such that $\Phi = T\varphi$ and it satisfies:

(1) $\Phi^*\Omega_{\mathcal{L}} = (T\varphi)^*\Omega_{\mathcal{L}} = \Omega_{\mathcal{L}}$.
(2) $\Phi^*E_{\mathcal{L}} = (T\varphi)^*E_{\mathcal{L}} = E_{\mathcal{L}} + c$ $(c \in \mathbb{R})$ (It is usual to write simply that $\Phi^*E_{\mathcal{L}} = E_{\mathcal{L}}$).

Definition 3.34. A vector field $Y \in \mathfrak{X}(TQ)$ is an **infinitesimal natural Lagrangian Noether symmetry** if there exists $Z \in \mathfrak{X}(Q)$ such that $Y = Z^C$ and it satisfies:

(1) $L(Y)\Omega_{\mathcal{L}} = L(Z^C)\Omega_{\mathcal{L}} = 0$.
(2) $L(Y)E_{\mathcal{L}} = L(Z^C)E_{\mathcal{L}} = 0$.

Every (infinitesimal) natural Lagrangian Noether symmetry is a (infinitesimal) natural Lagrangian dynamical symmetry.

Finally, as a particular case, we have:

Definition 3.35. A Lagrangian Noether symmetry is **exact** if $\Phi^* \Theta_\mathcal{L} = \Theta_\mathcal{L}$.

Definition 3.36. An infinitesimal Lagrangian Noether symmetry is **exact** if $L(Y)\Theta_\mathcal{L} = 0$.

In these circumstances, it is possible to state the Lagrangian geometric version of the Noether Theorem. First, observe that if $Y \in \mathfrak{X}(TQ)$ is a infinitesimal natural Lagrangian Noether symmetry, then Proposition 2.19 holds for these kinds of symmetries. Thus, for every $p \in TQ$, there exists an open set $U_p \ni p$ and $f_Y \in C^\infty(U_p)$, which is unique up to the sum constant functions, such that

$$i(Z^C)\Omega_\mathcal{L} = df_Y \qquad \text{(in } U_p\text{)}. \qquad (3.22)$$

Furthermore, there exists $\zeta_Y \in C^\infty(U_p)$, defined as $L(Z^C)\Theta_\mathcal{L} = d\zeta_Y$, in U_p, and such that

$$f_Y = \zeta_Y - i(Z^C)\Theta_\mathcal{L} = \zeta_Y - \Theta_\mathcal{L}(Z^C) = \zeta_Y - d\mathcal{L} \circ J(Z^C)$$
$$= \zeta_Y - d\mathcal{L}(Z^V) = \zeta_Y - i(Z^V)d\mathcal{L} = \zeta_Y - Z^V(\mathcal{L}) \qquad (3.23)$$

(up to the sum of constant functions in U_p). Then:

Theorem 3.11 (Lagrangian Noether). *If $Y = Z^C \in \mathfrak{X}(TQ)$ (with $Z \in \mathfrak{X}(Q)$) is an infinitesimal natural Lagrangian Noether symmetry, then $f_Y = \zeta_Y - Z^V(\mathcal{L})$ is a conserved quantity, that is, $L(X_\mathcal{L})f_Y = 0$.*

Proof. The proof is the same as in Theorem 2.8, considering (3.22) and (3.23). □

If the infinitesimal Noether symmetry is exact, then $Y = Z^C$, and we can take $f_Y = i(Y)\Theta_\mathcal{L} = Z^V(\mathcal{L})$.

3.5.3   *Equivalent Lagrangians and Noether's Theorem*

It is evident that if $\Phi \in \text{Diff}(TQ)$ (respectively, $Y \in \mathfrak{X}(TQ)$) is a canonical lift of a diffeomorphism (respectively, of a vector field) of Q to TQ, which, in addition, lets the Lagrangian function of the system be invariant, then the symplectic form $\Omega_\mathcal{L}$, the Lagrangian energy $E_\mathcal{L}$, and hence the Euler–Lagrange vector field $X_\mathcal{L}$ (i.e., the Euler–Lagrange equations) are also invariant by Φ. All of these assures that the conditions of Definitions

3.33 and 3.34 hold. Nevertheless, this requirement is too strong because there are Lagrangian functions that, being different, give the same form $\Omega_{\mathcal{L}}$ and the same Euler–Lagrange equations. This leads to the following:

Definition 3.37. Two Lagrangian functions $\mathcal{L}_1, \mathcal{L}_2 \in C^\infty(TQ)$ are **equivalent** if

$$\Omega_{\mathcal{L}_1} = \Omega_{\mathcal{L}_2} \quad \text{and} \quad X_{\mathcal{L}_1} = X_{\mathcal{L}_2}.$$

For regular Lagrangians, this definition is equivalent to the following:

Proposition 3.17. *Two regular Lagrangians* $\mathcal{L}_1, \mathcal{L}_2 \in C^\infty(TQ)$ *are equivalent if*

$$\Omega_{\mathcal{L}_1} = \Omega_{\mathcal{L}_2} \quad \text{and} \quad E_{\mathcal{L}_1} = E_{\mathcal{L}_2} + c \ (\ c \in \mathbb{R}).$$

Proof. We must prove that if $\Omega_{\mathcal{L}_1} = \Omega_{\mathcal{L}_2}$, then $X_{\mathcal{L}_1} = X_{\mathcal{L}_2}$ is equivalent to $E_{\mathcal{L}_1} = E_{\mathcal{L}_2} + c$.

If $X_{\mathcal{L}_1} = X_{\mathcal{L}_2}$, then

$$0 = i(X_{\mathcal{L}_1})\Omega_{\mathcal{L}_1} - \mathrm{d}E_{\mathcal{L}_1} = i(X_{\mathcal{L}_2})\Omega_{\mathcal{L}_2} - \mathrm{d}E_{\mathcal{L}_1},$$

which implies that $\mathrm{d}E_{\mathcal{L}_1} = \mathrm{d}E_{\mathcal{L}_2}$, and hence, $E_{\mathcal{L}_1} = E_{\mathcal{L}_2} + c$.

Conversely, if $E_{\mathcal{L}_1} = E_{\mathcal{L}_2} + c$, then

$$i(X_{\mathcal{L}_1})\Omega_{\mathcal{L}_1} = \mathrm{d}E_{\mathcal{L}_1} = \mathrm{d}E_{\mathcal{L}_2} = i(X_{\mathcal{L}_2})\Omega_{\mathcal{L}_2},$$

and as $\Omega_{\mathcal{L}_1} = \Omega_{\mathcal{L}_2}$, this implies that $X_{\mathcal{L}_1} = X_{\mathcal{L}_2}$, since $\mathcal{L}_1$ and $\mathcal{L}_2$ are regular Lagrangians and the solution is unique. $\qquad\square$

Next, we specify how equivalent Lagrangians are (see also [1]). First, we find the form of Lagrangian functions leading to vanishing Cartan forms:

Proposition 3.18. *A Lagrangian* $\mathcal{L} \in C^\infty(TQ)$ *satisfies* $\Omega_{\mathcal{L}} = 0$ *if, and only if, there exists a closed 1-form* $\alpha \in \Omega^1(Q)$ *in* Q *and a function* $f \in C^\infty(Q)$, *such that* $\mathcal{L} = \widehat{\alpha} + \tau_Q^* f$ *(up to a constant), where* $\widehat{\alpha} \in C^\infty(TQ)$ *is the function defined by*

$$\widehat{\alpha} : TQ \longrightarrow \mathbb{R}$$
$$(q, v) \mapsto \alpha_q(v)^.$$

Proof. Let $\Omega_{\mathcal{L}} = -\mathrm{d}\Theta_{\mathcal{L}}$; then $\Theta_{\mathcal{L}} = \mathrm{d}\mathcal{L} \circ J$ is a closed and semibasic form in TQ, and as a consequence, it is a basic form. Then, there exists

$\alpha \in \Omega^1(Q)$ such that

$$d\mathcal{L} \circ J = \tau_Q^* \alpha. \qquad (3.24)$$

Moreover, since $0 = d\Theta_{\mathcal{L}} = d(\tau_Q^* \alpha) = \tau_Q^*(d\alpha)$, then $d\alpha = 0$, that is, α is a closed 1-form in Q. Furthermore, a simple computation in local coordinates shows that $d\widehat{\alpha} \circ J = \tau_Q^* \alpha$, and from (3.24), we have

$$d\widehat{\alpha} \circ J = \tau_Q^* \alpha = d\mathcal{L} \circ J.$$

Then $d(\mathcal{L} - \widehat{\alpha}) \circ J = 0$, and therefore, the 1-form $d(\mathcal{L} - \widehat{\alpha})$ is closed and semi-basic. As a consequence, $d(\mathcal{L} - \widehat{\alpha})$ is a basic 1-form, that is, there exists $f \in C^\infty(Q)$ such that

$$d(\mathcal{L} - \widehat{\alpha}) = \tau_Q^* df = d(\tau_Q^* f)$$

and thus $\mathcal{L} = \widehat{\alpha} + \tau_Q^* f$ (up to a constant).

Conversely, let us suppose that $\mathcal{L} = \widehat{\alpha} + \tau_Q^* f$ (up to a constant). We have

$$\Theta_{\mathcal{L}} = d\mathcal{L} \circ J = d(\widehat{\alpha} + \tau_Q^* f) \circ J = d\widehat{\alpha} \circ J = \tau_Q^* \alpha,$$

since $d\tau_Q^* f$ vanishes on the vertical vector fields. As α is closed, $d\alpha = 0$, and we obtain

$$\Omega_{\mathcal{L}} = -d\Theta_{\mathcal{L}} = -d(\tau_Q^* \alpha) = -\tau_Q^*(d\alpha) = 0.$$

$\square$

In a local chart of natural coordinates of TQ, the local expression of a closed 1-form α is

$$\alpha = \alpha_i dq^i = \frac{\partial g}{\partial q^i} dq^i,$$

for some local function g; then, the local expression of the function $\widehat{\alpha}$ is

$$\widehat{\alpha} = \frac{\partial g}{\partial q^i} v^i.$$

Now, from this result, we obtain the explicit characterisation of the equivalent Lagrangians:

Proposition 3.19. *Two regular Lagrangians $\mathcal{L}_1, \mathcal{L}_2 \in C^\infty(TQ)$ are equivalent if, and only if, $\mathcal{L}_1 = \mathcal{L}_2 + \widehat{\alpha}$ (up to a constant).*

Proof. If $\mathcal{L}_1, \mathcal{L}_2$ are equivalent, as $\Omega_{\mathcal{L}_1} = \Omega_{\mathcal{L}_1}$, then $\Omega_{\mathcal{L}_1 - \mathcal{L}_2} = 0$. Therefore, by Proposition 3.18, there exist $\alpha \in Z^1(Q)$ and $f \in C^\infty(Q)$ such that $\mathcal{L}_1 - \mathcal{L}_2 = \widehat{\alpha} + \tau_Q^* f$ (up to a constant). From Proposition 3.17, we know that $E_{\mathcal{L}_1} = E_{\mathcal{L}_2}$, (up to a constant), or equivalently, $E_{\mathcal{L}_1} - E_{\mathcal{L}_2} = 0$ (up to a constant). Then,

$$0 = E_{\mathcal{L}_1} - E_{\mathcal{L}_2} = \Delta(\mathcal{L}_1) - \mathcal{L}_1 - \Delta(\mathcal{L}_2) + \mathcal{L}_2 = \Delta(\mathcal{L}_1 - \mathcal{L}_2) - (\mathcal{L}_1 - \mathcal{L}_2)$$
$$= \Delta(\widehat{\alpha} + \tau_Q^* f) - (\mathcal{L}_1 - \mathcal{L}_2) = \widehat{\alpha} - (\mathcal{L}_1 - \mathcal{L}_2) \quad \text{(up to a constant)}.$$

Conversely, suppose that $\mathcal{L}_1 = \mathcal{L}_2 + \widehat{\alpha}$ (up to a constant). First, a simple computation gives

$$\Omega_{\mathcal{L}_2} - \Omega_{\mathcal{L}_1} = d(\Theta_{\mathcal{L}_1} - \Theta_{\mathcal{L}_2})$$
$$= d(d(\mathcal{L}_1 - \mathcal{L}_2) \circ J) = d(d\widehat{\alpha} \circ J) = d(\tau_Q^* \alpha) = \tau_Q^*(d\alpha) = 0.$$

Thus, $\Omega_{\mathcal{L}_1} = \Omega_{\mathcal{L}_2}$. Furthermore, as $\Delta(\widehat{\alpha}) = \widehat{\alpha}$, we have, up to a constant,

$$E_{\mathcal{L}_1} = \Delta(\mathcal{L}_1) - \mathcal{L}_1 = \Delta(\mathcal{L}_2 + \widehat{\alpha}) - (\mathcal{L}_2 + \widehat{\alpha}) = E_{\mathcal{L}_2} + \widehat{\alpha} - \widehat{\alpha} = E_{\mathcal{L}_2}.$$

As $\Omega_{\mathcal{L}_1} = \Omega_{\mathcal{L}_2}$ and $E_{\mathcal{L}_1} = E_{\mathcal{L}_2}$ (up to a constant), then $\mathcal{L}_1$ and $\mathcal{L}_2$ are equivalent (Proposition 3.17). $\qquad\square$

Taking this into account, we have:

Definition 3.38. A **symmetry of the Lagrangian** is a diffeomorphism $\Phi: TQ \longrightarrow TQ$ such that $\mathcal{L}$ and $\Phi^* \mathcal{L}$ are equivalent Lagrangians, that is, $\Phi^* \mathcal{L} = \mathcal{L} + \widehat{\alpha}$ (up to constants), where $\widehat{\alpha} \in C^\infty(TQ)$ is the function defined in Proposition 3.18.

Definition 3.39. An **infinitesimal symmetry of the Lagrangian** is a vector field $Y \in \mathfrak{X}(TQ)$ such that the uniparametric groups of diffeomorphisms generated by its flux are symmetries of the Lagrangian, that is, $L(Y)\mathcal{L} = \widehat{\alpha}$.

A special case of these kinds of symmetries is:

Definition 3.40. A **strict symmetry of the Lagrangian** is a diffeomorphism $\Phi: TQ \longrightarrow TQ$ such that $\Phi^* \mathcal{L} = \mathcal{L}$.

Definition 3.41. An **infinitesimal strict symmetry of the Lagrangian** is a vector field $Y \in \mathfrak{X}(TQ)$ such that the uniparametric groups of diffeomorphisms generated by its flux are strict symmetries of the Lagrangian, that is, $L(Y)\mathcal{L} = 0$.

In particular, we have:

Definition 3.42. A (strict) symmetry of the Lagrangian $\Phi\colon TQ \longrightarrow TQ$ is said to be **natural** if there exists a diffeomorphism $\varphi\colon Q \longrightarrow Q$ such that $\Phi = T\varphi$.

Definition 3.43. An infinitesimal (strict) symmetry of the Lagrangian $Y \in \mathfrak{X}(TQ)$ is said to be **natural** if there exists a vector field $Z \in \mathfrak{X}(Q)$ such that $Y = Z^C$.

Remark 3.9. A symmetry of the Lagrangian $\Phi\colon TQ \longrightarrow TQ$ is not necessarily a Lagrangian Noether symmetry since, in general, $\Phi^*\Omega_{\mathcal{L}} \neq \Omega_{\Phi^*\mathcal{L}}$ and $\Phi^*E_{\mathcal{L}} \neq E_{\Phi^*\mathcal{L}}$, as a simple calculation in coordinates shows. In addition, it is not a Lagrangian dynamical symmetry. Nevertheless, the following relationship holds:

Proposition 3.20. *A diffeomorphism* $\Phi\colon T_k^1Q \longrightarrow T_k^1Q$ *is a natural Lagrangian Noether symmetry if, and only if, it is a natural symmetry of the Lagrangian.*

Proof. If $\Phi = T\varphi$, for some diffeomorphism $\varphi\colon Q \longrightarrow Q$, according to Lemma (3.16), we have

$$\Phi^*\Omega_{\mathcal{L}} = \Omega_{\Phi^*\mathcal{L}}, \quad \Phi^*E_{\mathcal{L}} = E_{\Phi^*\mathcal{L}},$$

and then

$$\left.\begin{array}{l} \Phi^*\Omega_{\mathcal{L}} = \Omega_{\mathcal{L}} \\ \Phi^*E_{\mathcal{L}} = E_{\mathcal{L}} \text{ (up to constants)} \end{array}\right\} \iff \left\{\begin{array}{l} \Omega_{\Phi^*\mathcal{L}} = \Omega_{\mathcal{L}} \\ E_{\Phi^*\mathcal{L}} = E_{\mathcal{L}} \text{ (up to constants)} \end{array}\right. ;$$

that is, Φ is a natural Lagrangian Noether symmetry if, and only if, $\mathcal{L}$ and $\Phi^*\mathcal{L}$ are equivalent Lagrangians, so Φ is a natural symmetry of the Lagrangian. $\square$

This result holds also for infinitesimal symmetries, as can be proven taking the flux of the vector fields that generate them. Thus, we have the following straightforward corollary:

Proposition 3.21. *A vector field $Y \in \mathfrak{X}(TQ)$ is an infinitesimal natural Lagrangian Noether symmetry if, and only if, it is an infinitesimal natural symmetry of the Lagrangian.*

Finally, a version of Noether's Theorem for infinitesimal natural strict symmetries of the Lagrangian can be established as follows:

Theorem 3.12 (Classical Noether for Lagrangian systems). *Let $Y = Z^C \in \mathfrak{X}(TQ)$, with $Z \in \mathfrak{X}(Q)$, be an infinitesimal natural strict symmetry of the Lagrangian. Then $\widetilde{f} = Z^V(\mathcal{L})$ is a conserved quantity, that is, $\mathrm{L}(X_{\mathcal{L}})\widetilde{f} = 0$.*

Proof. As every infinitesimal natural strict symmetries of the Lagrangian is a natural symmetry of the Lagrangian, and then it is a natural Noether Lagrangian symmetry, according to the above proposition; then the result is a straightforward consequence of Theorem 3.11, since $\zeta_Y = \mathrm{L}(Y)\Theta_{\mathcal{L}} = \mathrm{L}(Z^C)\Theta_{\mathcal{L}} = 0$. $\qquad\square$

The following table summarises the relationship among the different kinds of symmetries in the Lagrangian formalism of Lagrangian systems:

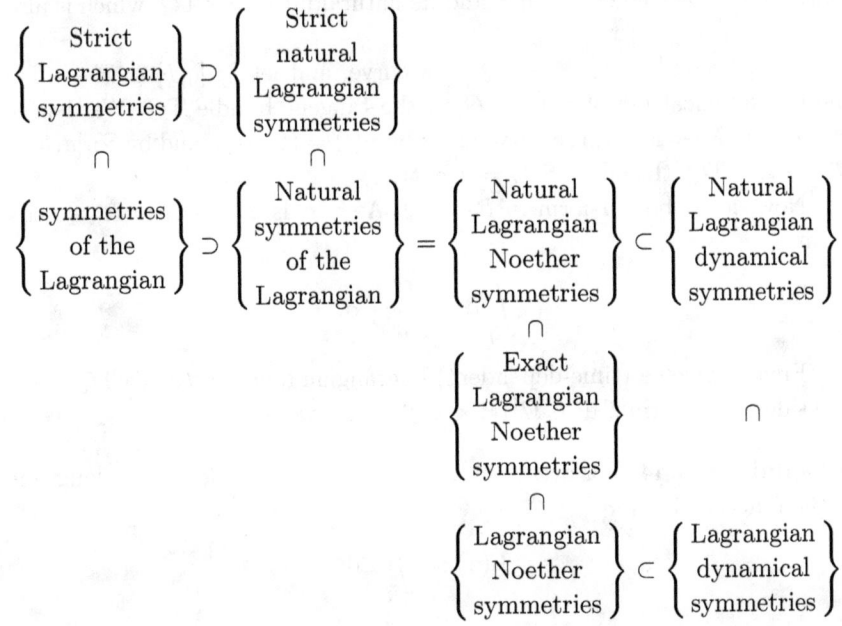

$$\left\{\begin{array}{c}\text{Strict}\\\text{Lagrangian}\\\text{symmetries}\end{array}\right\} \supset \left\{\begin{array}{c}\text{Strict}\\\text{natural}\\\text{Lagrangian}\\\text{symmetries}\end{array}\right\}$$

$$\cap \qquad\qquad \cap$$

$$\left\{\begin{array}{c}\text{symmetries}\\\text{of the}\\\text{Lagrangian}\end{array}\right\} \supset \left\{\begin{array}{c}\text{Natural}\\\text{symmetries}\\\text{of the}\\\text{Lagrangian}\end{array}\right\} = \left\{\begin{array}{c}\text{Natural}\\\text{Lagrangian}\\\text{Noether}\\\text{symmetries}\end{array}\right\} \subset \left\{\begin{array}{c}\text{Natural}\\\text{Lagrangian}\\\text{dynamical}\\\text{symmetries}\end{array}\right\}$$

$$\cap \qquad\qquad\qquad\qquad \cap$$

$$\left\{\begin{array}{c}\text{Exact}\\\text{Lagrangian}\\\text{Noether}\\\text{symmetries}\end{array}\right\}$$

$$\cap$$

$$\left\{\begin{array}{c}\text{Lagrangian}\\\text{Noether}\\\text{symmetries}\end{array}\right\} \subset \left\{\begin{array}{c}\text{Lagrangian}\\\text{dynamical}\\\text{symmetries}\end{array}\right\}$$

3.6   Variational Formulation for Lagrangian Systems

One of the main characteristics of the Lagrangian systems is that they
are variational, that is, their dynamical equations can be obtained from
a variational principle, and the same happens for Hamiltonian systems [5,
33, 34, 92, 147, 159, 173, 216, 234, 236, 246, 279, 286]. Next, we present the
variational formulation of Lagrangian systems. To study this topic, we need
to enlarge the phase space to include the time coordinate, so we have to
consider the trivial bundle $\mathbb{R} \times TQ$ (or $\mathbb{R} \times T^*Q$).

3.6.1   *Functional associated with a Lagrangian function*

Let $(TQ, \mathcal{L})$ be a Lagrangian dynamical system. We can associate to the
Lagrangian function a functional defined on a suitable space of curves,
which are considered to be possible trajectories of the system.

To obtain this, consider the manifolds $\mathbb{R} \times TQ$, $\mathbb{R} \times Q$, and $\mathbb{R}$ with the
canonical projections

$$\rho\colon \mathbb{R}\times Q \longrightarrow \mathbb{R}, \ \tau_{(1,0)} = id \times \tau_Q \colon \mathbb{R}\times TQ \longrightarrow \mathbb{R}\times Q, \ \tau_1 = \rho \circ \tau_{(1,0)} \colon \mathbb{R}\times TQ \longrightarrow \mathbb{R}.$$

The canonical projections ρ and τ_1 define a global coordinate in $\mathbb{R}$, which
is denoted as t and physically identified with "time". Then, we can take the
natural volume element dt in $\mathbb{R}$ and its natural lift to $\mathbb{R} \times TQ$, which is also
denoted by dt.

Now, let $\gamma\colon [a, b] \subset \mathbb{R} \longrightarrow Q$ be a curve, and let $\widetilde{\gamma}\colon [a, b] \subset \mathbb{R} \longrightarrow TQ$
be its canonical lift of γ from Q to the tangent bundle TQ. We denote
$\boldsymbol{\gamma}\colon [a, b] \subset \mathbb{R} \longrightarrow \mathbb{R} \times Q$, the curve given by $\boldsymbol{\gamma}(t) = (t, \gamma(t))$, and by $\widehat{\gamma}\colon [a, b] \subset \mathbb{R} \longrightarrow \mathbb{R} \times TQ$, the curve $\widehat{\gamma}(t) = (t, \widetilde{\gamma}(t))$.

Now, let α be a 1-form in $\mathbb{R} \times TQ$. As $\widehat{\gamma}^*\alpha$ is a 1-form in $\mathbb{R}$, we can
define

$$\int_{\widehat{\gamma}} \alpha = \int_a^b \widehat{\gamma}^*\alpha.$$

Finally, given a (time-dependent) Lagrangian function $\mathcal{L}\colon \mathbb{R} \times TQ \longrightarrow \mathbb{R}$,
consider the 1-form $\mathcal{L} \, dt \in \Omega^1(\mathbb{R} \times TQ)$. Then we define:

Definition 3.44. For every curve $\gamma\colon \mathbb{R} \longrightarrow Q$, the **action** of $\mathcal{L}$ along γ is
the functional

$$\mathbf{L}(\gamma) := \int_{\widehat{\gamma}} \mathcal{L} dt.$$

Remark 3.10. Usually, we take curves $\gamma\colon \mathbb{R} \longrightarrow Q$ defined on a closed interval $I = [a, b] \subset \mathbb{R}$, but we can extend the domain to the whole real numbers defining γ out of the interval as a constant function taking permanently the values on the two extreme points of the interval. In this case, observe that the extension may not be differentiable at the extreme points, but this is not a problem for the existence of the integral (from now on, it is only necessary that the curve is of class C^2).

If the Lagrangian function is time independent, it is enough to extend it from TQ to $\mathbb{R} \times TQ$.

3.6.2   *Hamilton variational principle*

The *variational problem* consists of optimising the above functional, that is, to choose the curves for which the functional takes the extremal value. As this is a very complicated problem, we restrict ourselves to look for the local extremal points of the functional; so we look for curves γ such that the value of $L(\gamma)$ does not change for small variations of the curve, but in first approximation. Next, we are going to clarify these ideas.

Definition 3.45. Let $\gamma\colon [a, b] \subset \mathbb{R} \longrightarrow Q$ be a curve such that $\gamma(a) = q_0$, $\gamma(b) = q_1$. A **variation** of γ is a map

$$\mu\colon (-\epsilon, \epsilon) \times [a, b] \subset \mathbb{R}^2 \longrightarrow Q$$

such that

(1) $\mu(0, t) = \gamma(t)$, $t \in [a, b]$.
(2) $\mu(s, a) = q_0$, $\mu(s, b) = q_1$, for $s \in (-\epsilon, \epsilon)$.

We denote μ_s the curve defined by $\mu_s\colon [a, b] \subset \mathbb{R} \longrightarrow Q$, with $\mu_s(t) = \mu(s, t)$.

Observe that given one of these curves μ_s, we can lift it to TQ and $\mathbb{R} \times TQ$ in the usual way. We denote these lifts as $\widetilde{\mu}_s$ and $\widehat{\mu}_s$, which are variations of $\widetilde{\gamma}$ and $\widehat{\gamma}$, respectively.

There is a natural way to construct different variations of that type. In fact, given $X \in \mathfrak{X}(Q)$ with $X(q_0) = 0$, $X(q_1) = 0$; if F_s is a local flux for X, a variation of the curve γ is obtained by

$$\mu(s, t) = (F_s \circ \gamma)(t) = F_s(\gamma(t)). \tag{3.25}$$

We know that $F_s\colon Q \longrightarrow Q$ is a diffeomorphism, and the corresponding lift $TF_s\colon TQ \longrightarrow TQ$ is the local flux corresponding to the canonical lift

$X^C \in \mathfrak{X}(TQ)$ of X to TQ. In the same way that (3.25) is a variation of γ, we have

$$\tilde{\mu}(s,t) = (\mathrm{T}F_s \circ \tilde{\gamma})(t) = \mathrm{T}F_s(\tilde{\gamma}(t)),$$

that is, a variation of $\tilde{\gamma}$. Observe that X^C is naturally a vector field in $\mathbb{R} \times TQ$ acting as the identity on the component $\mathbb{R}$. It is vertical with respect to the projection $\tau_1 \colon \mathbb{R} \times TQ \longrightarrow \mathbb{R}$.

Given γ and F_s as above, the following properties hold:

(1) $\mu_s(t) = (F_s \circ \gamma)(t) = F_s(\gamma(t))$.
(2) $\tilde{\mu}_s(t) = (\mathrm{T}F_s \circ \tilde{\gamma})(t) = \mathrm{T}F_s(\tilde{\gamma}(t))$.
(3) $\boldsymbol{\mu}_s(t) = (F_s \circ \boldsymbol{\gamma})(t) = F_s(\boldsymbol{\gamma})(t)) = (t, F_s(\gamma(t)))$.
(4) $\widehat{\boldsymbol{\mu}}_s(t) = (\mathrm{T}F_s \circ \widehat{\gamma})(t) = \mathrm{T}F_s(\widehat{\gamma}(t)) = (t, \mathrm{T}F_s(\tilde{\gamma}(t)))$.

These kinds of variations obtained from specific vector fields on the manifold Q are sufficient to obtain the local extreme of the functional $\mathbf{L}(\gamma)$ as we are going to see.

Definition 3.46. A curve γ is a **local extreme** for the functional **L** if for every variation $\mu_s = F_s \circ \gamma$, with F_s corresponding to a vector field X as above, the following condition holds:

$$\frac{d}{ds}\bigg|_{s=0} \int_{\widehat{\boldsymbol{\mu}}_s} \mathcal{L}\mathrm{d}t = 0.$$

This is the **Hamilton variational principle**.

To obtain a more useful way to manage with this definition, first we need the following results:

Proposition 3.22. *Let $X \in \mathfrak{X}(Q)$ and μ_s be the corresponding variations of a curve γ. Then*

(1) $\dfrac{d}{ds}\bigg|_{s=0} \displaystyle\int_{\widehat{\boldsymbol{\mu}}_s} \mathcal{L}\mathrm{d}t = \displaystyle\int_{\widehat{\gamma}} (\mathrm{L}(X^C)\mathcal{L}) \, \mathrm{d}t.$

(2) If $\alpha \in \Omega^1(TQ)$, then $\dfrac{d}{ds}\bigg|_{s=0} \displaystyle\int_{\tilde{\mu}_s} \alpha = \displaystyle\int_{\tilde{\gamma}} \mathrm{L}(X^C)\alpha.$

Proof. Considering that $\widetilde{\mu}_s = \widetilde{F}_s \circ \widetilde{\gamma}$ and $\widehat{\mu}_s = (TF_s \circ \widehat{\gamma})$, we have

(1) First,

$$\frac{d}{ds}\bigg|_{s=0} \int_{\widehat{\mu}_s} \mathcal{L} dt = \lim_{s \to 0} \frac{1}{s} \left(\int_{\widehat{\mu}_s} \mathcal{L} dt - \int_{\widehat{\mu}_0} \mathcal{L} dt \right)$$

$$= \lim_{s \to 0} \frac{1}{s} \left(\int_a^b \widetilde{\widehat{\mu}}_s^*(\mathcal{L}\, dt) - \int_a^b \widetilde{\widehat{\mu}}_0^*(\mathcal{L}\, dt) \right)$$

$$= \int_a^b \lim_{s \to 0} \frac{\widetilde{\widehat{\mu}}_s^*\mathcal{L} - \widetilde{\widehat{\mu}}_0^*\mathcal{L}}{s} dt$$

$$= \int_a^b \widehat{\gamma}^* \lim_{s \to 0} \frac{TF_s^*\mathcal{L} - TF_0^*\mathcal{L}}{s} dt$$

$$= \int_a^b \widehat{\gamma}^*(L(X^C)\mathcal{L}) dt = \int_{\widehat{\gamma}} (L(X^C)\mathcal{L}) dt,$$

because $\widehat{\mu}_s^* \alpha = (\widehat{\gamma}^* \circ TF_s^*)\alpha = \widehat{\gamma}^*(TF_s^*\alpha)$.

(2) Second,

$$\frac{d}{ds}\bigg|_{s=0} \int_{\widetilde{\mu}_s} \alpha = \lim_{s \to 0} \frac{1}{s} \left(\int_{\widetilde{\mu}_s} \alpha - \int_{\widetilde{\mu}_0} \alpha \right)$$

$$= \lim_{s \to 0} \frac{1}{s} \left(\int_a^b \widetilde{\mu}_s^* \alpha - \int_a^b \widetilde{\mu}_0^* \alpha \right)$$

$$= \int_a^b \lim_{s \to 0} \frac{\widetilde{\mu}_s^* \alpha - \widetilde{\mu}_0^* \alpha}{s} = \int_a^b \widetilde{\gamma}^* \lim_{s \to 0} \frac{T\widetilde{F}_s^* \alpha - T\widetilde{F}_0^* \alpha}{s}$$

$$= \int_a^b \widetilde{\gamma}^* L(X^C)\alpha = \int_{\widetilde{\gamma}} L(X^C)\alpha.$$

$\square$

Now, we must find the equation fulfilled by the curves $\gamma \colon [a, b] \subset \mathbb{R} \longrightarrow Q$ satisfying $\gamma(a) = q_0$, $\gamma(b) = q_1$, and such that

$$\frac{d}{ds}\bigg|_{s=0} \int_{\widetilde{\mu}_s} \mathcal{L} dt = \int_{\widetilde{\gamma}} L(X^C) dt = 0,$$

for every variation $\mu_s = F_s \circ \gamma$ of γ. Then, we need the following:

Lemma 3.1. *For every curve* $\gamma \colon [a, b] \subset \mathbb{R} \longrightarrow Q$, *we have*

$$\widetilde{\gamma}^*(A_{\mathcal{L}}\, dt) = \widetilde{\gamma}^*(\mathcal{L}\, dt) + \widetilde{\gamma}^*(E_{\mathcal{L}}\, dt) = \widetilde{\gamma}^* \Theta_{\mathcal{L}}.$$

Proof. If $t_0 \in [a, b]$, let (q^i, v^i) be a canonical local coordinate system in a neighbourhood of $\widetilde{\gamma}(t_0)$. On the one hand,

$$\left(\widetilde{\gamma}^*(A_{\mathcal{L}}\, dt)\right)_{t_0}\left(\frac{d}{dt}\Big|_{t_0}\right) = (\widetilde{\gamma}^* A_{\mathcal{L}})(t_0) = A_{\mathcal{L}}(\widetilde{\gamma}(t_0)) = \left(v^i \frac{\partial \mathcal{L}}{\partial v^i}\right)(\widetilde{\gamma}(t_0))$$

$$= \left(v^i \frac{\partial \mathcal{L}}{\partial v^i}\right)(\gamma(t_0), \dot{\gamma}(t_0)) = \dot{\gamma}(t_0)\frac{\partial \mathcal{L}}{\partial v^i}\Big|_{\widetilde{\gamma}(t_0)},$$

and on the other hand,

$$\left(\widetilde{\gamma}^* \Theta_{\mathcal{L}}\right)_{t_0}\left(\frac{d}{dt}\Big|_{t_0}\right) = (\Theta_{\mathcal{L}})_{t_0}\left(\mathrm{T}_{t_0}\widetilde{\gamma}\left(\frac{d}{dt}\Big|_{t_0}\right)\right)$$

$$= (\mathrm{d}\mathcal{L} \circ J)_{\widetilde{\gamma}(t_0)}\left(\mathrm{T}_{t_0}\widetilde{\gamma}\left(\frac{d}{dt}\Big|_{t_0}\right)\right)$$

$$= (\mathrm{d}\mathcal{L} \circ J)_{\widetilde{\gamma}(t_0)}\left(\dot{\gamma}^i(t_0)\frac{\partial}{\partial q^i}\Big|_{\widetilde{\gamma}(t_0)} + \ddot{\gamma}^i(t_0)\frac{\partial}{\partial v^i}\Big|_{\widetilde{\gamma}(t_0)}\right)$$

$$= (\mathrm{d}\mathcal{L})_{\widetilde{\gamma}(t_0)}\left(\dot{\gamma}^i(t_0)\frac{\partial}{\partial v^i}\Big|_{\widetilde{\gamma}(t_0)}\right) = \dot{\gamma}^i(t_0)\frac{\partial}{\partial v^i}\Big|_{\widetilde{\gamma}(t_0)},$$

and the result follows. $\square$

3.6.3   *Euler–Lagrange equations*

Now, we can obtain the equation for the curve γ solution to the variational problem.

Theorem 3.13. *The curves γ that are local extremes for the functional* **L** *are the solutions to the Euler–Lagrange equations.*

Proof. We have

$$0 = \frac{d}{ds}\Big|_{s=0}\int_{\widehat{\boldsymbol{\mu}}_s}(\mathcal{L}\, dt) = \frac{d}{ds}\Big|_{s=0}\int_a^b \widetilde{\mu}_s^* \mathcal{L}\, dt = \frac{d}{ds}\Big|_{s=0}\int_a^b \widehat{\mu}_s^* \Theta_{\mathcal{L}} - \widehat{\mu}_s^* E_{\mathcal{L}}\, dt$$

$$= \frac{d}{ds}\Big|_{s=0}\int_{\widehat{\boldsymbol{\mu}}_s}\Theta_{\mathcal{L}} - E_{\mathcal{L}}\, dt = \int_{\widetilde{\gamma}}\mathrm{L}(X^C)\Theta_{\mathcal{L}} - \mathrm{L}(X^C)(E_{\mathcal{L}})\, dt$$

$$= \int_{\widetilde{\gamma}}\mathrm{d}\, i(X^C)\Theta_{\mathcal{L}} + i(X^C)\mathrm{d}\Theta_{\mathcal{L}} - \mathrm{L}(X^C)(E_{\mathcal{L}})\, dt.$$

However, by the Stokes Theorem, being $X(q_0) = 0 = X(q_1)$ and $\Theta_{\mathcal{L}}$ a semibasic form, we have $\int_{\widetilde{\gamma}} d\, i(\widetilde{X}) d\Theta_{\mathcal{L}} = 0$; hence,

$$
\begin{aligned}
0 &= \int_{\widetilde{\gamma}} i(X^C) d\Theta_{\mathcal{L}} - \mathrm{L}(X^C)(E_{\mathcal{L}}) dt \\
&= \int_a^b \widetilde{\gamma}^*(i(X^C) d\Theta_{\mathcal{L}}) - \widetilde{\gamma}^*(\mathrm{L}(X^C)(E_{\mathcal{L}})) dt \\
&= \int_a^b \left[d\Theta_{\mathcal{L}}(X^C, (\mathrm{T}_t\widetilde{\gamma}) \frac{d}{dt}) - (\mathrm{L}(X^C)(E_{\mathcal{L}}))(\widetilde{\gamma}(t)) \right] dt,
\end{aligned}
$$

because

$$
\begin{aligned}
\left(\widetilde{\gamma}^*(i(X^C) d\Theta_{\mathcal{L}}) \frac{d}{dt}\right)\Big|_t &= (i(X^C) d\Theta_{\mathcal{L}}) \left(\mathrm{T}_t\widetilde{\gamma} \frac{d}{dt}\right) \\
&= (d\Theta_{\mathcal{L}})_{\widetilde{\gamma}(t)} \left(X^C, \mathrm{T}_t\gamma(\frac{d}{dt})\right),
\end{aligned}
$$

and we have

$$
0 = \int_a^b i(X^C_{\widetilde{\gamma}(t)}) \left[-i\left(\mathrm{T}_t\widetilde{\gamma}\frac{d}{dt}\right) d\Theta_{\mathcal{L}} - dE_{\mathcal{L}} \right] dt.
$$

However, X^C is an arbitrary vector field; then, observing $\mathrm{T}_t\widetilde{\gamma}\left(\dfrac{d}{dt}\right) = (\widetilde{\gamma}(t), \dot{\widetilde{\gamma}}(t))$, we have

$$
-i(\widetilde{\gamma}(t), \dot{\widetilde{\gamma}}(t)) d\Theta_{\mathcal{L}} = dE_{\mathcal{L}},
$$

and as $-d\Theta_{\mathcal{L}} = \Omega_{\mathcal{L}}$, we obtain

$$
i(\widetilde{\gamma}(t), \dot{\widetilde{\gamma}}(t)) \Omega_{\mathcal{L}} = dE_{\mathcal{L}}.
$$

(Recall that $(\widetilde{\gamma}(t), \dot{\widetilde{\gamma}}(t))$ is the tangent vector to the curve $\widetilde{\gamma}$ at the point $\gamma(t)$.)

If we consider that the curves $\widetilde{\gamma}(t)$, which are the solutions to the problem, are the integral curves of a vector field $X_{\mathcal{L}} \in \mathfrak{X}(TQ)$, observing that the curve $\widetilde{\gamma}$ is a natural lift to TQ of the curve γ in Q, then this vector field must verify the following conditions:

(1) $i(X_{\mathcal{L}})\Omega_{\mathcal{L}} = dE_{\mathcal{L}}$.
(2) $X_{\mathcal{L}}$ is a SODE ($J(X_{\mathcal{L}}) = \Delta$).

As we know, these are the geometrical expression of the Euler–Lagrange equations. $\square$

As a final remark, we have observed that the vector field X^C is arbitrary, and this is not exact. The real arbitrary vector field is X instead of X^C. However, the 1-form $i\left(T\widetilde{\gamma}\left(\dfrac{d}{dt}\right)\right)d\Theta_{\mathcal{L}}+dE_{\mathcal{L}}$ is semibasic, and hence, when we apply it to X^C, the obtained value depends only on X and not on its canonical lift X^C from Q to TQ.

3.6.4 Hamilton–Jacobi variational principle: Relationship with the canonical Hamiltonian formalism

When we studied the Hamiltonian formalism in the previous sections, we commented about the possibility of obtaining the dynamical equations from a variational principle. Following the discussion on the equivalence between the Lagrangian and the Hamiltonian formalisms, we can establish the relationship between the corresponding variational formalisms.

We have studied the above variational approach to the Lagrangian formalism, then using the corresponding geometrical and dynamical elements, we can relate it with the Hamiltonian one. Thus, let $\mathcal{FL}\colon TQ \longrightarrow T^*Q$ be the Legendre map associated with the Lagrangian $\mathcal{L}$. If $\gamma\colon [a,b] \subset \mathbb{R} \longrightarrow Q$ is a curve, we have

$$\mathbf{L}(\gamma) = \int_{\widehat{\gamma}} \mathcal{L}dt = \int_{\widehat{\gamma}} \Theta_{\mathcal{L}} - E_{\mathcal{L}}\,dt = \int_{\widehat{\gamma}} \mathcal{FL}^*\Theta - \mathcal{FL}^*\mathrm{h}\,dt$$

$$= \int_{\widehat{\gamma}} \mathcal{FL}^*(\Theta - \mathrm{h}\,dt) = \int_{\mathcal{FL}\circ\widehat{\gamma}} \Theta - \mathrm{h}\,dt;$$

where we have extended the Legendre transformation $\mathcal{FL}\colon TQ \longrightarrow T^*Q$ (with the same notation) to $\mathcal{FL}\colon \mathbb{R} \times TQ \longrightarrow \mathbb{R} \times T^*Q$, as the identity on $\mathbb{R}$ (see Definition 4.6). Note that if γ is a solution to the Hamilton variational problem, that is of the Lagrangian formalism, then $\mathcal{FL} \circ \widehat{\gamma}$ is the solution to the following variational problem:

> **Definition 3.47.** The **Hamilton–Jacobi variational principle** consists of finding the curves $\zeta\colon [a,b] \subset \mathbb{R} \longrightarrow T^*Q$, with fixed extremes, such that they are extremal for the functional
>
> $$\mathbf{H}(\zeta) := \int_{\zeta} \Theta - \mathrm{h}dt.$$

By a similar computation as above, although something simpler because there are no canonical lifts, we obtain

Theorem 3.14. *The curves ζ that are local extremes for the functional* **H** *are the solutions to the Hamilton equations.*

Proof. In fact, let $\zeta\colon [a,b] \subset \mathbb{R} \longrightarrow T^*Q$ be a curve with $\zeta(a) = A, \zeta(b) = B$, and $Z \in \mathfrak{X}(T^*Q)$ with $Z(A) = 0, Z(B) = 0$. Let F_s be the flux of Z and consider the variation $\eta_s = F_s \circ \zeta$. We have

$$\mathbf{H}(F_s \circ \zeta) = \int_a^b (F_s \circ \zeta)^* \Theta - (\mathrm{h} \circ F_s \circ \zeta) dt = \int_a^b \zeta^*(F_s^* \Theta) - \zeta^*(F_s^* \mathrm{h}) dt.$$

Hence,

$$\frac{d}{ds}\Big|_{s=0} \mathbf{H}(F_s \circ \zeta) = \int_a^b \zeta^*(\mathrm{L}(Z)\Theta) - \zeta^*(\mathrm{L}(Z)\mathrm{h}) dt = \int_\zeta \mathrm{L}(Z)\Theta - (\mathrm{L}(Z)\mathrm{h}) dt.$$

Following the same patterns as in Theorem 3.13, we conclude that the tangent vector to the curve ζ satisfies the Hamilton equations

$$i(\widetilde{\zeta})\Omega = \mathrm{dh}(\zeta(t)),$$

with $\Omega = -\mathrm{d}\Theta$, the symplectic form of the cotangent bundle. If we suppose that the curve is an integral curve of a vector field X_h, then

$$i(X_\mathrm{h})\Omega = \mathrm{dh}.$$

$\square$

Remark 3.11. The Hamilton–Jacobi variational principle can be stated for every Hamiltonian system (M, Ω, α) in general, taking a symplectic potential $\theta \in \Omega^1(M)$, such that $\Omega = -\mathrm{d}\theta$, and a (local) Hamiltonian function $h \in C^\infty(M)$, such that $\alpha = \mathrm{dh}$.

In summary, in the texts of classical mechanics, all the above results are collected stating:

Assertion 3.1 (Minimal action principles). *Given a Lagrangian system* $(TQ, \mathcal{L})$ *and the corresponding associated canonical Hamiltonian system* $(T^*Q, \Omega, \mathrm{h})$, *we have*

Hamilton's principle of minimal action: *The dynamics of the Lagrangian system* $(TQ, \mathcal{L})$ *is given by the curves* $\gamma\colon [a,b] \subset \mathbb{R} \longrightarrow Q$, *with fixed extremes, such that they minimise the functional*

$$\mathbf{L}(\gamma) := \int_{\widehat{\gamma}} \mathcal{L} \mathrm{dt}.$$

Hamilton–Jacobi principle of minimal action: *The dynamics of the Hamiltonian system* (T^*Q, Ω, dh) *is given by the curves* $\zeta \colon [a,b] \subset \mathbb{R} \longrightarrow T^*Q$, *with fixed extremes, such that they minimise the functional*

$$\mathbf{H}(\zeta) := \int_\zeta \Theta - h \, dt.$$

3.7 Examples

In this last section, we study two of the most typical and relevant mechanical systems using the symplectic Lagrangian, the Hamiltonian, and the unified formalisms.

3.7.1 *Harmonic oscillator*

The classical harmonic oscillator is a mechanical system made of a point-particle with mass m moving in $\mathbb{R}$, submitted to a recuperative force (*Hook's law*). The configuration bundle is $Q = \mathbb{R}$, with coordinate (q).

3.7.1.1 *Lagrangian formalism*

The Lagrangian formalism takes place in $TQ \simeq \mathbb{R}^2$, with coordinates (q, v), and the Lagrangian function is

$$\mathcal{L} = \frac{1}{2}(mv^2 - kq^2), \quad k \in \mathbb{R}^+.$$

The Lagrangian elements are

$$E_{\mathcal{L}} = \frac{1}{2}(mv^2 + kq^2), \quad \Theta_{\mathcal{L}} = mv dq, \quad \Omega_{\mathcal{L}} = m \, dq \wedge dv,$$

and the Lagrangian is regular. For $X_{\mathcal{L}} = f\dfrac{\partial}{\partial q} + g\dfrac{\partial}{\partial v}$, Eq. (3.4) gives

$$i(X_{\mathcal{L}})\Omega_{\mathcal{L}} = m \, (f \, dv - g \, dq) = mv \, dv + kq \, dq = dE_{\mathcal{L}},$$

which leads to

$$f = v, \quad mg = -kq.$$

So, the Euler–Lagrange vector field is

$$X_{\mathcal{L}} = v\frac{\partial}{\partial q} - \frac{k}{m}q\frac{\partial}{\partial v},$$

and its integral curves $(q(t), v(t))$ are the solutions to

$$\frac{dq}{dt} = v, \quad m\frac{dv}{dt} = -kq \quad \Longrightarrow \quad m\frac{d^2q}{dt^2} = -kq,$$

which is the Euler–Lagrange equation for the system.

3.7.1.2 *Hamiltonian formalism*

For the Hamiltonian formalism, $T^*Q \simeq \mathbb{R}^2$, with coordinates (q, p). First, the Legendre transformation is

$$\mathcal{FL}^* q = q, \quad \mathcal{FL}^* p = mv,$$

which is a diffeomorphism (the Lagrangian is hyperregular). The canonical Hamiltonian function is

$$h = \frac{p^2}{2m} + kq^2.$$

As $\Omega = dq \wedge dp$, for $X_h = F \dfrac{\partial}{\partial q} + G \dfrac{\partial}{\partial p}$, Eq. (3.11) gives

$$i(X_h)\Omega = F\, dp - G\, dq- = \frac{p}{m}\, dp + kq\, dq = dh,$$

which leads to

$$F = \frac{p}{m}, \quad G = -kq.$$

So, the Hamiltonian vector field is

$$X_h = \frac{p}{m} \frac{\partial}{\partial q} - kq \frac{\partial}{\partial p},$$

and its integral curves $(q(t), p(t))$ are the solutions to

$$m\frac{dq}{dt} = p, \quad \frac{dp}{dt} = -kq,$$

which are the Hamilton equations for the system.

Observe that using the Legendre map, the Hamilton and the Euler–Lagrange equations of the system are, in fact, equivalent. We have $\mathcal{FL}_* X_{\mathcal{L}} = X_h$.

3.7.1.3 *Unified Lagrangian–Hamiltonian formalism*

Let us consider the unified bundle $\mathcal{W} = TQ \times_Q T^*Q \simeq \mathbb{R}^3$ with coordinates (q, v, p). In this, we have the canonical presymplectic form

$$\Omega_{\mathcal{W}} = dq \wedge dp$$

and the Hamiltonian function

$$\mathcal{H} = pv - \frac{1}{2}(mv^2 - kq^2).$$

For $X_{\mathcal{H}} = f\dfrac{\partial}{\partial q} + g\dfrac{\partial}{\partial v} + G\dfrac{\partial}{\partial p}$, Eq. (3.16) gives

$$i(X_{\mathcal{H}})\Omega_{\mathcal{W}} = f\, dp - G\, dq = kq\, dq + (p - mv)\, dv + v\, dp = d\mathcal{H},$$

which leads to

$$f = v, \quad G = -kq, \quad p = mv.$$

The last equation is a constraint that defines the submanifold $\mathcal{W}_0 \hookrightarrow \mathcal{W}$ and gives the Legendre map. Therefore, the Hamiltonian vector field is

$$X_{\mathcal{H}}|_{\mathcal{W}_0} = v\frac{\partial}{\partial q} + g\frac{\partial}{\partial v} - kq\frac{\partial}{\partial p}.$$

Then, the tangency condition leads to

$$X_{\mathcal{H}}(p - mv) = -kq - gm = 0 \iff g = -\frac{kq}{m} \quad (\text{on } \mathcal{W}_0)$$

and then

$$X_{\mathcal{H}}|_{\mathcal{W}_0} = v\frac{\partial}{\partial q} - \frac{kq}{m}\frac{\partial}{\partial v} - kq\frac{\partial}{\partial p}.$$

Its integral curves $(q(t), v(t), p(t))$ are the solutions to

$$\frac{dq}{dt} = v, \quad m\frac{dv}{dt} = -kq, \quad \frac{dp}{dt} = -kq.$$

The first two equations are equivalent to

$$m\frac{d^2q}{dt^2} = -kq,$$

which is the Euler–Lagrange equation of the system. Furthermore, using the constraint $p = mv$ (the Legendre map), the first and third equations are

$$\frac{dq}{dt} = \frac{p}{m}, \quad \frac{dp}{dt} = -kq,$$

which are the Hamilton equations for the system.

3.7.2 *Central forces: The Kepler problem*

The *Kepler problem* consists of studying the motion of a particle of mass m under the action of *Newtonian central forces*. It is well known that the motion of such a particle is on a plane and hence $Q = \mathbb{R}^2$. We take polar coordinates (r, ϕ) in the plane (with origin at the center of the force).

3.7.2.1 *Lagrangian formalism*

The Lagrangian formalism takes place in $TQ \simeq \mathbb{R}^4$, with local coordinates (r, ϕ, v_r, v_ϕ). The Lagrangian function is

$$\mathcal{L} = \frac{1}{2}m(v_r^2 + r^2 v_\phi^2) - \frac{K}{r}, \quad K \neq 0 \,;$$

therefore

$$E_{\mathcal{L}} = \tfrac{1}{2}m(v_r^2 + r^2 v_\phi^2) + \tfrac{K}{r},$$
$$\Theta_{\mathcal{L}} = m(v_r\, dr + r^2 v_\phi\, d\phi),$$
$$\Omega_{\mathcal{L}} = m(dr \wedge dv_r + r^2 d\phi \wedge dv_\phi - 2r v_\phi\, dr \wedge d\phi),$$

and the Lagrangian is regular. For $X_{\mathcal{L}} = f_r \dfrac{\partial}{\partial r} + f_\phi \dfrac{\partial}{\partial \phi} + g_r \dfrac{\partial}{\partial v_r} + g_\phi \dfrac{\partial}{\partial v_\phi}$, Eq. (3.4) gives

$$i(X_{\mathcal{L}})\Omega_{\mathcal{L}} = m\left[f_r\, dv_r + f_\phi r^2\, dv_\phi - (g_r - 2r v_\phi f_\phi)\, dr - (g_\phi r^2 + 2r v_\phi f_r)\, d\phi \right]$$
$$= mv_r\, dv_r + mr^2 v_\phi\, dv_\phi + \left(mr v_\phi^2 - \frac{K}{r^2} \right) dr = dE_{\mathcal{L}},$$

which leads to

$$f_r = v_r, \quad f_\phi = v_\phi, \quad mg_r = 2mr v_\phi f_\phi - mr v_\phi^2 + \frac{K}{r^2}, \quad g_\phi = -\frac{2v_\phi f_r}{r},$$

and then the Euler–Lagrange vector field is

$$X_{\mathcal{L}} = v_r \frac{\partial}{\partial r} + v_\phi \frac{\partial}{\partial \phi} + \left(r v_\phi^2 + \frac{K}{mr^2} \right) \frac{\partial}{\partial v_r} - \frac{2v_\phi v_r}{r} \frac{\partial}{\partial v_\phi}.$$

Then, its integral curves $(r(t), \phi(t), v_r(t), v_\phi(t))$ are the solutions to

$$\frac{dr}{dt} = v_r, \quad \frac{d\phi}{dt} = v_\phi, \quad m\frac{dv_r}{dt} = mr v_\phi^2 + \frac{K}{r^2}, \quad \frac{dv_\phi}{dt} = -\frac{2v_\phi v_r}{r}$$

$$\implies \quad m\frac{d^2 r}{dt^2} = mr\left(\frac{d\phi}{dt} \right)^2 + \frac{K}{r^2}, \quad \frac{d^2\phi}{dt^2} = -\frac{2}{r}\frac{d\phi}{dt}\frac{dr}{dt}$$

$$\implies \quad \frac{d}{dt}\left(m\frac{dr}{dt} \right) = mr\left(\frac{d\phi}{dt} \right)^2 + \frac{K}{r^2}, \quad \frac{d}{dt}\left(mr^2 \frac{d\phi}{dt} \right) = 0,$$

which are the Euler–Lagrange equations for this system.

There is a Lagrangian exact Noether symmetry, which is generated by the vector field $Y = \dfrac{\partial}{\partial \phi}$, since

$$
\begin{aligned}
L(Y)\Theta_{\mathcal{L}} &= L\left(\frac{\partial}{\partial \phi}\right)(m(v_r\, dr + r^2 v_\phi\, d\phi)) \\
&= d\, i\left(\frac{\partial}{\partial \phi}\right)(m(v_r\, dr + r^2 v_\phi\, d\phi)) + i\left(\frac{\partial}{\partial \phi}\right) d\left(m(v_r\, dr + r^2 v_\phi\, d\phi)\right) \\
&= d(mr^2 v_\phi) + i\left(\frac{\partial}{\partial \phi}\right)(m(dv_r \wedge dr + r^2 dv_\phi \wedge d\phi - 2rv_\phi\, d\phi \wedge dr)) \\
&= m(2rv_\phi dr + r^2 dv_\phi - r^2 dv_\phi - 2rv_\phi dr) = 0,
\end{aligned}
\tag{3.26}
$$

$$
L(Y)E_{\mathcal{L}} L\left(\frac{\partial}{\partial \phi}\right)\left(\frac{1}{2}m(v_r^2 + r^2 v_\phi^2) + \frac{K}{r}\right) = 0,
\tag{3.27}
$$

and its associated conserved quantity is

$$
f_Y = i\left(\frac{\partial}{\partial \phi}\right)\Theta_{\mathcal{L}} = mr^2 v_\phi,
$$

that is, the angular momentum, as the last Euler–Lagrange equation shows.

3.7.2.2   *Hamiltonian formalism*

For the Hamiltonian formalism, $T^*Q \simeq \mathbb{R}^4$, with local coordinates (r, ϕ, p_r, p_ϕ). First, the Legendre transformation is,

$$
\mathcal{FL}^*r = r, \quad \mathcal{FL}^*\phi = \phi, \quad \mathcal{FL}^*p_r = mv_r, \quad \mathcal{FL}^*p_\phi = mr^2 v_\phi,
$$

which is a diffeomorphism (the Lagrangian is hyperregular). The canonical Hamiltonian function is

$$
h = \frac{p_r^2}{2m} + \frac{p_\phi^2}{2mr^2} + \frac{K}{r}.
$$

As $\Omega = dr \wedge dp_r + d\phi \wedge dp_\phi$, for $X_h = F_r \dfrac{\partial}{\partial r} + F_\phi \dfrac{\partial}{\partial \phi} + G_r \dfrac{\partial}{\partial p_r} + G_\phi \dfrac{\partial}{\partial p_\phi}$, Eq. (3.11) gives

$$
\begin{aligned}
i(X_h)\Omega &= F_r\, dp_r + F_\phi\, dp_\phi - G_r\, dr - G_\phi\, d\phi \\
&= \frac{p_r}{m}\, dp_r + \frac{p_\phi}{mr^2}\, dp_\phi - \left(\frac{p_\phi^2}{mr^3} + \frac{K}{r^2}\right) dr = dh,
\end{aligned}
$$

which leads to

$$
F_r = \frac{p_r}{m}, \quad F_\phi = \frac{p_\phi}{mr^2}, \quad G_r = \frac{p_\phi^2}{mr^3} + \frac{K}{r^2}, \quad G_\phi = 0.
$$

Then, the Hamiltonian vector field is

$$X_h = \frac{p_r}{m}\frac{\partial}{\partial r} + \frac{p_\phi}{mr^2}\frac{\partial}{\partial \phi} + \left(\frac{p_\phi^2}{mr^3} + \frac{K}{r^2}\right)\frac{\partial}{\partial p_r},$$

and its integral curves $(r(t), \phi(t), p_r(t), p_\phi(t))$ are the solutions to

$$m\frac{dr}{dt} = p_r, \quad mr^2\frac{d\phi}{dt} = p_\phi, \quad \frac{dp_r}{dt} = \frac{p_\phi^2}{mr^3} + \frac{K}{r^2}, \quad \frac{dp_\phi}{dt} = 0,$$

which are the Hamilton equations for this system.

As in the above example, the Legendre map is used to check that the Hamilton and the Euler–Lagrange equations of the system are, in fact, equivalent. Then, $\mathcal{FL}_*X_{\mathcal{L}} = X_h$.

The Hamiltonian exact Noether symmetry is again the vector field $Y = \frac{\partial}{\partial \phi}$, since

$$L(Y)\Theta = L\left(\frac{\partial}{\partial \phi}\right)(p_r\, dr + p_\phi\, d\phi)$$

$$= d\, i\left(\frac{\partial}{\partial \phi}\right)(p_r\, dr + p_\phi\, d\phi) + i\left(\frac{\partial}{\partial \phi}\right) d\,(p_r\, dr + p_\phi\, d\phi)$$

$$= dp_\phi - dp_\phi = 0,$$

$$L(Y)h = L\left(\frac{\partial}{\partial \phi}\right)\left(\frac{p_r^2}{2m} + \frac{p_\phi^2}{2mr^2} + \frac{K}{r}\right) = 0,$$

and as the last Hamilton equation shows, its associated conserved quantity is again the angular momentum

$$f_Y = i\left(\frac{\partial}{\partial \phi}\right)\Theta = p_\phi.$$

This Hamiltonian system is also a good and simple example to show how the geometric method of reduction by symmetries proceeds. The symmetry group G is the group of rotations on the orbit plane. The Lie algebra $\mathbf{g}$ is spanned by the vector field $\xi \equiv \frac{\partial}{\partial \phi}$, so $\mathbf{g}^* = \{d\phi\}$. Thus, the set of fundamental vector fields $\tilde{\mathbf{g}}$ is generated by the vector field $\tilde{\xi} \equiv Y = \frac{\partial}{\partial \phi} \in \mathfrak{X}(T^*Q)$. The action $\Phi\colon G \times T^*Q \longrightarrow T^*Q$ is effective, free, and proper, and it is a strongly symplectic action on the symplectic manifold (T^*Q, Ω), since it is exact, as we have seen. In this way, the momentum map is given by

$$(J(r, p_r, \phi, p_\phi))\left(\frac{\partial}{\partial \phi}\right) := p_\phi \quad \text{(for every } (r, p_r, \phi, p_\phi) \in T^*Q),$$

and for every weakly regular value $\mu = \mu_0 d\phi \in \mathbf{g}^*$, the level sets of this map are

$$\mathrm{J}^{-1}(\mu) := \{(r, p_r, \phi, p_\phi) \in \mathrm{T}^*Q \mid p_\phi = \mu_0\} \,.$$

They are defined by the constraints $\zeta := p_\phi - \mu_0$, that is, the hypersurfaces of constant angular momentum in T^*Q, and hence, $\dfrac{\partial}{\partial \phi}$ is tangent to all of them. On each one, we have the presymplectic Hamiltonian system $(\mathcal{J}^{-1}(\mu), \Omega_\mu, \mathrm{h}_\mu)$, where

$$\Omega_\mu := j_\mu^* \Omega = dp_r \wedge dr, \quad \mathrm{h}_\mu = \frac{p_r^2}{2m} + \frac{K}{r} \,,$$

so $\ker \Omega_\mu = \left\langle \dfrac{\partial}{\partial \phi} \right\rangle$. In this case, $G_\mu = G$, and applying the Marsden–Weinstein reduction theorem, this presymplectic system reduces to another symplectic one, $(\mathcal{J}^{-1}(\mu)/G, \widehat{\Omega}, \widehat{\mathrm{h}})$, where the local coordinates are (r, p_r), and

$$\widehat{\Omega} = dp_r \wedge dr, \quad \widehat{\mathrm{h}} = \frac{p_r^2}{2m} + \frac{K}{r}.$$

The Hamiltonian equation $i(\widehat{X})\widehat{\Omega} = d\widehat{\mathrm{h}}$, with $\widehat{X} \in \mathfrak{X}(\mathcal{J}^{-1}(\mu)/G)$, gives the Hamiltonian vector field

$$\widehat{X} = \frac{p_r}{m} \frac{\partial}{\partial r} + \left(\frac{p_\phi^2}{mr^3} + \frac{K}{r^2} \right) \frac{\partial}{\partial p_r} \,,$$

whose integral curves $(r(t), p_r(t))$ are the solutions to the Hamilton equations

$$\frac{dr}{dt} = \frac{p_r}{m}, \quad \frac{dp_r}{dt} = \frac{p_\phi^2}{mr^3} + \frac{K}{r^2} \,.$$

To obtain the complete set of Hamiltonian equations of the system, first remember that the Hamiltonian vector field $X_\mathrm{h} \in \mathfrak{X}(\mathrm{T}^*Q)$ is tangent to the level sets $p_\phi = ctn.$, and second, that by the Legendre map, $p_\phi = mr^2 v_\phi$. Therefore, for the integral curves of $X_\mathrm{h} \in \mathfrak{X}(\mathrm{T}^*Q)$, we have

$$\frac{d\phi}{dt} = \frac{p_\phi}{mr^2}, \quad \frac{dp_\phi}{dt} = 0.$$

3.7.2.3 Unified Lagrangian–Hamiltonian formalism

Consider the unified bundle $\mathcal{W} = \mathrm{T}Q \times_Q \mathrm{T}^*Q \simeq \mathbb{R}^6$ with coordinates $(r, \phi, v_r, v_\phi, p_r, p_\phi)$. The canonical presymplectic form is

$$\Omega_{\mathcal{W}} = dr \wedge dp_r + d\phi \wedge dp_\phi \,,$$

and the Hamiltonian function is

$$\mathcal{H} = p_r v_r + p_\phi v_\phi - \frac{1}{2}m(v_r^2 + r^2 v_\phi^2) + \frac{K}{r}.$$

For $X_\mathcal{H} = f_r \dfrac{\partial}{\partial r} + f_\phi \dfrac{\partial}{\partial \phi} + g_r \dfrac{\partial}{\partial v_r} + g_\phi \dfrac{\partial}{\partial v_\phi} + G_r \dfrac{\partial}{\partial p_r} + G_\phi \dfrac{\partial}{\partial p_\phi}$, Eq. (3.16)

gives

$$i(X_\mathcal{H})\Omega_\mathcal{W} = f_r\, dp_r + f_\phi\, dp_\phi - G_r\, dr - G_\phi\, d\phi$$
$$= -\left(\frac{K}{r^2} + mrv_\phi^2\right)dr + (p_r - mv_r)\, dv_r$$
$$+ (p_\phi - mr^2 v_\phi)\, dv_\phi + v_r\, dp_r + v_\phi\, dp_\phi$$
$$= d\mathcal{H},$$

which leads to

$$f_r = v_r, \quad f_\phi = v_\phi, \quad G_r = \frac{K}{r^2} + mrv_\phi^2, \quad G_\phi = 0,$$
$$p_r = mv_r, \quad p_\phi = mr^2 v_\phi.$$

The last two equations are constraints defining the submanifold $\mathcal{W}_0 \hookrightarrow \mathcal{W}$, which give the Legendre map. The Hamiltonian vector field is

$$X_\mathcal{H}|_{\mathcal{W}_0} = v_r \frac{\partial}{\partial r} + v_\phi \frac{\partial}{\partial \phi} + g_r \frac{\partial}{\partial v_r} + g_\phi \frac{\partial}{\partial v_\phi} + \left(\frac{K}{r^2} + mrv_\phi^2\right)\frac{\partial}{\partial p_r},$$

and the tangency condition leads to

$$X_\mathcal{H}(p_r - mv_r) = \frac{K}{r^2} + mrv_\phi^2 - g_r m = 0 \iff g_r = \frac{K}{mr^2} + rv_\phi^2 \ \text{(on } \mathcal{W}_0\text{)},$$
$$X_\mathcal{H}(p_\phi - mr^2 v_\phi) = -m(g_\phi r^2 + 2f_r rv_\phi) = 0 \iff g_\phi = -\frac{2v_r v_\phi}{r} \ \text{(on } \mathcal{W}_0\text{)};$$

therefore,

$$X_\mathcal{H}|_{\mathcal{W}_0} = v_r \frac{\partial}{\partial r} + v_\phi \frac{\partial}{\partial \phi} + \left(rv_\phi^2 + \frac{K}{mr^2}\right)\frac{\partial}{\partial v_r} - \frac{2v_r v_\phi}{r}\frac{\partial}{\partial v_\phi} + \left(\frac{K}{r^2} + mrv_\phi^2\right)\frac{\partial}{\partial p_r},$$

and its integral curves $(r(t), \phi(t), v_r(t), v_\phi(t), p_r(t), p_\phi(t))$ are the solutions to

$$\frac{dr}{dt} = v_r, \quad \frac{d\phi}{dt} = v_\phi, \quad \frac{dv_r}{dt} = \frac{K}{mr^2} + rv_\phi^2,$$
$$\frac{dv_\phi}{dt} = -\frac{2v_r v_\phi}{r}, \quad \frac{dp_r}{dt} = \frac{K}{r^2} + mrv_\phi^2, \quad \frac{dp_\phi}{dt} = 0.$$

The first four equations are equivalent to

$$m\frac{d^2 r}{dt^2} = mr\left(\frac{d\phi}{dt}\right)^2 + \frac{K}{r^2}, \quad \frac{d^2\phi}{dt^2} = -\frac{2}{r}\frac{d\phi}{dt}\frac{dr}{dt},$$

which are the Euler–Lagrange equations of the system. Furthermore, using the constraints $p_r = mv_r$ and $p_\phi = mr^2 v_\phi$ (i.e., the Legendre map), the first, second, fifth, and sixth equations are

$$\frac{dr}{dt} = \frac{p_r}{m}, \quad \frac{d\phi}{dt} = \frac{p_\phi}{mr^2}, \quad \frac{dp_r}{dt} = \frac{p_\phi^2}{mr^3} + \frac{K}{r^2}, \quad \frac{dp_\phi}{dt} = 0,$$

which are the Hamilton equations for the system.

3.8 Exercises

3.8.1 Geometric structures in tangent and cotangent bundles

Exercise 3.1

A 1-form α in TQ is *horizontal* or *semibasic* if it vanishes on the vertical tangent vectors. Give the local expression of any semibasic 1-form in a natural chart of TQ.

A 1-form α in TQ is *basic* if there exists another 1-form $\beta \in \Omega^1(Q)$ such that $\tau_Q^* \beta = \alpha$. Check that a basic form is semibasic. Give the local expression of any basic 1-form in a natural chart of TQ.

Exercise 3.2

A function $f \colon TQ \longrightarrow \mathbb{R}$ is *homogeneous* of degree $k \in \mathbb{N}$ if $f(p, \lambda u) = \lambda^k f(p, u)$, for every $(p, u) \in TQ$, $\lambda \in \mathbb{R}$. Prove that this condition for f is equivalent to $\mathrm{L}(\Delta) f = k f$, where Δ is the Liouville vector field in TQ.

Exercise 3.3

Let $X \in \mathfrak{X}(Q)$ and let X^C and X^V be the complete and vertical lifts of X to TQ, respectively. Let $\Gamma \in \mathfrak{X}(TQ)$ be a SODE vector field. Prove that

(1) $J(X^C) = X^V$ and $J(X^V) = 0$.
(2) $[X^C, Y^C] = [X, Y]^C$.
(3) $[X^V, Y^C] = [X, Y]^V$.
(4) $[X^C, \Delta] = 0$ and $[X^V, \Delta] = 0$.
(5) $J([X^C, \Gamma]) = [X^C, J(\Gamma)] = 0$.
(6) $J([X^V, \Gamma]) = [X^V, J(\Gamma)] = 0$.

Exercise 3.4

Let $\{X_1, \ldots, X_m\}$ be a basis of the vector fields in an open set $U \subset Q$. Check that $\{X_1^C, \ldots, X_m^C, X_1^V, \ldots, X_m^V\}$ is a basis of the vector fields in $\tau_Q^{-1}(U) \subset TQ$.

(**Comment:** *This means that to determine a covariant tensor field in TQ, it is enough to know how it acts on the canonical lifts and the vertical lifts.*)

Exercise 3.5

Let $W_1, W_2 \in \mathfrak{X}(TQ)$. Prove that

$$(\mathrm{L}(W_1) J)(W_2) = [W_1, J(W_2)] - J([W_1, W_2]).$$

Exercise 3.6

Given $c \in \mathbb{R}$, we denote $\Phi_c \colon TQ \longrightarrow TQ$ the homothety with ratio c on the fibres of TQ. Let $Z \in \mathfrak{X}(TQ)$ be a SODE. Prove that the following conditions are equivalent:

(1) For every curve $c \in \mathbb{R}$ and every solution $\xi \colon I \subset \mathbb{R} \longrightarrow Q$ to the second-order differential equation defined by Z, the curve $\gamma(t) = \xi(ct)$ is also a solution.

(2) For every $c \in \mathbb{R}$ and every $(q, v) \in TQ$,

$$Z(c(q, v)) = c\,T\Phi_c(Z(q, v)).$$

(3) In a natural chart, the local expression of Z is

$$Z(q, v) = v^i \frac{\partial}{\partial q^i} + h^i(q, v) \frac{\partial}{\partial v^i},$$

where $h^i(q, v)$ are homogeneous functions of degree 2 in v.

(4) $[\Delta, Z] = Z$.

(**Comment:** *A second-order vector field satisfying these equivalent conditions is called a* **spray**.)

Exercise 3.7

Let M and N be differentiable manifolds, $\Phi \colon M \longrightarrow N$ a differentiable map, and $Z \colon M \longrightarrow TN$ a vector field along Φ (i.e., Z satisfies that $\tau_N \circ Z = \Phi$). Given a function $h \colon N \longrightarrow \mathbb{R}$, we define the function $\mathrm{L}(Z)h \colon M \longrightarrow \mathbb{R}$ as

$$(\mathrm{L}(Z)h)(x) = Z(x) \cdot h = \langle \mathrm{d}\,h(f(x)), Z(x)\rangle.$$

Prove that $\mathrm{L}(Z) \colon \mathrm{C}^\infty(N) \longrightarrow \mathrm{C}^\infty(M)$ is a derivation.

Exercise 3.8

Prove that the canonical 1-form in T^*Q is the unique form satisfying $\beta^*(\Theta) = \beta$, for every 1-form $\beta \in \Omega^1(Q)$.

If Ω is the canonical 2-form in T^*Q, prove that $\beta^*(\Omega) = -\mathrm{d}\,\beta$, for every $\beta \in \Omega^1(Q)$.

Exercise 3.9

If $X \in \mathfrak{X}(Q)$ and X^* is its lift to T^*Q, prove that

(1) $[X^*, Y^*] = [X, Y]^*$.

(2) $[X^*, \Delta] = 0$, where $\Delta = p_i \dfrac{\partial}{\partial p_i}$ is the Liouville vector field of T^*Q.

Exercise 3.10

In $\mathbb{R}^3$, consider the vector fields

$$x\frac{\partial}{\partial x} + y\frac{\partial}{\partial y} + z\frac{\partial}{\partial z}, \quad -y\frac{\partial}{\partial x} + x\frac{\partial}{\partial y}, \quad -x\frac{\partial}{\partial z} + z\frac{\partial}{\partial x}, \quad -z\frac{\partial}{\partial y} + y\frac{\partial}{\partial z}.$$

(1) Obtain the expressions of their uniparametric group of diffeomorphisms and the canonical lifts of these diffeomorphisms to $T\mathbb{R}^3$ and $T^*\mathbb{R}^3$.

(2) Determine the canonical lifts of these vector fields to $T\mathbb{R}^3$ and $T^*\mathbb{R}^3$.

3.8.2 *Lagrangian and Hamiltonian formalisms*

Exercise 3.11

Given a Lagrangian function $\mathcal{L} \in C^\infty(TQ)$, prove that $\flat_{\Omega_{\mathcal{L}}} : T(TQ) \longrightarrow T^*(TQ)$ maps vertical vector fields into horizontal forms and that it is bijective if $\mathcal{L}$ is regular.

Exercise 3.12

Let $\mathcal{L} \colon TQ \longrightarrow \mathbb{R}$ be a Lagrangian and $X \in \mathfrak{X}(TQ)$. Check if the equation $\mathrm{L}(X)\Theta_{\mathcal{L}} = \mathrm{d}\,\mathcal{L}$ is equivalent to the Euler–Lagrange equation.

Exercise 3.13

Let $\mathcal{L} \in C^\infty(TQ)$ be a regular Lagrangian.

(1) For a smooth function f in TQ, we write $\mathrm{d}^J f = \mathrm{d}f \circ J$. Prove that the vector field $\widetilde{X}_f$, associated with $\mathrm{d}^J f$ by $\Omega_{\mathcal{L}}$, that is $i(\widetilde{X}_f)\Omega_{\mathcal{L}} = \mathrm{d}^J f$, is vertical.

(2) Let $Y = X_{\mathcal{L}} + \widetilde{X}_f$. Prove that Y is a SODE and $\mathrm{L}(\widetilde{X}_f)E_{\mathcal{L}} = -\Delta(f)$.
   (*Comment: We say that Y is the perturbed Lagrangian system given by $\mathcal{L}$ and that $\widetilde{X}_f$ is the perturbation. The perturbation is said to be dissipative if $\mathrm{L}(\widetilde{X}_f)E_{\mathcal{L}} < 0$.*)

(3) Give an interpretation of the *Van der Pol equation*,
$$\ddot{x} + \mu(x^2 - 1)\dot{x} + x = 0,$$
as the dynamical equation of a Lagrangian system perturbed by some $\widetilde{X}_f$ for a suitable function f. In which conditions is this perturbation dissipative?

Exercise 3.14

Let $\mathcal{L}$ be a hyperregular Lagrangian, with Hamiltonian function $h \in C^\infty(T^*Q)$. Prove that the derivative of h along the fibre, as an application $\mathcal{F}h \colon T^*Q \longrightarrow TQ$, is the inverse diffeomorphism of $\mathcal{F}\mathcal{L}$.

Exercise 3.15

Let $Q = \mathbb{R}$ and $h \in C^\infty(T^*Q)$ given by $h(q, p) = (1/2)(p^2 + \omega^2 q^2)$. Find X_h and calculate its flux. Which are the corresponding Lagrangian L and Lagrangian vector field $X_{\mathcal{L}}$?

Exercise 3.16

Consider the dynamical system of a particle in $\mathbb{R}^3$ with mass m, submitted to a central potential, whose Hamiltonian function is $h = \dfrac{m}{2} v_{\mathbf{r}}^2 + V(r)$, where $\mathbf{r} = (x, y, z)$, $r = \|\mathbf{r}\|$, and $v_{\mathbf{r}}^2 = v_x^2 + v_y^2 + v_z^2$.

(1) Let $\mathbf{p} = (p_x, p_y, p_z)$. Prove that the three components of the *angular momentum*, $\mathbf{L} := \mathbf{r} \times \mathbf{p} = (yp_z - zp_y, = zp_x - xp_z, xp_y - yp_x)$, are conserved quantities.

(2) Determine the vector fields that are associated with these conserved quantities. Prove that they are natural Noether symmetries which correspond to the rotations in the configuration space $\mathbb{R}^3$ of the system.
(**Hint:** *Prove that these Noether vector fields are the canonical lifts to* $T^*\mathbb{R}^3$ *of the vector fields that generate the rotations.*)

Exercise 3.17

For a point particle of mass m and electric charge e that moves in $\mathbb{R}^3$ subjected to the action of a magnetic field $\mathbf{B} \in \mathfrak{X}(\mathbb{R}^3)$, the classical Newton equation is

$$m \frac{d^2 \mathbf{x}}{dt^2} = \frac{e}{c} \frac{d\mathbf{x}}{dt} \times \mathbf{B}, \tag{3.28}$$

where c is the speed of light and "$\times$" indicates the vector product. This is because the acting force is the *Lorentz force* $\mathbf{F} = \dfrac{e}{c} \dfrac{d\mathbf{x}}{dt} \times \mathbf{B}$. Then, if $\mathbf{B} = B^1 \dfrac{\partial}{\partial x} + B^2 \dfrac{\partial}{\partial y} + B^3 \dfrac{\partial}{\partial z}$, consider the form $\beta \in \Omega^2(\mathbb{R}^3)$ given by

$$\beta := i(\mathbf{B})(dx \wedge dy \wedge dz) = B^1 dy \wedge dz + B^2 dz \wedge dx + B^3 dx \wedge dy,$$

which is closed if there are no magnetic monopoles present (and we suppose that it is so). If $\beta = d\mathcal{A}$, with $\mathcal{A} \in \Omega^1(\mathbb{R}^3)$, we say that $\mathcal{A}$ is a *vector potential* for $\mathbf{B}$.

(1) Prove that Eq. (3.28) is the Euler–Lagrange equation for the Lagrangian function $\mathcal{L}(q, u) = \dfrac{1}{2} m\|u\|^2 + \dfrac{e}{c} \mathcal{A}_p(u)$, $(q, u) \in T\mathbb{R}^3$.

(2) If $\mathcal{L}_0(q, u) = \dfrac{1}{2} m\|u\|^2$ is the Lagrangian function for a free particle in $\mathbb{R}^3$, find the relationship between the Lagrangian forms and the energy functions of $\mathcal{L}$ and $\mathcal{L}_0$.

Let $Q = \mathbb{R}^3 \times S^1$, with local coordinates (x, y, z, ϑ), and consider the Lagrangian function on TQ

$$\mathcal{L} = \frac{1}{2}[m(v_x^2 + v_y^2 + v_z^2) + (A_1 v_x + A_2 v_y + A_3 v_z + v_\vartheta)^2].$$

(3) Prove that the curves solution to the variational problem associated with $\mathcal{L}$ are the geodesics for a certain Riemannian metric g in Q. Find the matrix of g in the given local chart.

(4) Prove that $f = \mathcal{A}(\mathbf{v}) + v_\vartheta$, where $\mathbf{v} = (v_x, v_y, v_z)$, is a constant of motion.

(5) Show that doing $f = \dfrac{e}{c}$, we obtain the same solutions for $\mathcal{L}$ in Q than for $\mathcal{L}$ in $\mathbb{R}^3$ (and this allows us to determine the charge e).

(*Comment:* *The formulation described in items (3), (4), and (5) above is the Kaluza–Klein formulation, a particular case of the Yang-Mills field theory. Note that in this formulation, the electric charge e is an invariant associated with the particle, which is obtained from the geometrical formalism used. Note that the trajectories of the particles are geodesics of a metric in a certain manifold: we have changed the configuration space from $\mathbb{R}^3$ to $Q = \mathbb{R}^3 \times S^1$, and the metric has been changed by the action of the magnetic field to interpret the solution curves as geodesics of a certain metric.*)
(*Hint:* *For more details, see [265], p. 181.*)

For the systems described in Exercises 3.18 to 3.25:

(1) Give the configuration space Q of the system, and provide it with a system of generalised coordinates.
(2) Obtain the Lagrangian function, the Lagrangian energy, and the Lagrangian forms.
(3) Write the Lagrangian equations for the Lagrangian dynamical vector field $X_{\mathcal{L}}$, and obtain the expression of this vector field. Write the equations of the integral curves of $X_{\mathcal{L}}$, and obtain the Euler–Lagrange equations of the system and integrate them, as much as possible.
(4) Calculate the Legendre map for $\mathcal{L}$ specifying what its image is and if $\mathcal{FL}$ is a diffeomorphism. Obtain the Hamiltonian function h and the corresponding Hamilton equations.
(5) Write the Hamiltonian equations for the Hamiltonian dynamical vector field X_{h}, and obtain the expression of this vector field. Write the equations of the integral curves of X_{h}, and obtain the Hamilton equations of the system and integrate them, as much as possible. Prove that the

integral curves of $X_{\mathcal{L}}$ are in correspondence with those of X_h by means of the Legendre map.

(6) Can you give any conserved quantities of the system? Are they associated with infinitesimal symmetries of the Noether kind? In such cases, give the vector fields generating them. Are they natural symmetries?

Exercise 3.18 (Quadratic regular Lagrangians).

In the manifold Q, with coordinates (q^i), $i = 1, \ldots, n$, consider the system described by the Lagrangian function

$$\mathcal{L} = f_{ij}(q^k)\, v^i\, v^j + g(q^k),$$

where the matrix $(f_{ij}(q^k))$ is regular everywhere.

Exercise 3.19

Consider a particle with mass m, which moves without friction on the paraboloid S defined by the equation $z = k(x^2 + y^2), k > 0$, subjected to the force of gravity $F = (0, 0, -mg)$.

Study also if the change of coordinates in Q, from cylindrical to Cartesian ones, gives a canonical transformation in an open set of T^*Q, specifying the open set where it goes.

Exercise 3.20

Consider a pendulum whose mass is m_2 and length is l_2 in a vertical plane. Its suspension point has mass m_1 and moves horizontally in the same plane at the end of a spring (of negligible mass) of length-at-rest l_1 and elastic constant k.

Exercise 3.21

The *Atwood machine* consists of a pulley with radius R_1, a mass m_1 and another pulley with radius R_2 hanging from an inextensible wire of length l_1 (both pulleys have negligible masses), and two additional masses m_2 i m_3, which hang from the second pulley by means of another inextensible wire of length l_2.

Exercise 3.22

Let us consider a pulley, with radius R and mass M, with a wire of length l, with masses m_1 and m_2 in its extremes.

(**Hint:** *The kinetic energy of a rigid body spinning around a fixed axis with angular velocity ω is $\frac{1}{2}I\omega^2$, where I is the inertia moment of the body with respect to the axis.*)

Exercise 3.23

Let $Q = \mathbb{R}^d \times \mathbb{R}^*$, with coordinates (q^i, e), $i = 1, \ldots, d$, where the vector space $\mathbb{R}^d$ is endowed with a metric. Then, consider the system described by the Lagrangian function $\mathcal{L} = \dfrac{1}{2e} \displaystyle\sum_{i=1}^{d} (v^i)^2 + \dfrac{m^2 e}{2}$.

Exercise 3.24

A mechanical system consists of two particles that move in the space $\mathbb{R}^3$. Taking ordinary Cartesian coordinates, $\mathbf{r}_1 = (x_1, y_1, z_1)$ and $\mathbf{r}_2 = (x, y_2, z_2)$ give the position of the two particles. The first one has mass m_1 and is constrained to move along the OZ axis. The second one has mass m_2 and moves in the circle defined by $x^2 + y^2 = a^2$, $z = 0$. Both particles are joined by an elastic spring with elastic constant k and length b when it is at rest. It is supposed that gravitation is acting with acceleration g in the negative direction of the OZ axis.

Given a function f independent of the Hamiltonian, such that $\{f, \mathrm{h}\} = 0$; can you obtain another function g, independent of f and h, such that $\{g, \mathrm{h}\} = 0$?

Exercise 3.25

A mass m_1 is moving vertically, hanging from an elastic spring with elastic constant k. From m_1 hangs a stiff rod, of negligible mass and length l_2, at the end of which there is another mass m_2 that moves in a vertical plane. Let l_1 be the length of the spring when the mass m_1, the rod, and the mass m_2 are vertically aligned and at rest.

In the dynamical equations, study the case in which the angle θ of the rod with the vertical is small (then $\sin \theta \sim \theta$ and $\cos \theta \sim 1$) and solve them.

Exercise 3.26

Consider a particle of mass m moving on $Q = \mathbb{R}^2$, subjected to the action of constant gravity.

(1) Find the Lagrangian function $\mathcal{L}$ of the system in Cartesian coordinates. Which is the manifold $\mathcal{W} = TQ \times_Q T^*Q$ for this system?

Denoted by $\varrho_2 \colon \mathcal{W} \longrightarrow TQ$, the natural projection, determine the expression, in the natural coordinates of $\mathcal{W}$ of the function

$$\mathcal{H}(q, p_q, v_q) = \langle p_q, v_q \rangle - \varrho_2^* \mathcal{L} \quad ; \quad q \in Q, \ v_q \in T_q Q, \ p_q \in T_q^* Q.$$

(2) We define the 2-form $\Omega_{\mathcal{W}} = \varrho_1^* \Omega \in \Omega^2(\mathcal{W})$, where $\varrho_1 \colon M \longrightarrow T^*Q$ is the natural projection and Ω is the canonical 2-form in T^*Q. Obtain

the vector field $X_{\mathcal{H}} \in \mathfrak{X}(\mathcal{W})$ solution to the equation $i(X_{\mathcal{H}})\Omega_{\mathcal{W}} = d\mathcal{H}$. Which is the domain of X?

(3) Determine the integral curves of $X\mathcal{H}$, and from them, the curves $\phi: \mathbb{R} \longrightarrow Q$ and $\psi: \mathbb{R} \longrightarrow T^*Q$, which are solutions to the Euler–Lagrange equations and the Hamilton equations of the system, respectively.

(4) Check that the vector field $X_h \in \mathfrak{X}(T^*Q)$ whose integral curves are the curves ψ is the solution to the equation $i(X_h)\Omega = dh$, where $h \in C^\infty(T^*Q)$ is the Hamiltonian function of the system associated with the Lagrangian $\mathcal{L}$.

Also, check that the vector field $X_{\mathcal{L}} \in \mathfrak{X}(TQ)$ whose integral curves are the canonical lifts to TQ of curves ϕ is a solution to the equation $i(X_{\mathcal{L}})\Omega_{\mathcal{L}} = dE_{\mathcal{L}}$, where $\Omega_{\mathcal{L}} \in \Omega^2(TQ)$ and $E_{\mathcal{L}} \in C^\infty(TQ)$ are the Lagrangian 2-form and the Lagrangian energy defined by $\mathcal{L}$.

(**Comment:** *This is the Skinner–Rusk unified Lagrangian–Hamiltonian formalism for this system.*)

Exercise 3.27

Do the same analysis as in the above Exercise 3.26 for the systems of Exercises 3.18 to 3.25.

For the systems described in Exercises 3.28 to 3.31:

(1) Give the Legendre map $\mathcal{FL}$, determine its image $\mathcal{FL}(TQ) \equiv P_0$, and describe the fibres $\mathcal{FL}^{-1}(\mathcal{FL}(p))$ for every $p \in TQ$. Prove that the Lagrangian is almost-regular.

(2) Obtain the expressions of the Lagrangian forms $\Theta_{\mathcal{L}}$ and $\Omega_{\mathcal{L}}$, and the Lagrangian energy $E_{\mathcal{L}}$.

(3) Analyse if there exists $X_{\mathcal{L}} \in \mathfrak{X}(TQ)$ which is the solution to the Lagrangian equation $i(X_{\mathcal{L}})\Omega_{\mathcal{L}} - dE_{\mathcal{L}} = 0$. If it does not exist, determine if there is any submanifold $M \hookrightarrow TQ$ such that $(i(X_{\mathcal{L}})\Omega_{\mathcal{L}} - dE_{\mathcal{L}})|_M = 0$, and in that case, determine the general solution $X_{\mathcal{L}}|_M$.
(**Hint:** *Write the equation $i(X_{\mathcal{L}})\Omega_{\mathcal{L}} - dE_{\mathcal{L}} = 0$ in coordinates, and check whether the resulting equations include constraints or not.*)

(4) Figure out if there exists any SODE $\Gamma_{\mathcal{L}} \in \mathfrak{X}(TQ)$ which is a solution to the Lagrangian equation on M. If it does not exist, find if there is a submanifold $S \hookrightarrow M$ such that there exists a SODE, $\Gamma_{\mathcal{L}} \in \mathfrak{X}(TQ)$, solution to the Lagrangian equation on S.
(**Hint:** *If $X_{\mathcal{L}} = A^i \dfrac{\partial}{\partial q^i} + B^i \dfrac{\partial}{\partial v^i}$, check whether the SODE condition $A^i = v^i$ originates new constraints or not.*)

(5) Discover if the SODE $\Gamma_\mathcal{L}|_M$ is tangent to S. If it is not tangent, determine if there is any submanifold $S_f \hookrightarrow S$ such that $(i(\Gamma_\mathcal{L})\Omega_\mathcal{L} - dE_\mathcal{L})|_{S_f} = 0$, calculate the general solution $\Gamma_\mathcal{L}|_{S_f}$, and obtain the integral curves, whether it is feasible.

(**Hint:** *Check whether the Lie derivative by $\Gamma_\mathcal{L}$ of the constraints defining S vanishes or not over S. When these tangency conditions originate new constraints, repeat the procedure under no new constraints appear.*)

(6) Obtain the Hamiltonian system (P_0, Ω_0, h_0) associated with the Lagrangian system $(TQ, \Omega_\mathcal{L}, E_\mathcal{L})$. Without doing any calculation, what can be said about the existence and unicity of vector fields $X_0 \in \mathfrak{X}(P_0)$, which are solutions to the equation $i(X_0)\Omega_0 - dh_0 = 0$?

(7) Calculate if there exists $X_0 \in \mathfrak{X}(P_0)$, that is the solution to the Hamiltonian equation $i(X_0)\Omega_0 - dh_0 = 0$. If not, find if there exists some manifold $P_1 \hookrightarrow P_0$ such that there are $X_0 \in \mathfrak{X}(P_0)$ solutions to the equation $(i(X_0)\Omega_0 - dh_0)|_{P_1} = 0$.

(8) Check if the vector fields $X_0|_{P_f}$ are tangent to P_1, and if they are not, find if there exists a submanifold $P_f \hookrightarrow P_1$ such that there are $X_0 \in \mathfrak{X}(P_0)$ solutions to the equation $(i(X_0)\Omega_0 - dh_0)|_{P_f} = 0$. In that case, determine the general solution $X_0|_{P_f}$, and obtain the integral curves, if it is feasible.

(9) Analyse the relationships between the vector fields $X_\mathcal{L}$, $\Gamma_\mathcal{L}$, and X_0. Answer the same question for the manifolds M, P_1, S_f, and P_f.

Exercise 3.28 (Affine Lagrangians).
In the manifold Q, with coordinates (q^i), $i = 1, \ldots, n$, consider the system described by the Lagrangian function

$$\mathcal{L} = f_i(q^j)\, v^i - g(q^j).$$

As a particular case, in $TQ = \mathbb{R}^2 \times \mathbb{R}^2$, with coordinates (x, y, v_x, v_y), use the Lagrangian function $\mathcal{L} = yv_x - xv_y - x^2 - y^2$.

Exercise 3.29 (Quadratic singular Lagrangians).
In the manifold Q, with coordinates (q^i), $i = 1, \ldots, n$, consider the system described by the Lagrangian function

$$\mathcal{L} = f_{ij}(q^k)\, v^i\, v^j - g(q^k),$$

where the matrix $(f_{ij}(q^k))$ has a constant rank equal to $m < n$.

As a particular case, in $T\mathbb{R}^3$, consider a local chart of coordinates $(r, \varphi, z; v_r, v_\varphi, v_z)$, with $r \in \mathbb{R}^+$, $\varphi \in (0, 2\pi)$, $z \in \mathbb{R}$, $v_r \in \mathbb{R}$, $v_\varphi \in \mathbb{R}$, $v_z \in \mathbb{R}$, and the Lagrangian function $\mathcal{L} = \dfrac{1}{2}\left(v_r^2 + r^2(v_\varphi - z)^2\right) - V(r)$.

Exercise 3.30

In $TQ = \mathbb{R}^4 \times \mathbb{R}^4$, consider the system described by the Lagrangian function

$$\mathcal{L} = \frac{1}{2}\left(v_1^2 - q_1^2 - 2q_1 q_3 + (v_4 - q_2)^2\right).$$

Exercise 3.31

Taking $Q = (\mathbb{R}^2 - \{(0, 0)\}) \times \mathbb{R}$ endowed with cylindrical coordinates (r, θ, z), consider the system described by the Lagrangian in TQ given by

$$\mathcal{L} = \frac{1}{2}(v_r^2 + r^2(v_\theta - z)^2) - V(r).$$

Exercise 3.32

For the *Kepler problem* analysed in Section 3.7.2, develop the Marsden–Weinstein reduction procedure in the Lagrangian formalism, following the same patterns as in the Hamiltonian formalism.

Exercise 3.33 (Lagrangian Hamilton–Jacobi theory).

Let $(TQ, \Omega_{\mathcal{L}}, E_{\mathcal{L}})$ be a regular Lagrangian system and $\Gamma_{\mathcal{L}} \in C^\infty(TQ)$ be the corresponding Euler–Lagrange vector field. The **generalised Lagrangian Hamilton–Jacobi problem** consists of finding a vector field $X \in \mathfrak{X}(Q)$ such that if $\gamma \colon \mathbb{R} \longrightarrow Q$ is an integral curve of X, then $X \circ \gamma = \dot{\gamma} \colon \mathbb{R} \longrightarrow TQ$ is an integral curve of $\Gamma_{\mathcal{L}}$, that is, if $X \circ \gamma = \dot{\gamma}$, then $\Gamma_{\mathcal{L}} \circ \dot{\gamma} = \overline{X \circ \gamma}$ (and X is said to be a **solution to the generalised Lagrangian Hamilton–Jacobi problem**).

Prove that the above definition is equivalent to any one of the following statements:

(1) The vector fields X and $\Gamma_{\mathcal{L}}$ are X-related, that is, $\Gamma_{\mathcal{L}} \circ X = TX \circ X$. As a consequence, the integral curves of $\Gamma_{\mathcal{L}}$ with initial conditions in $\operatorname{Im} X$ project onto the integral curves of X.

(2) The submanifold $\operatorname{Im} X$ of TQ is invariant by $\Gamma_{\mathcal{L}}$ (i.e., $\Gamma_{\mathcal{L}}$ is tangent to $\operatorname{Im} X$).

(3) Prove the equation $i(X)(X^*\Omega_{\mathcal{L}}) = \mathrm{d}(X^* E_{\mathcal{L}})$.

$$
\begin{array}{ccc}
TQ & \xleftarrow{\;\;T\tau_Q\;\;} & T(TQ) \\
\end{array}
$$

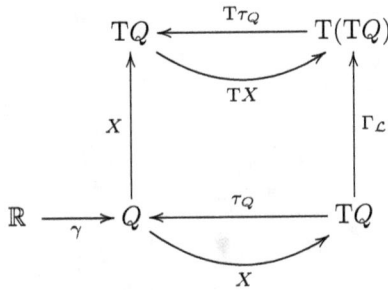

The **Lagrangian Hamilton–Jacobi problem** consists of finding a vector field $X \in \mathfrak{X}(Q)$ solution to the generalised Lagrangian Hamilton–Jacobi problem, which satisfies $X^*\Omega_{\mathcal{L}} = 0$ (and X is said to be a **solution to the Lagrangian Hamilton–Jacobi problem**). Then, as

$$
0 = X^*\Omega_{\mathcal{L}} = -X^*\mathrm{d}\Theta_{\mathcal{L}} = -\mathrm{d}(X^*\Theta_{\mathcal{L}}),
$$

for every point of Q, there is a neighbourhood $U \subset Q$ and a function S such that $X^*\Theta_{\mathcal{L}} = \mathrm{d}S$ (in U).

Prove that this definition is equivalent to any of the following statements:

(1) $\operatorname{Im} X$ is a Lagrangian submanifold of TQ, which is invariant by $\Gamma_{\mathcal{L}}$.
(2) $\mathrm{d}(X^*E_{\mathcal{L}}) = 0$ (equivalently, the function $E_{\mathcal{L}} \circ \mathrm{d}S \colon Q \longrightarrow \mathbb{R}$ is locally constant).
(3) In the natural coordinates of TQ, the following equation holds:

$$
\frac{\partial S}{\partial q^i} = \frac{\partial \mathcal{L}}{\partial v^i}(q^i, X^i).
$$

(**Solution:** See [65].)

Chapter 4

Cosymplectic Mechanics: Nonautonomous Dynamical Systems

In the previous chapters, we have studied autonomous Hamiltonian and Lagrangian systems, that is, dynamical systems described by Hamiltonian or Lagrangian functions which are independent of time. Now, we are going to analyse the case of *nonautonomous dynamical systems*, which are described by time-dependent Hamiltonian or Lagrangian functions.

The geometrical description of nonautonomous Hamiltonian and Lagrangian systems can be provided using different approaches. For instance, one can use the *contact formalism* [1, 109, 154, 155, 314] or its generalisation, the *jet bundle formalism*, using *jet* and *fibre bundles* [106, 154, 255, 275, 317]. However, these kinds of systems can also be described as symplectic Hamiltonian systems by means of the *extended formalisms* [13, 154, 233, 301, 302, 331], as singular (presymplectic) dynamical systems [64, 154], or using the Lagrangian–Hamiltonian *unified formalism* [25, 101]. Nevertheless, one of the most elegant and simplest geometric descriptions of time-dependent systems is the *cosymplectic formulation* [54, 89], and this is where we focus our attention in this chapter.

As mentioned in the previous chapters, first we state the geometrical foundations on which this formulation is based, which are the *cosymplectic manifolds* and their properties. Next, we introduce the concept of *cosymplectic Hamiltonian system*, and we describe, in particular, the Lagrangian and Hamiltonian formalisms of nonautonomous Lagrangian dynamical systems using this formulation. Symmetries, conserved quantities, and Noether's Theorem are also discussed in this context. The chapter is completed with a brief presentation of two other very common formulations of time-dependent mechanics, the *contact* and the *extended symplectic formulations*, and by showing their equivalence with the cosymplectic picture.

135

Finally, the examples from the previous chapter are analysed when the Lagrangians include time dependence.

4.1 Notions on Cosymplectic Geometry

First, we establish the basic foundations of cosymplectic geometry (see, for instance, [54, 57, 89] for details).

4.1.1 *Cosymplectic manifolds*

Definition 4.1. Let M be a differentiable manifold of dimension $2n + 1$. A **cosymplectic structure** on M is a couple (η, ω), where $\eta \in \Omega^1(M)$ and $\omega \in \Omega^2(M)$ are closed forms such that $\eta \wedge \omega^n$ is a volume form. Then, (M, η, ω) is called a **cosymplectic manifold**.

If $\eta \wedge \omega^n$ is not a volume form and $\dim M$ is arbitrary, then we say that (η, ω) is a **precosymplectic structure** on M and (M, η, ω) is a **precosymplectic manifold**.

The (pre)cosymplectic structure is said to be **exact** if $\omega = d\theta$ for some $\theta \in \Omega^1(M)$.

Proposition 4.1. *If (η, ω) is a cosymplectic structure on M, then there exists a unique vector field $R \in \mathfrak{X}(M)$, called the **Reeb vector field**, which is characterised by the conditions*

$$i(R)\eta = 1, \quad i(R)\omega = 0. \tag{4.1}$$

Proof. By the second condition, observe that $R \in \ker \omega$. From the condition that $\eta \wedge \omega^n$ is a volume form, we have $\operatorname{rank} \omega = 2n$, and hence, $\ker \omega$ is a 1-dimensional $C^\infty(M)$ module. Therefore, the first condition allows us to select one generator of this module. $\qquad\square$

The local structure of cosymplectic manifolds is given by the following extension of Darboux Theorem [138, 143]:

Theorem 4.1 (Darboux). *Let (M, η, ω) be a cosymplectic manifold. Then, for every $p \in M$, there exists an open neighbourhood $U \subset M$, $p \in U$, which is the domain of a local chart of coordinates (t, x^i, y_i),*

$1 \leq i \leq n$, *such that*

$$\eta|_U = \mathrm{d}t, \quad \omega|_U = \mathrm{d}x^i \wedge \mathrm{d}y_i, \quad R|_U = \frac{\partial}{\partial t}.$$

These are called **Darboux** *or* **canonical coordinates** *of the cosymplectic manifold.*

Proof. In a cosymplectic manifold, there is a symplectic foliation, which is made of the leaves of the distribution generated by ker η (which are $2n$-dimensional submanifolds, and on each one of them, the restriction of ω is a symplectic form since it has maximal rank $2n$ by the condition of the volume form). Then, we take coordinates adapted to the foliation, and on each leaf, apply the symplectic Darboux Theorem. In this way, we have local coordinates $(x^i, y_i, \tilde{z})$ such that $\omega = \mathrm{d}x^i \wedge \mathrm{d}y_i$. Finally, we write η as a combination of all of these coordinates, $\eta = f(q, p, \tilde{z})\mathrm{d}\tilde{z}$, with f a nonvanishing function, and then we can redefine the coordinate z. In these coordinates, the Reeb vector field has the expression given by the theorem. □

Remark 4.1. For precosymplectic manifolds, there is a similar result [138]. In fact, if (M, η, ω) is a precosymplectic manifold with rank $\omega = 2r < \dim M - 1 \equiv m - 1$, then, for every point on M, there exists a local chart $(U; t, x^i, y_i, z^j)$, where $1 \leq i \leq r$, $1 \leq j \leq m - 2r - 1$, such that

$$\eta|_U = \mathrm{d}t, \quad \omega|_U = \mathrm{d}x^i \wedge \mathrm{d}y_i.$$

These local charts are the *precosymplectic charts*, and their coordinates are the *canonical coordinates* or *Darboux coordinates* of the precosymplectic manifold in this chart [143]. In addition, for precosymplectic manifolds, the solution to Eq. (4.1) is not unique and Reeb vector fields are not uniquely defined.

The fundamental examples of cosymplectic manifolds are the following:
Canonical model: The canonical model for cosymplectic manifolds is the following: consider the manifold $\mathbb{R} \times \mathrm{T}^*Q$ with canonical projections

$$\pi_1 \colon \mathbb{R} \times \mathrm{T}^*Q \to \mathbb{R}, \ \pi_2 \colon \mathbb{R} \times \mathrm{T}^*Q \to \mathrm{T}^*Q,$$
$$\pi_0 \colon \mathbb{R} \times \mathrm{T}^*Q \to Q, \ \pi_{1,0} \colon \mathbb{R} \times \mathrm{T}^*Q \to \mathbb{R} \times Q.$$

If (q^i) are local coordinates on $U \subseteq Q$, the induced local coordinates (t, q^i, p_i) on $\pi_0^{-1}(U) = \mathbb{R} \times T^*U$ are given by

$$t(\mathrm{t}, \alpha_q) = \mathrm{t}, \quad q^i(\mathrm{t}, \alpha_q) = x^i(q), \quad p_i(\mathrm{t}, \alpha_q) = \alpha_q \left(\frac{\partial}{\partial q^i} \Big|_q \right),$$

for $\mathrm{t} \in \mathbb{R}$, $q \in Q$, and $\alpha_q \in T_q^*Q$. We define the differential forms on $\mathbb{R} \times T^*Q$,

$$\eta = \pi_1^* dt, \quad \theta = \pi_2^* \Theta, \quad \omega = \pi_2^* \Omega,$$

where Θ and Ω are the canonical forms on T^*Q. In local coordinates,

$$\eta = dt, \quad \theta = p_i dq^i, \quad \omega = dq^i \wedge dp_i.$$

Hence, $(\mathbb{R} \times T^*Q, \eta, \omega)$ is a cosymplectic manifold, and the natural coordinates of $\mathbb{R} \times T^*Q$ are Darboux coordinates for this canonical cosymplectic structure. Furthermore, $\dfrac{\partial}{\partial t}$ is its Reeb vector field.

Almost-canonical cosymplectic manifolds: There is another kind of cosymplectic manifolds which are specially relevant: those of the form $M = \mathbb{R} \times N$, where (N, Ω) is a symplectic manifold. Then, if

$$\pi_\mathbb{R} \colon \mathbb{R} \times N \to \mathbb{R}, \quad \pi_N \colon \mathbb{R} \times N \to N$$

are the canonical projections, we have the differential forms

$$\eta = \pi_\mathbb{R}^* dt, \quad \omega = \pi_M^* \Omega.$$

The conditions given in Definition 4.1 are verified, so $\mathbb{R} \times N$ is a cosymplectic manifold. From the Darboux Theorem 4.1, we have local coordinates (t, x^i, y_i) on $\mathbb{R} \times N$. These kinds of cosymplectic manifolds are sometimes called **almost-canonical cosymplectic manifolds**. Observe that the standard model is a particular class of this kind of k-cosymplectic manifold, where $N = T^*Q$.

Every cosymplectic manifold (M, η, ω) is endowed with the natural vector bundle isomorphism

$$\flat_{(\eta,\omega)} \colon TM \longrightarrow T^*M$$

$$(\mathrm{p}, X_\mathrm{p}) \mapsto (\mathrm{p}, i(X_\mathrm{p})\omega_\mathrm{p} + ((i(X_\mathrm{p})\eta_\mathrm{p})\eta_\mathrm{p})$$

and its inverse $\sharp_{(\eta,\omega)} = \flat_{(\eta,\omega)}^{-1} \colon T^*M \to TM$. Their natural extensions, which are denoted with the same notation, are the $C^\infty(M)$-module isomorphisms

$$\flat_{(\eta,\omega)} \colon \mathfrak{X}(M) \longrightarrow \Omega^1(M)$$
$$X \longmapsto i(X)\omega + (i(X)\boldsymbol{\eta})\boldsymbol{\eta}$$

and its inverse $\sharp_{(\eta,\omega)} = \flat_{(\eta,\omega)}^{-1} \colon \Omega^1(M) \to \mathfrak{X}(M)$. In particular, for the Reeb vector field, we have $\flat_{(\eta,\omega)}(R) = \eta$.

4.1.2   Hamiltonian, gradient, and evolution vector fields

Using the natural $C^\infty(M)$-module isomorphism $\flat_{(\eta,\omega)}$, one can associate some particular vector fields to every function $f \in C^\infty(M)$:

> **Definition 4.2.** Let (M, η, ω) be a cosymplectic manifold and a function $f \in C^\infty(M)$.
>
> The **Hamiltonian vector field** associated with f is the vector field $X_f \in \mathfrak{X}(M)$ defined by $\flat_{(\eta,\omega)}(X_f) := \mathrm{d}f - R(f)\eta$.
>
> The **gradient vector field** associated with f is the vector field $\mathrm{grad}\, f \in \mathfrak{X}(M)$ defined by $\flat_{(\eta,\omega)}(\mathrm{grad}\, f) := \mathrm{d}f$.
>
> The **evolution vector field** associated with f is the vector field $\mathcal{E}_f \in \mathfrak{X}(M)$ defined by $\flat_{(\eta,\omega)}(\mathcal{E}_f) = \mathrm{d}f - (R(f) - 1)\eta$, that is, $\mathcal{E}_f := R + X_f$.

These vector fields can be equivalently characterised as follows:

> **Proposition 4.2.** *The Hamiltonian vector field associated with f is determined by the equations:*
>
> $$i(X_f)\eta = 0, \quad i(X_f)\omega = \mathrm{d}f - R(f)\eta. \tag{4.2}$$
>
> *The gradient vector field associated with f is determined by:*
>
> $$i(\mathrm{grad}\, f)\eta = R(f), \quad i(\mathrm{grad}\, f)\omega = \mathrm{d}f - R(f)\eta. \tag{4.3}$$
>
> *The evolution vector field associated with f is determined by:*
>
> $$i(\mathcal{E}_f)\eta = 1, \quad i(\mathcal{E}_f)\omega = \mathrm{d}f - R(f)\eta. \tag{4.4}$$

Proof. For every $f \in C^\infty(M)$, if X_f is the Hamiltonian vector field associated with f, using the definitions of the isomorphism $\flat_{(\eta,\omega)}$ and of X_f, first we have

$$i(X_f)\mathrm{d}\eta + (i(X_f)\eta)\eta = \flat_{(\eta,\omega)}(X_f) = \mathrm{d}f - R(f)\,\eta \tag{4.5}$$

and contracting both members with R and using (4.1), we get

$$(i(X_f)\eta)\, i(R)\eta = i(R)\mathrm{d}f - R(f)\, i(R)\eta = R(f) - R(f) = 0 \iff i(X_f)\eta = 0$$

since this holds for every f. Now, going to (4.5), we obtain

$$i(X_f)\mathrm{d}\eta = \mathrm{d}f - R(f)\,\eta.$$

Repeating the same procedure for the gradient and the evolution vector fields, we obtain the corresponding equations for $\mathrm{grad}\, f$ and $\mathcal{E}_f$. $\qquad\square$

From these results, it is straightforward to obtain:

Proposition 4.3. *The equations for the integral curves $c \colon I \subset \mathbb{R} \longrightarrow M$ of the Hamiltonian, gradient, and evolution vector fields associated with $f \in C^\infty(M)$ are, respectively:*

$$i(\widetilde{c})(\eta \circ c) = 0, \quad i(\widetilde{c})(\omega \circ c) = (\mathrm{d}f - R(f)\eta) \circ c,$$

$$i(\widetilde{c})(\eta \circ c)) = R(f) \circ c, \quad i(\widetilde{c})(\omega \circ c)) = (\mathrm{d}f - R(f)\eta) \circ c,$$

$$i(\widetilde{c})(\eta \circ c)) = 1, \quad i(\widetilde{c})(\omega \circ c)) = (\mathrm{d}f - R(f)\eta) \circ c. \qquad (4.6)$$

Local expressions: In Darboux coordinates on M, we have $R(f) = \dfrac{\partial f}{\partial t}$; therefore, $\mathrm{d}f - R(f)\eta = \dfrac{\partial f}{\partial x^i}\mathrm{d}x^i + \dfrac{\partial f}{\partial y_i}\mathrm{d}y_i$, and from (4.2), (4.3), and (4.4), we obtain[1]

$$X_f = \frac{\partial f}{\partial y_i}\frac{\partial}{\partial x^i} - \frac{\partial f}{\partial x^i}\frac{\partial}{\partial y_i},$$

$$\operatorname{grad} f = \frac{\partial f}{\partial t}\frac{\partial}{\partial t} + \frac{\partial f}{\partial y_i}\frac{\partial}{\partial x^i} - \frac{\partial f}{\partial x^i}\frac{\partial}{\partial y_i},$$

$$\mathcal{E}_f = \frac{\partial}{\partial t} + \frac{\partial f}{\partial y_i}\frac{\partial}{\partial x^i} - \frac{\partial f}{\partial x^i}\frac{\partial}{\partial y_i}. \qquad (4.7)$$

Therefore, if $c(s) = (t(s), x^i(s), y_i(s))$ is an integral curve of some of these vector fields, these expressions imply that $c(s)$ should satisfy the following systems of differential equations:

$$\frac{dt}{ds} = 0, \quad \frac{dx^i}{ds} = \frac{\partial f}{\partial y_i}, \quad \frac{dy_i}{ds} = -\frac{\partial f}{\partial x^i},$$

$$\frac{dt}{ds} = \frac{\partial f}{\partial t}, \quad \frac{dx^i}{ds} = \frac{\partial f}{\partial y_i}, \quad \frac{dy_i}{ds} = -\frac{\partial f}{\partial x^i},$$

$$\frac{dt}{ds} = 1, \quad \frac{dx^i}{ds} = \frac{\partial f}{\partial y_i}, \quad \frac{dy_i}{ds} = -\frac{\partial f}{\partial x^i}. \qquad (4.8)$$

As in the symplectic case, a Poisson bracket can be defined using gradient vector fields. Indeed, it is straightforward to prove that:

---

[1]Observe that the local expression of the Hamiltonian vector field in the cosymplectic formulation is the same as the Hamiltonian vector field in the symplectic formulation, which justifies the terminology.

Proposition 4.4. *Every cosymplectic manifold* (M, η, ω) *is a Poisson manifold, with the Poisson bracket defined by*

$$\{f, q\} := \omega(\operatorname{grad} f, \operatorname{grad} g); \quad f, g \in C^\infty(M).$$

The expression of this Poisson bracket in Darboux coordinates is the usual one:

$$\{f, g\} = \frac{\partial f}{\partial y_i} \frac{\partial g}{\partial x^i} - \frac{\partial f}{\partial x^i} \frac{\partial g}{\partial y_i}. \tag{4.9}$$

As a consequence of these definitions and properties, many of the results stated in Section 2.1 about Poisson brackets and canonical transformations for symplectic manifolds can also be extended to this case.

4.2 Nonautonomous Hamiltonian Dynamical Systems

4.2.1 *Nonautonomous Hamiltonian systems*

Using the above considerations and the set of postulates for autonomous dynamical systems, we can state the following postulates for the geometric study of nonautonomous Hamiltonian dynamical systems.

Postulate 4.1 (First postulate of nonautonomous Hamiltonian mechanics). The phase space of a regular (respectively, singular) nonautonomous dynamical system is a differentiable manifold M endowed with a cosymplectic (respectively, precosymplectic) structure (η, ω).

Postulate 4.2 (Second postulate of nonautonomous Hamiltonian mechanics). The observables or physical magnitudes of a nonautonomous dynamical system are functions of $C^\infty(M)$.

Postulate 4.3 (Third postulate of nonautonomous Hamiltonian mechanics). The dynamics of a nonautonomous dynamical system is given by a function $h \in C^\infty(M)$ (or, in general, a closed 1-form $\alpha \in Z^1(M)$, such that $\alpha = dh$, locally), which is called the **Hamiltonian function** (or the **Hamiltonian 1-form**) of the system. This function represents the energy of the system.

Postulate 4.4 (Fourth postulate of nonautonomous Hamiltonian mechanics). The dynamical trajectories of a nonautonomous dynamical system are the integral curves of the evolution vector field $\mathcal{E}_h \in \mathfrak{X}(M)$ associated with h, that is, of the vector field solution to Eq. (4.4). Thus, they are the solutions to Eq. (4.6).

Then, we have:

Definition 4.3. A *regular nonautonomous* or *cosymplectic Hamiltonian dynamical system* is a set $(M, \eta, \omega; h)$, where (M, η, ω) is a cosymplectic manifold and $h \in C^\infty(M)$ is the Hamiltonian function of the system. If (M, η, ω) is a precosymplectic manifold, then $(M, \eta, \omega; h)$ is said to be a *singular nonautonomous* or *precosymplectic Hamiltonian dynamical system*.

Definition 4.4. Given a nonautonomous Hamiltonian dynamical system $(M, \eta, \omega; h)$, the **Hamiltonian problem** posed by the system consists of finding the evolution vector field $\mathcal{E}_h \in \mathfrak{X}(M)$ associated with h (if it exists).

In addition, we have:

Proposition 4.5. *If $(M, \eta, \omega; h)$ is a regular nonautonomous Hamiltonian system, then there exists a unique evolution vector field $\mathcal{E}_h \in \mathfrak{X}(M)$, that is, a unique vector field, which is the solution to Eq. (4.4).*

Proof. It is straightforward because if the system is regular, the existence of the isomorphism $\sharp_{(\eta,\omega)}$ is assured. $\qquad\square$

Remark 4.2. If (M, η, ω, h) is a precosymplectic Hamiltonian system, Eq. (4.4) is not necessarily compatible everywhere on M and a suitable *constraint algorithm* must be implemented to find a *final constraint submanifold* $P_f \hookrightarrow M$ (if it exists) where there are evolution vector fields $\mathcal{E}_h \in \mathfrak{X}(M)$, tangent to P_f which is a solution to Eq. (4.4) on P_f (they are not necessarily unique). Singular nonautonomous systems and the corresponding constraint algorithms are studied in [62, 90, 123–125, 345].

4.3   Nonautonomous Lagrangian Dynamical Systems

4.3.1   *Geometric elements*

To develop the cosymplectic Lagrangian formalism, we must first extend some canonical structures of TQ to $T(\mathbb{R} \times TQ)$. Note that as $\mathbb{R} \times TQ$ is a product manifold, we can write

$$T(\mathbb{R} \times TQ) = T\mathbb{R} \oplus_{\mathbb{R} \times TQ} T(TQ),$$

and this splitting extends naturally to vector fields. Thus, any operation on tangent vectors to TQ acts on tangent vectors to $\mathbb{R} \times TQ$. In particular, the canonical endomorphism J of $T(TQ)$ yields a *canonical endomorphism* $\mathcal{J} \colon T(\mathbb{R} \times TQ) \longrightarrow T(\mathbb{R} \times TQ)$, and similarly, the Liouville vector field on TQ yields a *Liouville vector field* $\Delta \in \mathfrak{X}(\mathbb{R} \times TQ)$, which is the Liouville vector field of the vector bundle structure $\pi_{1,0} \colon \mathbb{R} \times TQ \longrightarrow \mathbb{R} \times Q$. In natural coordinates, their local expressions are

$$\mathcal{J} = \frac{\partial}{\partial v^i} \otimes dq^i, \quad \Delta = v^i \frac{\partial}{\partial v^i}.$$

Definition 4.5. Let $\mathbf{c} \colon \mathbb{R} \to \mathbb{R} \times Q$ be a curve. We can write $\mathbf{c} = (c_1, c_2)$, where $c_1 \colon \mathbb{R} \to \mathbb{R}$ and $c_2 \colon \mathbb{R} \to Q$. The **lift** of $\mathbf{c}$ to $\mathbb{R} \times TQ$ is the curve

$$\widehat{\mathbf{c}} = (c_1, \widetilde{c}_2) \colon \mathbb{R} \longrightarrow \mathbb{R} \times TQ,$$

where $\widetilde{c}_2$ is the canonical lift of c_2 to TQ. The curve $\widehat{\mathbf{c}}$ is said to be **holonomic** on $\mathbb{R} \times TQ$.

A vector field $\Gamma \in \mathfrak{X}(\mathbb{R} \times TQ)$ is said to be a SODE on $\mathbb{R} \times TQ$ if its integral curves are holonomic.

As in the autonomous case, the last definition can be equivalently expressed in terms of the canonical structures of $\mathbb{R} \times TQ$, and so it is straightforward to prove:

Proposition 4.6. *A vector field* $\Gamma \in \mathfrak{X}(\mathbb{R} \times TQ)$ *is a SODE if, and only if,* $\mathcal{J}(\Gamma) = \Delta$.

In coordinates, if $\mathbf{c}(t) = (s(t), q^i(t))$, then

$$\widehat{\mathbf{c}}(t) = \left(s(t), q^i(t), \frac{dq^i}{dt}(t) \right),$$

and the local expression of a SODE is

$$\Gamma = g \frac{\partial}{\partial s} + v^i \frac{\partial}{\partial q^i} + f^i \frac{\partial}{\partial v^i},$$

so, in coordinates, a SODE defines the following system of differential equations:

$$\frac{ds}{dt} = g(q, \dot{q}, s), \quad \frac{d^2 q^i}{dt^2} = f^i(q, \dot{q}, s).$$

4.3.2 *Lagrangian formalism: Nonautonomous Lagrangian systems*

The foundations of the Lagrangian formulation of (first-order) nonautonomous dynamical system are analogous to those given in Section 3.2.1 for autonomous Lagrangian systems, and they can be stated as follows:

Postulate 4.5 (First postulate of nonautonomous Lagrangian mechanics). The configuration space of a system with n degrees of freedom is $\mathbb{R} \times Q$, where Q is an n-dimensional differentiable manifold. The phase space is the bundle $\mathbb{R} \times TQ$.

Postulate 4.6 (Second postulate of nonautonomous Lagrangian mechanics). The observables or physical magnitudes of the system are functions of $C^\infty(\mathbb{R} \times TQ)$.

Postulate 4.7 (Third postulate of nonautonomous Lagrangian mechanics). There is a function $\mathcal{L} \in C^\infty(\mathbb{R} \times TQ)$, called the **Lagrangian function**, which contains the dynamical information of the system.

Let $\mathcal{L} \in C^\infty(\mathbb{R} \times TQ)$ be a Lagrangian function. As in the autonomous case, using the canonical structures in $\mathbb{R} \times TQ$, we can introduce the *Lagrangian forms* $\theta_{\mathcal{L}} \in \Omega^1(\mathbb{R} \times TQ)$ and $\omega_{\mathcal{L}} \in \Omega^2(\mathbb{R} \times TQ)$ associated with $\mathcal{L}$, which are defined as follows: $\theta_{\mathcal{L}} = \mathcal{J}(d\mathcal{L})$ and $\omega_{\mathcal{L}} = -d\theta_{\mathcal{L}}$. They have the local expressions

$$\theta_{\mathcal{L}} = \frac{\partial \mathcal{L}}{\partial v^i} \, dq^i, \quad \omega_{\mathcal{L}} = dq^i \wedge d\left(\frac{\partial \mathcal{L}}{\partial v^i}\right).$$

In the same way, we define the *energy Lagrangian function* associated with $\mathcal{L}$ as $E_{\mathcal{L}} := \Delta(\mathcal{L}) - \mathcal{L}$, whose local expression is

$$E_{\mathcal{L}} = v^i \frac{\partial \mathcal{L}}{\partial v^i} - \mathcal{L}.$$

Definition 4.6. For a Lagrangian $\mathcal{L} \in C^\infty(\mathbb{R} \times TQ)$, the **Legendre map** associated with $\mathcal{L}$ is the fibre derivative of $\mathcal{L}$, considered as a function on the vector bundle $\pi_0 \colon \mathbb{R} \times TQ \longrightarrow Q \times \mathbb{R}$, that is, $F\mathcal{L} \colon \mathbb{R} \times TQ \longrightarrow \mathbb{R} \times T^*Q$ given by

$$F\mathcal{L}(t, q, v_q) = (t, F\mathcal{L}_t(q, v_q)),$$

where $\mathcal{L}_t \colon TQ \longrightarrow \mathbb{R}$ denotes the restriction of $\mathcal{L}$ to each fibre of the bundle $\pi_1 \colon \mathbb{R} \times TQ \longrightarrow \mathbb{R}$ (i.e., the Lagrangian $\mathcal{L}$ with t "frozen").

In natural coordinates, we have

$$t \circ F\mathcal{L} = t, \quad q^i \circ F\mathcal{L} = q^i, \quad p_i \circ F\mathcal{L} = \frac{\partial \mathcal{L}}{\partial v^i}.$$

Considering the canonical cosymplectic manifold $(\mathbb{R} \times T^*Q, \eta, \omega)$, where $\omega = -d\theta$, the Lagrangian forms can also be defined as

$$\theta_\mathcal{L} = F\mathcal{L}^*\theta, \quad \omega_\mathcal{L} = F\mathcal{L}^*\omega. \tag{4.10}$$

As in the autonomous case, we also have:

Definition 4.7. A Lagrangian function $\mathcal{L} \in C^\infty(\mathbb{R} \times TQ)$ is said to be **regular** if the corresponding Legendre map $F\mathcal{L}$ is a local diffeomorphism and is hyperregular if FL is a global diffeomorphism. Elsewhere, $\mathcal{L}$ is called a **singular** Lagrangian.

A singular Lagrangian function $\mathcal{L} \in C^\infty(\mathbb{R} \times TQ)$ is called **almost-regular** if $\mathcal{P} := F\mathcal{L}(\mathbb{R} \times TQ)$ is a closed submanifold of $\mathbb{R} \times T^*Q$ (we will denote the natural embedding by $\jmath_0 \colon \mathcal{P} \hookrightarrow \mathbb{R} \times T^*Q$), $F\mathcal{L}$ is a submersion onto its image, and the fibres $F\mathcal{L}^{-1}(FL(\mathrm{p}))$ for every $\mathrm{p} \in \mathbb{R} \times TQ$ are connected submanifolds of $\mathbb{R} \times TQ$.

Once again, as in the autonomous case, it is straightforward to prove that $\mathcal{L}$ is regular if, and only if, the matrix $\left(\dfrac{\partial^2 \mathcal{L}}{\partial v^i \partial v^j} \right)$ is regular everywhere. Therefore, the following equivalences also follow logically:

Proposition 4.7. *The following conditions are equivalent:*

(1) The Lagrangian $\mathcal{L}$ is regular.

(2) The Legendre map $F\mathcal{L}$ is a local diffeomorphism.

(3) The pair $(dt, \omega_\mathcal{L})$ is a cosymplectic structure on $\mathbb{R} \times TQ$.

If $\mathcal{L}$ is not regular, then $(dt, \omega_\mathcal{L})$ is a precosymplectic structure on $\mathbb{R} \times TQ$. Thus, if $\mathcal{L}$ is a regular (or alternatively, singular) Lagrangian, we have $(\mathbb{R} \times TQ, dt, \omega_{\mathcal{L}}; E_{\mathcal{L}})$ that is a cosymplectic (or alternatively, pre-cosymplectic) Hamiltonian system. Then we have:

Definition 4.8. A **nonautonomous** or **(pre)cosymplectic La-grangian dynamical system** is a pair $(\mathbb{R} \times TQ, \mathcal{L})$, where Q is an n-dimensional manifold, and $\mathcal{L} \in C^{\infty}(\mathbb{R} \times TQ)$ is the Lagrangian function of the system.

Now, we can state the following:

Postulate 4.8 (Fourth postulate of nonautonomous Lagrangian mechanics). The dynamical trajectories of a nonautonomous La-grangian system are the integral curves of a vector field $\Gamma_{\mathcal{L}} \in \mathfrak{X}(\mathbb{R} \times TQ)$ satisfying the conditions:

(1) $\Gamma_{\mathcal{L}}$ is the evolution vector field associated with $E_{\mathcal{L}}$, that is, it is a solution to the equations

$$i(\Gamma_{\mathcal{L}})dt = 1, \quad i(\Gamma_{\mathcal{L}})\omega_{\mathcal{L}} = dE_{\mathcal{L}} - R_{\mathcal{L}}(E_{\mathcal{L}})dt. \qquad (4.11)$$

(2) $\Gamma_{\mathcal{L}}$ is a SODE: $\mathcal{J}(\Gamma_{\mathcal{L}}) = \Delta$.

Therefore, these trajectories are the holonomic curves $\mathbf{c} \colon I \subset \mathbb{R} \longrightarrow \mathbb{R} \times TQ$, which are the solutions to equations

$$i(\widetilde{\mathbf{c}})(dt \circ \mathbf{c}) = 1, \quad i(\widetilde{\mathbf{c}})(\omega_{\mathcal{L}} \circ \mathbf{c}) = (dE_{\mathcal{L}} - R_{\mathcal{L}}(E_{\mathcal{L}})dt) \circ \mathbf{c}. \qquad (4.12)$$

Equations (4.12) are the **nonautonomous Euler–Lagrange equations for curves**. Equations (4.11) are called the **nonautonomous La-grangian equations for vector fields**, and a vector field solution to them (if it exists) is a **nonautonomous Lagrangian dynamical vector field**. If, in addition, $\Gamma_{\mathcal{L}}$ is a SODE, then it is called a **nonautonomous Euler–Lagrange vector field** of the system.

In this postulate, $R_{\mathcal{L}}$ is a Reeb vector field of the (pre)cosymplectic structure $(\mathbb{R} \times TQ, dt, \omega_{\mathcal{L}})$, determined by the corresponding equations to (4.1), which are

$$i(R_{\mathcal{L}})dt = 1, \quad i(R_{\mathcal{L}})\omega_{\mathcal{L}} = 0.$$

Definition 4.9. Given a nonautonomous Lagrangian dynamical system $(\mathbb{R} \times TQ, \mathcal{L})$, the **Lagrangian problem** posed by the system consists of finding a SODE vector field $\Gamma_{\mathcal{L}} \in \mathfrak{X}(\mathbb{R} \times TQ)$ solutions to (4.11).

Local expressions: Consider a natural chart $(U; t, q^i, v^i)$ of $\mathbb{R} \times TQ$. Considering the local expressions of the several geometric elements appearing in the nonautonomous dynamical equations, first we obtain the Reeb vector field is given by

$$
R_{\mathcal{L}} = \frac{\partial}{\partial t} + R^i \frac{\partial}{\partial v^i},
$$

where the functions R^i are obtained from

$$
\frac{\partial^2 \mathcal{L}}{\partial t \partial v^j} + \frac{\partial^2 \mathcal{L}}{\partial v^j \partial v^i} R^i = 0,
$$

and then we have

$$
R_{\mathcal{L}}(E_{\mathcal{L}}) = -\frac{\partial \mathcal{L}}{\partial t}.
$$

Therefore, if $\Gamma_{\mathcal{L}} = f \dfrac{\partial}{\partial t} + A^i \dfrac{\partial}{\partial q^i} + B^i \dfrac{\partial}{\partial v^i}$, Eq. (4.11), written in coordinates, leads to $f = 1$ and

$$
0 = \frac{\partial^2 \mathcal{L}}{\partial v^j \partial v^i} B^i + \left(\frac{\partial^2 \mathcal{L}}{\partial q^j \partial v^i} + \frac{\partial^2 \mathcal{L}}{\partial v^j \partial q^i} \right) A^i + \frac{\partial^2 \mathcal{L}}{\partial v^i \partial q^j} v^i + \frac{\partial^2 \mathcal{L}}{\partial v^j \partial t} - \frac{\partial \mathcal{L}}{\partial q^j},
$$

$$
0 = \frac{\partial^2 \mathcal{L}}{\partial v^j \partial v^i} (A^i - v^i).
$$

To demand that $\Gamma_{\mathcal{L}}$ is a SODE is locally equivalent to demanding that $A^i = v^i$. Furthermore, the integral curves of $\Gamma_{\mathcal{L}}$ are holonomic, that is, they are of the form $\mathbf{c}(t) = (t, q^i(t), \dot{q}^i(t))$ and

$$
A^i = v^i = \frac{dq^i}{dt}, \quad B^i = \frac{d^2 q^i}{dt^2},
$$

and the combination of these expressions with the above equations leads to the equation of the integral curves of $\Gamma_{\mathcal{L}}$, which is

$$
\left(\frac{\partial^2 \mathcal{L}}{\partial v^j \partial v^i} \circ \mathbf{c} \right) \frac{d^2 q^i}{dt^2} + \left(\frac{\partial^2 \mathcal{L}}{\partial v^j \partial q^i} \circ \mathbf{c} \right) \frac{dq^i}{dt} + \left(\frac{\partial^2 \mathcal{L}}{\partial v^j \partial t} \circ \mathbf{c} \right) - \frac{\partial \mathcal{L}}{\partial q^j} \circ \mathbf{c} = 0,
$$

or also in an equivalent form as in (3.9), which is the classical coordinate expression of the Euler–Lagrange equations.

Remark 4.3. If the Lagrangian $\mathcal{L}$ is singular (in particular almost-regular), then the existence of solutions to Eq. (4.11) is not assured except, perhaps, in a submanifold of $\mathbb{R} \times TQ$. Furthermore, when these solutions exist, they are not SODE, in general. Thus, to recover the Euler–Lagrange equations (4.12) for the integral curves of $\Gamma_{\mathcal{L}}$, the condition $\mathcal{J}(\Gamma_{\mathcal{L}}) = \Delta$ must be added to the Lagrangian equations (4.11). Then, as in the autonomous case, a constraint algorithm must be implemented to find a submanifold $S \hookrightarrow \mathbb{R} \times TQ$ where the existence of SODE vector field solutions to the Lagrange equations on S and tangent to S is assured. Furthermore, Reeb vector fields $R_{\mathcal{L}}$ are not uniquely defined by Eq. (4.1) for this case; nevertheless, the dynamics and the constraint algorithm are independent of the selected Reeb vector field $R_{\mathcal{L}}$ [90].

4.3.3 Canonical Hamiltonian formalism

As in the autonomous case analysed in Section 3.3.2, we study the case of hyperregular systems, although all the results hold also for the case of regular systems, changing $\mathbb{R} \times T^*Q$ by $F\mathcal{L}(\mathbb{R} \times TQ) \subset \mathbb{R} \times T^*Q$. First, as $F\mathcal{L}$ is a diffeomorphism, we have:

Proposition 4.8. *Let $(\mathbb{R} \times TQ, \mathcal{L})$ be a hyperregular nonautonomous Lagrangian system. Then there exists a unique function $h \in C^\infty(\mathbb{R} \times T^*Q)$ such that $F\mathcal{L}^*h = E_{\mathcal{L}}$, which is the* **Hamiltonian function** *associated with the system $(\mathbb{R} \times TQ, \mathcal{L})$. The triple $(\mathbb{R} \times T^*Q, dt, \omega, h)$ is the* **canonical nonautonomous Hamiltonian system** *associated with $(\mathbb{R} \times TQ, \mathcal{L})$, where $\omega \in \Omega^2(\mathbb{R} \times T^*Q)$ is given in (4.10).*

Then, we have the cosymplectic Hamiltonian system $(\mathbb{R} \times T^*Q, dt, \omega; h)$, fulfilling the postulates and results established in Section 4.2.1, and hence, the Hamiltonian equations for vector fields $\mathcal{E}_h \in \mathfrak{X}(\mathbb{R} \times T^*Q)$ and their integral curves $\mathbf{c} \colon \mathbb{R} \longrightarrow \mathbb{R} \times T^*Q$ read

$$i(\mathcal{E}_h)dt = 1, \quad i(\mathcal{E}_h)\omega = dh - R(h)dt, \tag{4.13}$$

$$i(\widetilde{\mathbf{c}})(dt \circ \mathbf{c}) = 1, \quad i(\widetilde{\mathbf{c}})(\omega \circ \mathbf{c}) = (dh - R(h)dt) \circ \mathbf{c}. \tag{4.14}$$

In canonical coordinates, the local expression of the dynamical vector field $\mathcal{E}_h$ solution to Eq. (4.13) is given by (4.7) and the equations of its integral curves (4.14) are Eq. (4.8), with $f = h$. Then, since $\dfrac{dt}{ds} = 1$ implies

$t(s) = s + const$ (t is an affine transformation of s), we deduce that

$$\frac{dx^i}{dt} = \frac{\partial f}{\partial y_i}, \quad \frac{dy_i}{dt} = -\frac{\partial f}{\partial x^i}$$

which are the *cosymplectic Hamilton equations*.

The relationship between the Lagrangian and the canonical Hamiltonian formalisms of a (hyper)regular Lagrangian system is stated as in the autonomous case as follows:

Theorem 4.2 (Equivalence Theorem). *Let* $(\mathbb{R} \times TQ, \mathcal{L})$ *be a (hyper)regular Lagrangian system.*

(1) If $\Gamma_{\mathcal{L}}$ *is the Lagrangian vector field solution to Eq. (4.11), then there exists a unique vector field* $F\mathcal{L}_* \Gamma_{\mathcal{L}} \equiv \mathcal{E}_h \in \mathfrak{X}(\mathbb{R} \times T^*Q)$*, which is the solution to Eq. (4.13).*

Conversely, if $\mathcal{E}_h$ *is the evolution vector field solution to Eq. (4.13), then there exists a unique vector field* $F\mathcal{L}_*^{-1}\mathcal{E}_h \equiv \Gamma_{\mathcal{L}} \in \mathfrak{X}(\mathbb{R} \times TQ)$*, which is the solution to Eqs. (4.11) and (3.5).*

(2) Equivalently, if $\gamma\colon I \subset \mathbb{R} \longrightarrow \mathbb{R} \times Q$ *is a curve and its canonical lift* $\widehat{\gamma}\colon I \subset \mathbb{R} \longrightarrow \mathbb{R} \times TQ$ *is a solution to Eq. (4.12), then* $\zeta = F\mathcal{L} \circ \widehat{\gamma}$ *is a curve solution to Eq. (4.14).*

Conversely, if $\zeta\colon I \subset \mathbb{R} \longrightarrow \mathbb{R} \times T^*Q$ *is a curve solution to Eq. (4.14), then* $\widehat{\gamma} = \widehat{\pi_{(1,0)}} \circ \zeta\colon I \subset \mathbb{R} \longrightarrow \mathbb{R} \times TQ$ *is a curve solution to Eq. (4.12).*

Proof. The proof follows the same pattern as in Theorem 3.5. □

Remark 4.4. If the Lagrangian is almost regular, then there exists $h_0 \in C^\infty(\mathcal{P})$ such that $F\mathcal{L}_0^* h_0 = E_{\mathcal{L}}$, where $F\mathcal{L}_0\colon \mathbb{R} \times TQ \longrightarrow \mathcal{P}$ is defined by $\jmath_0 \circ F\mathcal{L}_0 = FL$. Now, taking $\omega_0 = \jmath_0^* \omega$, the set $(\mathcal{P}, dt, \omega_0; h_0)$ is a precosymplectic Hamiltonian system, which is called the *canonical Hamiltonian system* associated with $(\mathbb{R} \times TQ, \mathcal{L})$. In particular, the equations equivalent to (4.13) are

$$i(X_0)dt = 1, \quad i(X_0)(\jmath_0^*\omega_0) = dh_0 - R_0(h_0)dt,$$

where $X_0 \in \mathfrak{X}(\mathcal{P})$. The existence of a vector field X_0 solution to the above equations is not assured, except on a submanifold $\mathcal{P}_f \hookrightarrow \mathcal{P}$ to which it is tangent. Moreover, the solution is not unique, and although Reeb vector fields R are not uniquely, defined by Eq. (4.1) for this case, the dynamics and the constraint algorithm are independent of the selected Reeb vector field R_0. (Details on the construction of the canonical Hamiltonian formalism for almost-regular nonautonomous Lagrangians and a deeper study on the constraint algorithms and the equivalence between both formalisms can be found, for instance, in [62, 90, 123, 124, 211, 345].)

4.4 Unified Lagrangian–Hamiltonian Formalism for Nonautonomous Systems

This section is devoted to explaining the extension of the Skinner–Rusk unified formalism, which was presented in Section 3.4, to the case of nonautonomous dynamical systems, using the cosymplectic setting (see [25, 52, 194] for different but equivalent approaches). This extended formalism is quite similar to the autonomous case.

4.4.1 *Extended unified bundle: Unified nonautonomous formalism*

Definition 4.10. Let Q be an n-dimensional differentiable manifold. The **extended unified bundle** or **extended Pontryagin bundle** is $\mathcal{M} := \mathbb{R} \times TQ \times_Q T^*Q$ and has natural projections

$$\kappa_1 \colon \mathcal{M} \longrightarrow \mathbb{R} \times TQ, \quad \kappa_2 \colon \mathcal{M} \longrightarrow \mathbb{R} \times T^*Q, \quad \kappa_0 \colon \mathcal{M} \longrightarrow \mathbb{R} \times Q.$$

Its natural coordinates are (t, q^i, v^i, p_i).

Definition 4.11. A curve $\mathbf{c} \colon \mathbb{R} \to \mathcal{M}$ is **holonomic** on $\mathcal{M}$ if $\kappa_1 \circ \mathbf{c} \colon \mathbb{R} \longrightarrow \mathbb{R} \times TQ$ is holonomic.

A vector field $\Gamma \in \mathfrak{X}(\mathcal{M})$ is a **holonomic vector field** on $\mathcal{M}$ if its integral curves are holonomic on $\mathcal{M}$.

The coordinate expressions of holonomic curves and vector fields on $\mathcal{M}$ are the following:

$$\mathbf{c}(t) = \left(s(t), q^i(t), \frac{dq^i}{dt}(t), p_i(t), \right),$$

$$\Gamma = g \frac{\partial}{\partial s} + v^i \frac{\partial}{\partial q^i} + F^i \frac{\partial}{\partial v^i} + G_i \frac{\partial}{\partial p_i}.$$

The extended unified bundle $\mathcal{M}$ is endowed with the following canonical structures:

Definition 4.12.

(1) The **coupling function** on $\mathcal{M}$ is the map $C \colon \mathcal{M} \longrightarrow \mathbb{R}$, defined by:

$$C \colon \quad \mathcal{M} = \mathbb{R} \times TQ \times_Q T^*Q \quad \longrightarrow \mathbb{R}$$
$$(t, q, v_q, \xi_q) = (t, q^i, v^i, p_i) \mapsto \langle v_q, \xi_q \rangle = v^i p_i.$$

(2) If $\theta \in \Omega^1(\mathbb{R} \times T^*Q)$ and $\omega \in \Omega^2(\mathbb{R} \times T^*Q)$ are the canonical forms on $\mathbb{R} \times T^*Q$, then the canonical forms on $\mathcal{M}$ are

$$\Theta_{\mathcal{M}} := \kappa_2^* \theta \in \Omega^1(\mathcal{M}), \quad \Omega_{\mathcal{M}} := \kappa_2^* \omega = -d\Theta_{\mathcal{M}} \in \Omega^2(\mathcal{M}).$$

We denote by dt the pull-back to $\mathcal{M}$ of the canonical 1-form on $\mathbb{R}$.

Definition 4.13. Given a Lagrangian function $\mathcal{L} \in C^\infty(\mathbb{R} \times TQ)$, denoting $\mathfrak{L} = \kappa_1^* \mathcal{L}$, the **Hamiltonian function** is defined as

$$H = C - \mathfrak{L} \in C^\infty(\mathcal{M}).$$

In coordinates, we have

$$\Theta_{\mathcal{M}} = p_i dq^i, \quad \Omega_{\mathcal{M}} = dq^i \wedge dp_i, \quad H = v^i p_i - \mathfrak{L}(t, q^i, v^i).$$

The triple $(\mathcal{M}, dt, \Omega_{\mathcal{M}})$ is a precosymplectic manifold, since $\ker \Omega_{\mathcal{M}} = \left\langle \dfrac{\partial}{\partial t}, \dfrac{\partial}{\partial v^i} \right\rangle$, where the Reeb vector field can be taken to be $R = \dfrac{\partial}{\partial t}$, so $(\mathcal{M}, dt, \Omega_{\mathcal{M}}, H)$ is a precosymplectic Hamiltonian system. The **dynamical problem** for this system consists of finding an evolution vector field $X_H \in \mathfrak{X}(\mathcal{M})$, which is a solution to the equations

$$i(X_H)dt = 1, \quad i(X_H)\Omega_{\mathcal{M}} = dH - R(H)dt, \tag{4.15}$$

and then the integral curves $\mathbf{c} \colon \mathbb{R} \longrightarrow W$ of X_H are solutions to the equations

$$i(\widetilde{\mathbf{c}})(dt \circ \mathbf{c}) = 1, \quad i(\widetilde{\mathbf{c}})(\Omega_W \circ \mathbf{c})) = (dH - R(H)dt) \circ \mathbf{c}. \tag{4.16}$$

However, as $(\mathcal{M}, \Omega_{\mathcal{M}}, H)$ is a precosymplectic Hamiltonian system, these equations are not compatible on $\mathcal{M}$. In fact, for an arbitrary vector field in $\mathfrak{X}(\mathcal{M})$,

$$X_H = g\frac{\partial}{\partial s} + f^i\frac{\partial}{\partial q^i} + F^i\frac{\partial}{\partial v^i} + G_i\frac{\partial}{\partial p_i}.$$

Equation (4.15) leads to

$$g = 1, \quad f^i = v^i, \quad G_i = \frac{\partial \mathcal{L}}{\partial q^i}, \quad p_i = \frac{\partial \mathcal{L}}{\partial v^i}, \tag{4.17}$$

which gives different kinds of information:

- The first equation fixes the evolution parameter $s = t$.
- The second equation assures that X_H is a holonomic vector field on $\mathcal{M}$ (regardless of the regularity of the Lagrangian function).
- The third equation allows us to determine the component functions G_i.
- The fourth equation is about compatibility conditions, that is, *compatibility constraints* defining a submanifold $\mathcal{M}_0 \hookrightarrow \mathcal{M}$ where vector fields X_H solution to (4.15) are defined. As in the autonomous case, these constraints give the Legendre map and $\mathcal{M}_0 = \mathrm{graph}(\mathrm{F}\mathcal{L})$.

In this way, we have obtained

$$X_H|_{\mathcal{M}_0} = \frac{\partial}{\partial t} + v^i \frac{\partial}{\partial q^i} + F^i \frac{\partial}{\partial v^i} + \frac{\partial \mathcal{L}}{\partial q^i} \frac{\partial}{\partial p_i},$$

where the functions F^i are still undetermined. Now, the constraint algorithm continues by demanding the tangency of X_H to the submanifold $\mathcal{M}_0$, that is, we have to impose that $X_H \left(p_i - \dfrac{\partial \mathcal{L}}{\partial v^i} \right)\bigg|_{\mathcal{M}_0} = 0$, and then we obtain the equations for the functions F^i:

$$\frac{\partial^2 L}{\partial v^i \partial v^j} F^j + \frac{\partial^2 L}{\partial q^j \partial v^i} v^j + \frac{\partial^2 \mathcal{L}}{\partial t \partial v^i} - \frac{\partial \mathcal{L}}{\partial q^i} = 0 \quad (\text{on } \mathcal{M}_0). \qquad (4.18)$$

If $\mathcal{L}$ is a regular Lagrangian, these equations are compatible, and they have a unique vector field X_H, which is the solution to (4.15) on $\mathcal{M}_0$, and the last system of equations gives the dynamical trajectories (i.e., the solutions to Eq. (4.16) on $\mathcal{M}_0$). If $\mathcal{L}$ is singular, Eq. (4.18) can be compatible or not, and eventually, new compatibility constraints can appear, defining a new submanifold $\mathcal{M}_1 \hookrightarrow \mathcal{M}_0$. Then, the constraint algorithm continues by demanding the tangency of solutions to $\mathcal{M}_1$. In the most favorable cases, there is a submanifold $\mathcal{M}_f \hookrightarrow \mathcal{M}_0$ such that there exist holonomic vector fields $X_H \in \mathfrak{X}(\mathcal{M})$ defined on $\mathcal{M}_0$ and tangent to $\mathcal{M}_f$, which are solutions to Eq. (4.15) at support on $\mathcal{M}_f$ (they are not unique necessarily).

4.4.2 Recovering the nonautonomous Lagrangian and Hamiltonian formalisms

The analysis of the equivalence among the unified, the Lagrangian, and Hamiltonian formalisms is performed again for the hyperregular case (the regular case is locally the same; see [25] for details about the singular case).

As in the autonomous case, denoted by $\jmath_0 \colon \mathcal{M}_0 \hookrightarrow \mathcal{M}$ the natural embedding, we have the projections

$$(\kappa_1 \circ \jmath_0) \colon \mathcal{M}_0 \longrightarrow \mathbb{R} \times \mathrm{T}Q, \quad (\kappa_2 \circ \jmath_0) \colon \mathcal{M}_0 \longrightarrow \mathbb{R} \times \mathrm{T}^*Q,$$

where $\kappa_1 \circ \jmath_0$ is a diffeomorphism. The diagram that summarises the situation is

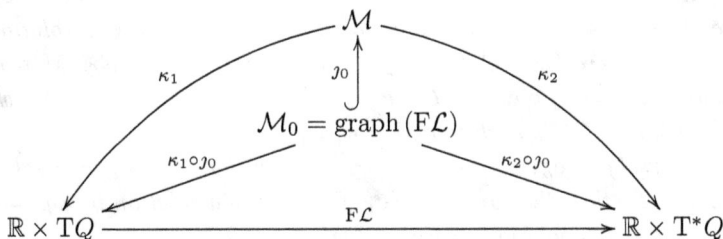

Hence, functions, differential forms, and vector fields on $\mathcal{M}$ tangent to $\mathcal{M}_0$ can be restricted to $\mathcal{M}_0$, and then they can be translated to the Lagrangian side by using the diffeomorphism $\kappa_1 \circ \jmath_0$ and to the Hamiltonian side using the Legendre map and the projection κ_2.

Therefore, if $\mathbf{c}(t) = (t, q^i(t), v^i(t), p_i(t))$ is a solution to Eq. (4.15), and hence, it is an integral curve of the vector field X_H solution to Eq. (4.15), then (4.18) leads to

$$\frac{d}{dt}\left(\frac{\partial \mathcal{L}}{\partial v^i} \circ \mathbf{c}\right) = \frac{\partial \mathcal{L}}{\partial q^i} \circ \mathbf{c}, \tag{4.19}$$

and from Eq. (4.17), we obtain

$$\frac{dq^i}{dt} = v^i, \quad \frac{dp_i}{dt} = \frac{\partial \mathcal{L}}{\partial q^i} \circ \mathbf{c} = -\frac{\partial H}{\partial q^i} \circ \mathbf{c}, \quad p_i = \frac{\partial \mathcal{L}}{\partial v^i} \circ \mathbf{c}. \tag{4.20}$$

From the first group of these equations, together with (4.19), we recover the Euler–Lagrange equations for the curves $\mathbf{c}_\mathcal{L}(t) = (t, q^i(t), v^i(t))$. In addition, $\dfrac{\partial \mathcal{L}}{\partial q^i} = -\dfrac{\partial H}{\partial q^i}$, and hence, the second group of equations (4.20) is

$$\frac{dp_i}{dt} = -\frac{\partial H}{\partial q^i} \circ \mathbf{c},$$

and using the local expression of H and the first group of Eq. (4.20), we get

$$\frac{\partial H}{\partial q^i} \circ \mathbf{c} = v^i = \frac{dq^i}{dt},$$

but using the Legendre map (i.e., the third group of equations (4.17)), we have $H = F\mathcal{L}^*h$, and these last equations become the *Hamilton equations* for the curves $\mathbf{c}_h(t) = (t, q^i(t), p_i(t))$.

For every curve $\mathbf{c} \colon \mathbb{R} \longrightarrow \mathcal{M}$, taking values on $\mathcal{M}_0$ can be viewed as $\mathbf{c} = (\mathbf{c}_\mathcal{L}, \mathbf{c}_h)$, where $\mathbf{c}_\mathcal{L} = \kappa_1 \circ \mathbf{c} \colon \mathbb{R} \longrightarrow \mathbb{R} \times TQ$ and $\mathbf{c}_h = F\mathcal{L} \circ \mathbf{c}_\mathcal{L} \colon \mathbb{R} \longrightarrow \mathbb{R} \times T^*Q$. Thus, we have proven that

> **Theorem 4.3.** *If* $\mathbf{c}\colon \mathbb{R} \longrightarrow \mathcal{M}$, *with* $\mathrm{Im}\, \mathbf{c} \subset \mathcal{M}_0$, *is a curve fulfilling Eq. (4.16), then* $\mathbf{c}_{\mathcal{L}}$ *is the lift to* $\mathbb{R} \times TQ$ *of the projected curve* $\mathbf{c}_o = \kappa_0 \circ \mathbf{c}\colon \mathbb{R} \longrightarrow \mathbb{R} \times Q$ *(i.e.,* $\mathbf{c}_{\mathcal{L}}$ *is a holonomic curve), and it is a solution to Eq. (4.12), where* $E_{\mathcal{L}} \in \mathrm{C}^{\infty}(\mathbb{R} \times TQ)$ *is such that* $\mathrm{H} = \kappa_1^* E_{\mathcal{L}}$. *Moreover, the curve* $\mathbf{c}_{\mathrm{h}} = \kappa_2 \circ \mathbf{c} = F\mathcal{L} \circ \mathbf{c}_{\mathcal{L}}$ *is a solution to Eq. (4.14), where* $\mathrm{h} \in \mathrm{C}^{\infty}(\mathbb{R} \times T^*Q)$ *is such that* $\mathrm{H} = \kappa_1^* \mathrm{h}$.
>
> *Conversely, if* $\mathbf{c}_o\colon \mathbb{R} \longrightarrow Q$ *is a curve such that* $\widetilde{\mathbf{c}}_o \equiv \mathbf{c}_{\mathcal{L}}$ *is a solution to Eq. (4.12), then the curve* $\mathbf{c} = (\mathbf{c}_{\mathcal{L}}, F\mathcal{L} \circ \mathbf{c}_{\mathcal{L}})$ *is a solution to Eq. (4.16), and* $F\mathcal{L} \circ \mathbf{c}_{\mathcal{L}}$ *is a solution to Eq. (4.14).*

The curves $\mathbf{c}\colon \mathbb{R} \longrightarrow \mathcal{M}$ that are solutions to Eq. (4.16) are the integral curves of the holonomic vector field $X_{\mathrm{H}} \in \mathfrak{X}(\mathcal{M})$, which is the solution to Eq. (4.15), and the curves $\mathbf{c}_{\mathcal{L}}\colon \mathbb{R} \longrightarrow \mathbb{R} \times TQ$ are the integral curves of the holonomic vector field $\Gamma_{\mathcal{L}} \in \mathfrak{X}(\mathbb{R} \times TQ)$, which is the solution to Eq. (4.11), and the curves $\mathbf{c}_{\mathrm{h}}\colon \mathbb{R} \longrightarrow \mathbb{R} \times T^*Q$ are the integral curves of the vector field $\mathcal{E}_{\mathrm{h}} \in \mathfrak{X}(\mathbb{R} \times T^*Q)$, which is the solution to Eq. (4.13). Therefore:

> **Theorem 4.4.** *Let* $X_{\mathrm{H}} \in \mathfrak{X}(\mathcal{M})$ *be the solution to Eq. (4.15) (on* $\mathcal{M}_0$*), which is tangent to* $\mathcal{M}_0$. *Then:*
>
> *The vector field* $\Gamma_{\mathcal{L}} \in \mathfrak{X}(\mathbb{R} \times TQ)$, *defined by* $\Gamma_{\mathcal{L}} \circ \kappa_1 = T\kappa_1 \circ X_{\mathrm{H}}$, *is the SODE solution to Eq. (4.11), where* $E_{\mathcal{L}} \in \mathrm{C}^{\infty}(TQ)$ *is such that* $\mathrm{H} = \kappa_1^* E_{\mathcal{L}}$.
>
> *The vector field* $\mathcal{E}_{\mathrm{h}} \in \mathfrak{X}(\mathbb{R} \times T^*Q)$, *defined by* $\mathcal{E}_{\mathrm{h}} \circ \kappa_2 = T\kappa_2 \circ X_{\mathrm{H}}$, *is the solution to Eq. (4.13), where* $\mathrm{h} \in \mathrm{C}^{\infty}(\mathbb{R} \times T^*Q)$ *is such that* $\mathrm{H} = \kappa_1^* \mathrm{h}$. *Furthermore,* $F\mathcal{L}_* \Gamma_{\mathcal{L}} = \mathcal{E}_{\mathrm{h}}$.

4.5 Symmetries of Regular Nonautonomous Dynamical Systems

We restrict the study of symmetries of regular nonautonomous dynamical systems to the case in which the phase space is an almost-canonical cosymplectic manifold, that is, $\mathbb{R} \times M$, with M being a symplectic manifold, and in particular, to the canonical case $M = T^*Q$. Recall that in this case, the natural projection $\pi_{\mathbb{R}}\colon \mathbb{R} \times M \longrightarrow \mathbb{R}$ defines a global canonical coordinate t.

This study is inspired by the ideas introduced mainly in [4] and follows the scheme given in Section 2.3 (see also [314] for a very interesting description of symmetries in the Lagrangian formalism).

4.5.1 Symmetries of nonautonomous Hamiltonian systems: Noether Theorem

Let $(\mathbb{R} \times M, \eta, \omega; h)$ be a regular nonautonomous (almost-canonical) Hamiltonian system and $\mathcal{E}_h \in \mathfrak{X}(\mathbb{R} \times M)$ be its evolution vector field.

Definition 4.14. A function $f \in C^\infty(\mathbb{R} \times M)$ is a **conserved quantity** or a **constant of motion** if

$$L(\mathcal{E}_h)f = 0.$$

Definition 4.15. A **dynamical symmetry** of the system $(\mathbb{R} \times M, \eta, \omega; h)$ is a diffeomorphism $\Phi \colon \mathbb{R} \times M \longrightarrow \mathbb{R} \times M$ satisfying

$$\Phi_* \mathcal{E}_h = \mathcal{E}_h.$$

Definition 4.16. An **infinitesimal dynamical symmetry** of the system $(\mathbb{R} \times M, \eta, \omega; h)$ is a vector field $Y \in \mathfrak{X}(M)$ such that the local diffeomorphisms generated by its flux is dynamical symmetries of the system, that is,

$$L(Y)\mathcal{E}_h = [Y, \mathcal{E}_h] = 0.$$

It is straightforward to check that the results obtained in Section 2.3.3 for the dynamical symmetries of autonomous Hamiltonian systems also hold in this case.

Definition 4.17. A **cosymplectic Noether symmetry** of the system $(\mathbb{R} \times M, \eta, \omega; h)$ is a diffeomorphism $\Phi \colon \mathbb{R} \times M \longrightarrow \mathbb{R} \times M$ satisfying the following conditions:

$$\text{(a)} \quad \Phi^*\omega = \omega, \quad \text{(b)} \quad \Phi^*t = t, \quad \text{(c)} \quad \Phi^*h = h.$$

If the cosymplectic structure is exact, and $\omega = -d\theta$, then a cosymplectic Noether symmetry is said to be **exact** if $\Phi^*\theta = \theta$.

Definition 4.18. An **infinitesimal cosymplectic Noether symmetry** of the system $(\mathbb{R} \times M, \eta, \omega; h)$ is a vector field $Y \in \mathfrak{X}(\mathbb{R} \times M)$ whose local fluxes are local cosymplectic Noether symmetries, that is, it satisfies

$$\text{(a)} \quad L(Y)\omega = 0, \quad \text{(b)} \quad L(Y)t = i(Y)\eta = 0, \quad \text{(c)} \quad L(Y)h = 0.$$

If the cosymplectic structure is exact, an infinitesimal cosymplectic Noether symmetry is said to be **exact** if $L(Y)\theta = 0$.

For the Reeb vector field, we have the following property:

Proposition 4.9.

(1) If $\Phi \in \mathrm{Diff}\,(\mathbb{R} \times M)$ is a cosymplectic Noether symmetry, then $\Phi_* R = R$.

(2) If $Y \in \mathfrak{X}(\mathbb{R} \times M)$ is an infinitesimal cosymplectic Noether symmetry, then $[Y, R] = 0$.

Proof. (1) As Φ is a cosymplectic Noether symmetry, then $\Phi^* \omega = \omega$ and $\Phi^* \eta = \eta$; then, considering the conditions (4.1) that characterise the Reeb vector field, we have

$$i(\Phi_* R)\eta = i(R)(\Phi^{-1})^* \eta = i(R)\eta = 1,$$
$$i(\Phi_* R)\omega = i(R)(\Phi^{-1})^* \omega = i(R)\omega = 0,$$

and the result is a consequence of the unicity of the Reeb vector field, so $\Phi_* R = R$.

(2) We have

$$i([Y, R])\omega = \mathrm{L}(Y)\,i(R)\omega - i(R)\,\mathrm{L}(Y)\omega = 0 \iff [Y, R] \in \ker \omega,$$
$$i([Y, R])\eta = \mathrm{L}(Y)\,i(R)\eta - i(R)\,\mathrm{L}(Y)\eta = 0 \iff [Y, R] \in \ker \eta,$$

and then $[Y, R] \in \ker \omega \cap \ker \eta = \{0\}$ (because $\eta \wedge \omega^n$ is a volume form). $\qquad\square$

Remark 4.5.

(1) The condition $\Phi^* t = t$ means that cosymplectic Noether symmetries generate transformations along the fibres of the projection $\pi_{\mathbb{R}} : \mathbb{R} \times M \longrightarrow \mathbb{R}$, that is, they leave the fibres of the projection $\pi_{\mathbb{R}} : \mathbb{R} \times M \longrightarrow \mathbb{R}$ invariant, or what is the equivalent, $\pi_{\mathbb{R}} \circ \Phi = \pi_{\mathbb{R}}$.

(2) In the case of infinitesimal cosymplectic Noether symmetries, the analogous condition is $i(Y)dt = 0$, which means that Y has the local expression $Y = f^i \dfrac{\partial}{\partial x^i} + g_i \dfrac{\partial}{\partial y_i}$. Y is tangent to the fibres of the projection $\pi_{\mathbb{R}}$. Thus, these infinitesimal symmetries only generate transformations along these fibres, the local flux of the generators Y leaves the fibres of the projection $\pi_{\mathbb{R}}$ invariant. Furthermore, as a consequence of the above proposition, and considering that $R = \dfrac{\partial}{\partial t}$,

in this local expression for Y, the component functions f^i, g_i do not depend on the coordinate t.

Proposition 4.10.

(1) If $\Phi \in \mathrm{Diff}\,(\mathbb{R} \times M)$ is a cosymplectic Noether symmetry, then it is a dynamical symmetry.

(2) If $Y \in \mathfrak{X}(\mathbb{R} \times M)$ is an infinitesimal cosymplectic Noether symmetry, then it is an infinitesimal dynamical symmetry.

Proof. (1) As Φ is a cosymplectic Noether symmetry, then $\Phi^*\omega = \omega$, $\Phi^*\eta = \eta$, $\Phi^*h = h$, and $\Phi_* R = R$; therefore, from the equation $0 = i(\mathcal{E}_h)\omega - \mathrm{d}h + (R(h))\eta$, we obtain

$$
\begin{aligned}
0 &= \Phi^*[i(\mathcal{E}_h)\omega - \mathrm{d}h + (R(h))\eta]\\
&= i(\Phi_*^{-1}\mathcal{E}_h)(\Phi^*\omega) - \Phi^*\mathrm{d}h + ((\Phi_*^{-1}R)(\Phi^*h))(\Phi^*\eta)]\\
&= i(\Phi_*^{-1}\mathcal{E}_h)\omega - \mathrm{d}h + (R(h))\eta,
\end{aligned}
$$

and from $0 = i(\mathcal{E}_h)\eta - 1$, we get

$$
0 = \Phi^*(i(\mathcal{E}_h)\eta - 1) = i(\Phi_*^{-1}\mathcal{E}_h)(\Phi^*\eta) - 1 = i(\Phi_*^{-1}\mathcal{E}_h)\eta - 1,
$$

but as the system is regular, the evolution vector field solution is unique, so $\Phi_*^{-1}\mathcal{E}_h = \mathcal{E}_h$, and hence, the result holds.

(2) The result for infinitesimal cosymplectic Noether symmetries follows from the above item using the local uniparametric group of diffeomorphisms associated with their fluxes (or also, using the properties (4.1) that characterise the Reeb vector field and reasoning as in the first item). $\qquad\qquad\square$

In addition, it is straightforward to prove that if $Y_1, Y_2 \in \mathfrak{X}(\mathbb{R} \times M)$ are infinitesimal cosymplectic Noether symmetries, then so is $[Y_1, Y_2]$.

As in the previous chapters, to establish Noether's Theorem for nonautonomous systems, we will stick to the case of infinitesimal Noether symmetries:

Proposition 4.11. *Let $Y \in \mathfrak{X}(\mathbb{R} \times M)$ be an infinitesimal cosymplectic Noether symmetry. Then, for every $p \in \mathbb{R} \times M$, there is an open neighbourhood $U \ni p$, such that:*

(1) There exists $\mathcal{F} \in C^\infty(U)$, which is unique up to a constant function, such that

$$i(Y)\omega = d\mathcal{F} \quad \text{(on } U\text{)} . \tag{4.21}$$

(2) There exists $\zeta \in C^\infty(U)$, verifying that $L(Y)\theta = d\zeta$, on U, and then

$$\mathcal{F} = i(Y)\theta - \zeta \quad \text{(up to a constant function on } U\text{)} .$$

Then, for exact infinitesimal cosymplectic Noether symmetries, we have $\mathcal{F} = i(Y)\theta$ (up to a constant function).

Proof. Recalling the conditions set out in Definition 4.18:

(1) It is a consequence of the Poincaré Lemma and the condition

$$0 = L(Y)\omega = i(Y)d\omega + d\,i(Y)\omega = d\,i(Y)\omega.$$

(2) We have

$$d\,L(Y)\theta = L(Y)d\theta = -\,L(Y)\omega = 0,$$

and hence, $L(Y)\theta$ are closed forms. Therefore, by the Poincaré Lemma, there exists $\zeta \in C^\infty(U)$, verifying that $L(Y)\theta = d\zeta$ on U. Furthermore, as (4.21) holds on U, we obtain

$$d\zeta = L(Y)\theta = d\,i(Y)\theta + i(Y)d\theta = d\,i(Y)\theta - i(Y)\omega = d\{i(Y)\theta - \mathcal{F}\},$$

and the result holds.   □

Remark 4.6. Observe that using Darboux coordinates on $\mathbb{R} \times M$, item (2) of Proposition 4.11 tells us that the conserved quantities associated with infinitesimal cosymplectic Noether symmetries do not depend on the coordinate t (since the generators of these symmetries, the vector fields Y, do not depend on them either).

Finally, the classical Noether's Theorem can be stated for these kinds of symmetries as follows:

Theorem 4.5 (Noether's Theorem). *If $Y \in \mathfrak{X}(\mathbb{R} \times M)$ is an infinitesimal cosymplectic Noether symmetry, then, for every $p \in \mathbb{R} \times M$, there is an open neighbourhood $U \ni p$ such that the function $\mathcal{F} = i(Y)\theta - \zeta$ is a conserved quantity.*

Proof. From (4.21), we obtain

$$L(\mathcal{E}_h)\mathcal{F} = i(\mathcal{E}_h)\mathrm{d}\mathcal{F} = i(\mathcal{E}_h)\,i(Y)\omega = -\,i(Y)\,i(\mathcal{E}_h)\omega$$
$$= -\,i(Y)\mathrm{d}h + i(Y)((R(h))\eta) = -\,L(Y)h + (R(h))\,i(Y)\eta = 0,$$

because Y is an infinitesimal cosymplectic Noether symmetry. $\square$

As a final situation, consider the case of the canonical model $\mathbb{R} \times T^*Q$. If $\varphi\colon Q \longrightarrow Q$ is a diffeomorphism, we construct the diffeomorphism $\Phi :=$ $(\mathrm{Id}_{\mathbb{R}}, T^*\varphi)\colon \mathbb{R}\times T^*Q \longrightarrow \mathbb{R}\times T^*Q$, where $T^*\varphi\colon T^*Q \longrightarrow T^*Q$ is the canonical lift of φ to T^*Q. Then Φ is said to be the **canonical lift** of φ to $\mathbb{R} \times T^*Q$. Any transformation Φ of this kind is called a **natural transformation** of $\mathbb{R} \times T^*Q$.

In the same way, given a vector field $Z \in \mathfrak{X}(Q)$ we can define its **complete lift** to $\mathbb{R} \times T^*Q$ as the vector field $Y \in \mathfrak{X}(\mathbb{R} \times T^*Q)$ whose local flux is the canonical lift of the local flux of Z to $\mathbb{R} \times T^*Q$, that is, $Y = Z^*$, where Z^* denotes the complete lift of Z to $\mathbb{R} \times T^*Q$. Any vector field Y of this kind is called a **natural infinitesimal transformation** of $\mathbb{R} \times T^*Q$.

Then we have:

Definition 4.19. Let $(\mathbb{R} \times T^*Q, \eta, \omega; h)$ be a regular nonautonomous Hamiltonian system.

A symmetry (respectively, cosymplectic Noether symmetry) $\Phi \in C^\infty(\mathbb{R} \times T^*Q)$ is said to be **natural** if Φ is a natural transformation of $\mathbb{R} \times T^*Q$.

An infinitesimal symmetry (respectively, infinitesimal cosymplectic Noether symmetry) $Y \in \mathfrak{X}(\mathbb{R} \times T^*Q)$ is said to be **natural** if Y is a natural infinitesimal transformation of $\mathbb{R} \times T^*Q$.

Remark 4.7. The study of symmetries and conserved quantities in the Lagrangian formalism of nonautonomous dynamical systems follows the same patterns as in the canonical Hamiltonian formalism, considering the system $(\mathbb{R} \times TQ, \eta_{\mathcal{L}}, \omega_{\mathcal{L}}, E_{\mathcal{L}})$ and the development made in Section 3.5.2 (see also [102, 314]).

4.6 Other Geometrical Settings for Nonautonomous Dynamical Systems

As mentioned in the beginning of this chapter, there are other ways to describe geometrically time-dependent dynamical systems. Next, we briefly review two of the best known methods.

4.6.1   *Contact formulation*

(See [1, 90, 109, 154, 155, 314, 316] for more details.)

The manifold where this formalism is developed is the same as in the cosymplectic case, that is $\mathbb{R} \times M$, where (M, Ω) is a symplectic manifold. In particular, consider the canonical case, where $M = \mathrm{T}^*Q$ (Hamiltonian formalism), or $M = \mathrm{T}Q$ (Lagrangian formalism), and $\mathbb{R} \times Q$ is the configuration space of the system.

Thus, in the Lagrangian description of this contact formulation, the dynamics takes place on the manifold $\mathbb{R} \times \mathrm{T}Q$. Then, given an admissible time-dependent Lagrangian function $\mathcal{L} \in C^\infty(\mathbb{R} \times \mathrm{T}Q)$ (then $\operatorname{rank} \omega_\mathcal{L} = 2n$), we define the *Poincaré-Cartan forms* associated with $\mathcal{L}$ as

$$\boldsymbol{\Theta}_\mathcal{L} := \theta_\mathcal{L} - E_\mathcal{L}\, \mathrm{d}t, \quad \boldsymbol{\Omega}_\mathcal{L} := -\mathrm{d}\boldsymbol{\Theta}_\mathcal{L} = \omega_\mathcal{L} + \mathrm{d}E_\mathcal{L} \wedge \mathrm{d}t.$$

The expression in coordinates of $\boldsymbol{\Theta}_\mathcal{L}$ is

$$\boldsymbol{\Theta}_\mathcal{L} = \frac{\partial \mathcal{L}}{\partial v^i}\mathrm{d}q^i - \left(v^i \frac{\partial \mathcal{L}}{\partial v^i} - \mathcal{L} \right) \mathrm{d}t.$$

If $\mathcal{L}$ is a regular Lagrangian, then the pair $(\mathbb{R} \times \mathrm{T}Q, \boldsymbol{\Omega}_\mathcal{L})$ is said to be a **regular time-dependent Lagrangian system**. The dynamical Lagrangian equations for a vector field $X_\mathcal{L} \in \mathfrak{X}(\mathbb{R} \times \mathrm{T}Q)$ are

$$i(X_\mathcal{L})\boldsymbol{\Omega}_\mathcal{L} = 0, \quad i(X_\mathcal{L})\mathrm{d}t = 1, \tag{4.22}$$

and the equations for the integral curves $\mathbf{c} \colon \mathbb{R} \longrightarrow \mathbb{R} \times \mathrm{T}Q$ of $X_\mathcal{L}$ are

$$i(\widetilde{\mathbf{c}})(\boldsymbol{\Omega}_\mathcal{L} \circ \mathbf{c}) = 0, \quad i(\widetilde{\mathbf{c}})(\mathrm{d}t \circ \mathbf{c}) = 1.$$

These equations are compatible and determinate, and in a local chart of coordinates on $\mathbb{R} \times \mathrm{T}Q$, its unique solution is

$$X_\mathcal{L} = \frac{\partial}{\partial t} + v^i \frac{\partial}{\partial q^i} + \left(\frac{\partial^2 \mathcal{L}}{\partial v^i \partial v^j} \right)^{-1} \left(\frac{\partial \mathcal{L}}{\partial q^j} - v^k \frac{\partial^2 \mathcal{L}}{\partial v^k \partial v^j} \right) \frac{\partial}{\partial v^i}.$$

Note that $X_\mathcal{L}$ is a SODE, so its integral curves are solutions to the Euler–Lagrange equations. Moreover, we obtain the additional information

$$\frac{dE_\mathcal{L}}{dt} \circ \mathbf{c} = \frac{\partial \mathcal{L}}{\partial t} \circ \mathbf{c},$$

on every integral curve of $X_\mathcal{L}$, $\mathbf{c} \colon I \subseteq \mathbb{R} \longrightarrow \mathbb{R} \times \mathrm{T}Q$. This gives the non-conservation law of the energy.

In the Hamiltonian formalism, as in the cosymplectic formulation, the dynamical information is given by the *time-dependent Hamiltonian function* $\mathrm{h} \in C^\infty(\mathbb{R} \times \mathrm{T}^*Q)$, which can only be defined locally. Then, if $\omega = \pi_2^*\Omega$, where Ω is the canonical 2-form on T^*Q, we define the form (only locally if h is a locally Hamiltonian function),

$$\Omega_h := \omega + dh \wedge dt \in Z^2(\mathbb{R} \times T^*Q),$$

which is called the *Hamilton–Cartan 2-form* associated with h. Observe that $\Omega_h = -d\Theta_h$, with $\Theta_h = \theta - h\,dt$, and $\theta = \pi_2^*\Theta$, where Θ is the Liouville's 1-form on T^*Q. Thus, $\omega = -d\theta$. Then, the pair $(\mathbb{R} \times T^*Q, \Omega_h)$ is said to be a **time-dependent regular (locally) Hamiltonian system**. Its dynamical equations for a vector field $X_h \in \mathfrak{X}(\mathbb{R} \times T^*Q)$ are

$$i(X_h)\Omega_h = 0; \quad i(X_h)dt = 1,$$

and for the integral curves $\mathbf{c} \colon \mathbb{R} \longrightarrow \mathbb{R} \times T^*Q$ of X_h, the equations are

$$i(\widetilde{\mathbf{c}})(\Omega_h \circ \mathbf{c}) = 0, \quad i(\widetilde{\mathbf{c}})(dt \circ \mathbf{c})) = 1.$$

This system of equations is compatible and determinate. In fact, in a chart of canonical coordinates on $\mathbb{R} \times T^*Q$, the unique solution is the vector field

$$X_h = \frac{\partial}{\partial t} + \frac{\partial h}{\partial p_i}\frac{\partial}{\partial q^i} - \frac{\partial h}{\partial q^i}\frac{\partial}{\partial p_i},$$

and if $\mathbf{c}(s) = (t(s), q^i(s), p_i(s))$ is an integral curve of X_h, then the second equation implies that $t = s + ctn$ and $c(t)$ is a solution to the Hamilton equations

$$\frac{dq^i}{dt} = \frac{\partial h}{\partial p_i} \circ \mathbf{c}, \quad \frac{dp_i}{dt} = -\frac{\partial h}{\partial q^i} \circ \mathbf{c}.$$

The connection between the Lagrangian and the Hamiltonian contact formalisms for nonautonomous systems is performed by the Legendre map $F\mathcal{L} \colon \mathbb{R} \times TQ \longrightarrow \mathbb{R} \times T^*Q$ described in Definition 4.6. Then, as $F\mathcal{L}$ is a (local) diffeomorphism, all the geometrical objects in $\mathbb{R} \times TQ$ are $F\mathcal{L}$-projectable; in particular, we have

$$F\mathcal{L}^*h = E_{\mathcal{L}}; \quad F\mathcal{L}^*\Omega_h = \Omega_{\mathcal{L}},$$

and therefore, for the solutions,

$$F\mathcal{L}_* X_{\mathcal{L}} = X_h.$$

Finally, the equivalence between the cosymplectic and the contact (Lagrangian) formulations is stated in the following:

Proposition 4.12. Let $(\mathbb{R} \times TQ, \mathcal{L})$ be a time-dependent Lagrangian system. Then Eqs. (4.22) and (4.11) are equivalents.

Proof. From (4.22), we have

$$0 = i(X_\mathcal{L})\Omega_\mathcal{L} = i(X_\mathcal{L})\omega_\mathcal{L} + i(X_\mathcal{L})dE_\mathcal{L}\,dt - dE_\mathcal{L}\,i(X_\mathcal{L})dt$$
$$= i(X_\mathcal{L})\omega_\mathcal{L} + i(X_\mathcal{L})dE_\mathcal{L}\,dt - dE_\mathcal{L}.$$

Now, taking the Reeb vector field $R_\mathcal{L}$, we obtain

$$0 = i(R_\mathcal{L})\,i(X_\mathcal{L})\omega_\mathcal{L} + i(X_\mathcal{L})dE_\mathcal{L}\,i(R_\mathcal{L})dt - i(R_\mathcal{L})dE_\mathcal{L}$$
$$= i(X_\mathcal{L})dE_\mathcal{L} - i(R_\mathcal{L})dE_\mathcal{L} \iff i(X_\mathcal{L})dE_\mathcal{L} = i(R_\mathcal{L})dE_\mathcal{L}\,,$$

and coming to the first equation, we obtain

$$i(X_\mathcal{L})\omega_\mathcal{L} = dE_\mathcal{L} - i(R_\mathcal{L})dE_\mathcal{L}\,dt\,,$$

which together with $i(X_\mathcal{L})dt = 1$ are Eq. (4.11). $\qquad\square$

Thus, the Lagrangian contact and cosymplectic formulations are equivalent. The equivalence between the corresponding Hamiltonian equations is proven in the same way and leads to the same Hamilton equations.

Remark 4.8 (Cartan's Theorem). It can be proven (see [1] and Exercise 4.5) that if $(\mathbb{R} \times T^*Q, \Omega_h)$ is a regular time-dependent Hamiltonian system, then $\Theta_h \wedge \Omega_h{}^n$ is a volume form on $\mathbb{R} \times T^*Q$ if, and only if, the condition $X_h(h) - h \neq 0$ holds everywhere. Then, Θ_h is said to be a *contact form*, and the pair $(\mathbb{R} \times T^*Q, \Theta_h)$ is called a *contact manifold* (see Definition 6.1).

In the same way, if $(\mathbb{R} \times TQ, \Omega_\mathcal{L})$ is a regular time-dependent Lagrangian system, then $\Theta_\mathcal{L} \wedge \Omega_\mathcal{L}{}^n$ is a volume form on $\mathbb{R} \times TQ$ if the condition $X_\mathcal{L}(E_\mathcal{L}) - E_\mathcal{L} \equiv -\mathcal{L} \neq 0$ everywhere. Then $\Theta_\mathcal{L}$ is a contact form, and $(\mathbb{R} \times TQ, \Theta_\mathcal{L})$ is a contact manifold.

All these justify the name *contact formulation*.

4.6.2 *Extended symplectic formulation*

(See [13, 154, 233, 301, 302] for details.)

In this formulation, the Lagrangian formalism is done starting from the extended configuration space $\mathbb{R} \times Q$ and then taking its tangent bundle

$T(\mathbb{R} \times Q) \cong T\mathbb{R} \times TQ \cong \mathbb{R} \times \mathbb{R} \times TQ$, which is called the *extended velocity phase space* and has the canonical projections

$$\nu: T(\mathbb{R} \times Q) \longrightarrow \mathbb{R} \times TQ; \quad \varpi: T(\mathbb{R} \times Q) \longrightarrow \mathbb{R}.$$

Then, we have the following extensions of the geometrical objects that are defined in $\mathbb{R} \times TQ$,

$$E_{\mathcal{L}_{ext}} := \nu^* E_{\mathcal{L}} + \varpi,$$

$$\Theta_{\mathcal{L}_{ext}} := \nu^* \Theta_{\mathcal{L}} + E_{\mathcal{L}_{ext}} \wedge dt,$$

$$\Omega_{\mathcal{L}_{ext}} := -d\Theta_{\mathcal{L}_{ext}} = \nu^* \Omega_{\mathcal{L}} - dE_{\mathcal{L}_{ext}} \wedge dt.$$

Another way to obtain these geometric elements is constructing the *extended Lagrangian*

$$\mathcal{L}_{ext} := \nu^* \mathcal{L} + \frac{1}{2}\varpi^2 \in C^\infty(T(\mathbb{R} \times Q)), \tag{4.23}$$

using the canonical structures (the vertical endomorphism and the Liouville vector field) of the tangent bundle $T(\mathbb{R} \times Q)$. It is clear that the Lagrangian $\mathcal{L}$ is regular if, and only if, $\Omega_{\mathcal{L}_{ext}}$ is a symplectic form, and $(T(\mathbb{R} \times Q), \Omega_{\mathcal{L}_{ext}}, E_{\mathcal{L}_{ext}})$ is a regular Hamiltonian system. This means that there exists a unique vector field $X_{\mathcal{L}_{ext}} \in \mathfrak{X}(T(\mathbb{R} \times Q))$ that is the solution to the Lagrangian equation of motion in this formalism, which is

$$i(X_{\mathcal{L}_{ext}})\Omega_{\mathcal{L}_{ext}} = dE_{\mathcal{L}_{ext}},$$

and whose integral curves $\mathbf{c}: \mathbb{R} \longrightarrow T(\mathbb{R} \times Q)$ are the solutions to the equation

$$i(\widetilde{\mathbf{c}})(\Omega_{\mathcal{L}_{ext}} \circ \mathbf{c}) = dE_{\mathcal{L}_{ext}} \circ \mathbf{c}.$$

In coordinates, this solution is

$$X_{\mathcal{L}_{ext}} = \frac{\partial}{\partial t} - \frac{\partial E_{\mathcal{L}_{ext}}}{\partial t} \frac{\partial}{\partial \varpi} + v^i \frac{\partial}{\partial q^i} + \left(\frac{\partial^2 \mathcal{L}}{\partial v^i \partial v^j}\right)^{-1} \left(\frac{\partial \mathcal{L}}{\partial q^j} - v^k \frac{\partial^2 \mathcal{L}}{\partial v^k \partial v^j}\right) \frac{\partial}{\partial v^i}.$$

Nevertheless, the real manifold of physical states is not $T(\mathbb{R} \times Q)$ but $\mathbb{R} \times TQ$. However, as $\ker \rho_* = \left\langle \frac{\partial}{\partial \varpi} \right\rangle$ and $\left[\frac{\partial}{\partial \varpi}, X_{\mathcal{L}_{ext}}\right] = 0$, this vector field $X_{\mathcal{L}_{ext}}$ is ρ-projectable, and actually, we have

$$\rho_* X_{\mathcal{L}_{ext}} = X_{\mathcal{L}}.$$

Note that $X_{\mathcal{L}}$ is a SODE on $T(\mathbb{R} \times Q)$, although $X_{\mathcal{L}_{ext}}$ is not a SODE on $T(\mathbb{R} \times Q)$, and the integral curves of $X_{\mathcal{L}}$ are solutions to the Euler–Lagrange equations. Thus, the Lagrangian contact (or the cosymplectic) formulation is recovered in this way.

For the Hamiltonian formalism, we take the *extended momentum phase space* $T^*(\mathbb{R} \times Q) \cong T^*\mathbb{R} \times T^*Q \cong \mathbb{R} \times \mathbb{R}^* \times T^*Q$ (or, generally, $M \times \mathbb{R} \times \mathbb{R}^*$, where M is any symplectic manifold), and its canonical projections

$$pr_1 \colon T^*(\mathbb{R} \times Q) \longrightarrow T^*\mathbb{R} \cong \mathbb{R} \times \mathbb{R}^* \; ; \; \mu \colon T^*(\mathbb{R} \times Q) \longrightarrow \mathbb{R} \times T^*Q$$

$$pr_2 \colon T^*(\mathbb{R} \times Q) \longrightarrow T^*Q \; ; \; u \colon T^*(\mathbb{R} \times Q) \longrightarrow \mathbb{R}.$$

If $\Omega \in \Omega^2(T^*Q)$ and $\Omega_\mathbb{R} \in \Omega^2(T^*\mathbb{R})$ are the canonical symplectic forms of T^*Q and $T^*\mathbb{R}$, respectively, then $T^*(\mathbb{R} \times Q)$ can be endowed with the following symplectic structure:

$$\Omega_{ext} := pr_2^*\Omega + pr_1^*\Omega_\mathbb{R}.$$

The dynamical information is still given by the time-dependent Hamiltonian function $h \in C^\infty(T^*(\mathbb{R} \times Q))$. Then, we can introduce the *extended time-dependent Hamiltonian function*

$$h_{ext} := \mu^*h + u \in C^\infty(T^*(\mathbb{R} \times Q)).$$

Thus, we have the symplectic Hamiltonian system $(T^*(\mathbb{R} \times Q), \Omega_{ext}, h_{ext})$. The Hamiltonian dynamical equation for $X_{ext} \in \mathfrak{X}(T^*(\mathbb{R} \times Q))$ is

$$i(X_{ext})\Omega_{ext} = dh_{ext},$$

and for its integral curves $\mathbf{c} \colon \mathbb{R} \longrightarrow T^*(\mathbb{R} \times Q)$

$$i(\widetilde{\mathbf{c}})(\Omega_{ext} \circ \mathbf{c}) = dh_{ext} \circ \mathbf{c}.$$

The unique solution to the above equation, in a chart of natural coordinates (t, u, q^i, p_i) of $T^*(\mathbb{R} \times Q)$, is

$$
\begin{aligned}
X_{ext} &= \frac{\partial}{\partial t} - \frac{\partial h_{ext}}{\partial t}\frac{\partial}{\partial u} + \frac{\partial h_{ext}}{\partial p_i}\frac{\partial}{\partial q^i} - \frac{\partial h_{ext}}{\partial q^i}\frac{\partial}{\partial p_i} \\
&= \frac{\partial}{\partial t} - \frac{\partial(\mu^*h)}{\partial t}\frac{\partial}{\partial u} + \frac{\partial(\mu^*h)}{\partial p_i}\frac{\partial}{\partial q^i} - \frac{\partial(\mu^*h)}{\partial q^i}\frac{\partial}{\partial p_i}.
\end{aligned}
$$

Then, the integral curves of this vector field are solutions to the Hamilton equations,

$$\frac{dq^i}{dt} = \frac{\partial \mu^*h}{\partial p_i}, \quad \frac{dp_i}{dt} = -\frac{\partial \mu^*h}{\partial q^i}, \quad \frac{du}{dt} = -\frac{\partial \mu^*h}{\partial t},$$

where the last equation shows that the energy is not conserved.

As the physical phase space is $\mathbb{R} \times T^*Q$ and not $T^*(\mathbb{R} \times Q)$), the real dynamical vector field cannot be X_{ext}, but another one on $\mathbb{R} \times T^*Q$, which must be related to it. However, as in the Lagrangian formalism, the vector field X_{ext} is μ-projectable, and in fact, we have

$$\mu_* X_{ext} = X_h.$$

In this way, the equivalence with the Hamiltonian contact (or the cosymplectic) formulation is proven.

To establish the equivalence between the Lagrangian and Hamiltonian formalisms in this extended symplectic picture, we need to connect the extended velocity and momentum phase spaces through a suitable "Legendre map", that is, the fibre derivative of some Lagrangian function on $T(\mathbb{R} \times Q)$. This function cannot be $\rho^* \mathcal{L}$ because its fibre derivative $F(\rho^* \mathcal{L})$ is not a (local) diffeomorphism since $\rho^* \mathcal{L}$ is not a regular Lagrangian on $T(\mathbb{R} \times Q)$. The correct choice consists of taking the extended Lagrangian (4.23) and its fibre derivative

$$F\mathcal{L}_{ext} : T(\mathbb{R} \times Q) \longrightarrow T^*(\mathbb{R} \times Q),$$

which in coordinates is

$$\mathcal{F}\mathcal{L}_{ext}^* q^i = q^i, \ \mathcal{F}\mathcal{L}_{ext}^* p_i = \frac{\partial \rho^* \mathcal{L}}{\partial v^i}, \ \mathcal{F}\mathcal{L}_{ext}^* t = t, \ \mathcal{F}\mathcal{L}_{ext}^* u = \varpi.$$

Observe that $F\mathcal{L}_{ext}$ is just an extension of $\mathcal{F}\mathcal{L}$ such that the following diagram commutes:

$$
\begin{array}{ccc}
T(\mathbb{R} \times Q) & \xrightarrow{\mathcal{F}\mathcal{L}_{ext}} & T^*(\mathbb{R} \times Q) \\
\nu \downarrow & & \downarrow \mu \\
\mathbb{R} \times TQ & \xrightarrow{\mathcal{F}\mathcal{L}} & \mathbb{R} \times T^*Q
\end{array}
$$

and $\mathcal{F}\mathcal{L}_{ext}$ is a (local) diffeomorphism if, and only if, $\mathcal{F}\mathcal{L}$ is a (local) diffeomorphism or if $\mathcal{L}$ is regular. Therefore,

$$F\mathcal{L}_{ext}^* \Omega = \Omega_{\mathcal{L}_{ext}}, \quad F\mathcal{L}_{ext}^* h_{ext} = E_{\mathcal{L}_{ext}},$$

and

$$F\mathcal{L}_{ext \ *} X_{\mathcal{L}_{ext}} = X_{ext}.$$

4.7 Examples

In this section, let us consider the examples studied at the end of the previous chapter, in which the Lagrangian functions of the systems have been modified to include time dependence.

4.7.1 *The forced harmonic oscillator with periodic perturbation*

We return to the system analysed in Section 3.7.1, but now the oscillator is also submitted to an external periodic force. In the cosymplectic formalism, the configuration bundle is $\mathbb{R} \times Q = \mathbb{R}^2$, with coordinates (t, q).

4.7.1.1 *Lagrangian formalism*

The cosymplectic Lagrangian formalism takes place on $\mathbb{R} \times TQ \simeq \mathbb{R}^3$, with coordinates (t, q, v), and the dynamics can be described by the Lagrangian function

$$\mathcal{L} = \frac{1}{2}(mv^2 - kq^2) + Aq \cos wt; \quad k, w \in \mathbb{R}^+.$$

Now, we have

$$E_{\mathcal{L}} = \frac{1}{2}(mv^2 + kq^2) - Aq \cos wt, \quad \theta_{\mathcal{L}} = mv \, dq, \quad \omega_{\mathcal{L}} = m \, dq \wedge dv,$$

and $\mathcal{L}$ is a regular Lagrangian. The Reeb vector field for this Lagrangian is just $R_{\mathcal{L}} = \dfrac{\partial}{\partial t}$. Then, for $\Gamma_{\mathcal{L}} = \lambda \dfrac{\partial}{\partial t} + f \dfrac{\partial}{\partial q} + g \dfrac{\partial}{\partial v}$, Eq. (4.11) gives $\lambda = 1$ and

$$i(\Gamma_{\mathcal{L}})\omega_{\mathcal{L}} = m\,(f \, dv - g \, dq) = mv \, dv + (kq - A \cos wt) \, dq = dE_{\mathcal{L}} + \frac{\partial \mathcal{L}}{\partial t} \, dt,$$

which leads to

$$f = v, \quad mg = -kq + A \cos wt.$$

So, the Euler–Lagrange vector field is

$$\Gamma_{\mathcal{L}} = \frac{\partial}{\partial t} + v \frac{\partial}{\partial q} + \frac{-kq + A \cos wt}{m} \frac{\partial}{\partial v},$$

and its integral curves $(q(t), v(t))$ are the solutions to

$$\frac{dq}{dt} = v, \quad m\frac{dv}{dt} = -kq + A \cos wt \quad \Longrightarrow \quad m\frac{d^2q}{dt^2} = -kq + A \cos wt,$$

which is the Euler–Lagrange equation for this time-dependent forced harmonic oscillator.

4.7.1.2 *Hamiltonian formalism*

For the Hamiltonian formalism, $\mathbb{R} \times T^*Q \simeq \mathbb{R}^3$ with coordinates (t, q, p). First, the Legendre transformation is

$$\mathcal{FL}^* t = t, \quad \mathcal{FL}^* q = q, \quad \mathcal{FL}^* p = mv,$$

which is a diffeomorphism (the Lagrangian is hyperregular). The canonical Hamiltonian function is

$$h = \frac{p^2}{2m} + kq^2 - Aq \cos wt.$$

As $\omega = dq \wedge dp$, the Reeb vector field is $R = \dfrac{\partial}{\partial t}$, and for the vector field $\mathcal{E}_h = \Lambda \dfrac{\partial}{\partial t} + F \dfrac{\partial}{\partial q} + G \dfrac{\partial}{\partial p}$, Eq. (4.13) give $\Lambda = 1$ and

$$i(\mathcal{E}_h)\omega = F\,dp - G\,dq = \frac{p}{m}\,dp + (kq - A\cos wt)\,dq = dh - R\frac{\partial h}{\partial t}\,dt\,,$$

which leads to

$$F = \frac{p}{m}, \quad G = -kq + A\cos wt.$$

So, the evolution vector field is

$$\mathcal{E}_h = \frac{\partial}{\partial t} + \frac{p}{m}\frac{\partial}{\partial q} - (kq - A\cos wt)\frac{\partial}{\partial p}\,,$$

and its integral curves $(q(t), p(t))$ are the solutions to

$$m\frac{dq}{dt} = p, \quad \frac{dp}{dt} = -kq + A\cos wt\,,$$

which are the Hamilton equations for this time-dependent forced harmonic oscillator.

Using the Legendre map, we check again that the Hamilton and the Euler–Lagrange equations of the system are equivalent. In particular, we have $\mathcal{FL}_*\Gamma_\mathcal{L} = \mathcal{E}_h$.

4.7.1.3 *Unified Lagrangian–Hamiltonian formalism*

The extended unified bundle is $\mathcal{M} = \mathbb{R} \times TQ \times_Q T^*Q \simeq \mathbb{R}^4$ with coordinates (t, q, v, p). The Reeb vector field is $R = \dfrac{\partial}{\partial t}$, the canonical forms are

$$\Theta_\mathcal{M} = p\,dq, \quad \Omega_\mathcal{M} = -d\Theta_\mathcal{M} = dq \wedge dp,$$

and the Hamiltonian function is

$$\mathrm{H} = pv - \frac{1}{2}(mv^2 - kq^2) - Aq\cos wt.$$

For $X_\mathrm{H} = g\dfrac{\partial}{\partial t} + f\dfrac{\partial}{\partial q} + F\dfrac{\partial}{\partial v} + G\dfrac{\partial}{\partial p}$, Eq. (4.15) gives $g = 1$ and

$$i(X_\mathrm{H})\Omega_\mathcal{W} = f\,dp - G\,dq = (kq - A\cos wt)\,dq + (p - mv)\,dv + v\,dp$$
$$= d\mathrm{H} - R(\mathrm{H})dt\,,$$

which leads to

$$f = v, \quad G = -(kq - A\cos wt), \quad p = mv.$$

The last equation is the constraint defining the submanifold $\mathcal{M}_0 \hookrightarrow \mathcal{M}$, which gives the Legendre map. The evolution vector field is

$$X_{\mathrm{H}}|_{\mathcal{M}_0} = \frac{\partial}{\partial t} + v\frac{\partial}{\partial q} + G\frac{\partial}{\partial v} - (kq - A\cos wt)\frac{\partial}{\partial p},$$

and the tangency condition leads to

$$X_{\mathrm{H}}(p - mv) = -(kq - A\cos wt) - Gm = 0$$
$$\Longleftrightarrow \; G = -\frac{kq - A\cos wt}{m} \quad (\text{on } \mathcal{M}_0).$$

Therefore,

$$X_{\mathrm{H}}|_{\mathcal{M}_0} = \frac{\partial}{\partial t} + v\frac{\partial}{\partial q} - \frac{kq - A\cos wt}{m}\frac{\partial}{\partial v} - (kq - A\cos wt)\frac{\partial}{\partial p}.$$

Its integral curves $(t, q(t), v(t), p(t))$ are the solutions to

$$\frac{dq}{dt} = v, \quad m\frac{dv}{dt} = -(kq - A\cos wt), \quad \frac{dp}{dt} = -(kq - A\cos wt).$$

The first two equations are equivalent to

$$m\frac{d^2q}{dt^2} = -kq + A\cos wt,$$

which is the Euler–Lagrange equation of the system. Furthermore, using the constraint $p = mv$ (the Legendre map), the first and third equations are

$$\frac{dq}{dt} = \frac{p}{m}, \quad \frac{dp}{dt} = -kq + A\cos wt,$$

which are the Hamilton equations for the system.

4.7.2 *A nonautonomous system with central forces*

Consider the *Kepler problem* analysed in Section 3.7.2 in the case where the mass of the particle subjected to the central field is not constant but a (strictly positive) function of time $m(t)$. As in the standard case, the motion is on a plane; hence, the configuration bundle is $\mathbb{R} \times Q = \mathbb{R}^3$, and the coordinates are (t, r, ϕ), where (r, ϕ) are polar coordinates on the adequate open set $U \subset Q = \mathbb{R}^2$.

4.7.2.1 *Lagrangian formalism*

The velocity phase space for the cosymplectic Lagrangian formalism is $\mathbb{R} \times TQ \simeq \mathbb{R}^5$, with local coordinates $(t, r, \phi, v_r, v_\phi)$, and the Lagrangian function is

$$\mathcal{L} = \frac{1}{2}m(t)(v_r^2 + r^2 v_\phi^2) - \frac{K}{r}, \quad K \neq 0.$$

Therefore,

$$E_\mathcal{L} = \frac{1}{2}m(t)(v_r^2 + r^2 v_\phi^2) + \frac{K}{r},$$

$$\theta_\mathcal{L} = m(t)(v_r \, dr + r^2 v_\phi \, d\phi),$$

$$\omega_\mathcal{L} = m(t)(dr \wedge dv_r + r^2 d\phi \wedge dv_\phi - 2r v_\phi \, dr \wedge d\phi)$$
$$-\dot{m}(t)(v_r \, dt \wedge dr + r^2 v_\phi \, dt \wedge d\phi),$$

where $\dot{m}(t) = \dfrac{dm(t)}{dt}$. The Lagrangian is regular and the Reeb vector field for this system is

$$R_\mathcal{L} = \frac{\partial}{\partial t} + \frac{\dot{m}(t)}{m(t)}\left(v_r \frac{\partial}{\partial v_r} + v_\phi \frac{\partial}{\partial v_\phi}\right).$$

If $\Gamma_\mathcal{L} = \lambda \dfrac{\partial}{\partial t} + f_r \dfrac{\partial}{\partial r} + f_\phi \dfrac{\partial}{\partial \phi} + g_r \dfrac{\partial}{\partial v_r} + g_\phi \dfrac{\partial}{\partial v_\phi}$, Eq. (4.11) give $\lambda = 1$ and

$$i(\Gamma_\mathcal{L})\omega_\mathcal{L} = \dot{m}(t)\left[-v_r \, dr - r^2 v_\phi \, d\phi + (v_r f_r + r^2 v_\phi f_\phi)\, dt\right] +$$
$$m(t)\left[f_r \, dv_r + f_\phi r^2 \, dv_\phi - (g_r - 2r v_\phi f_\phi)\, dr \right.$$
$$\left. -(g_\phi r^2 + 2r v_\phi f_r)\, d\phi\right]$$
$$= \dot{m}(t)(v_r^2 + r^2 v_\phi^2)\, dt + m(t)\, v_r \, dv_r$$
$$+ m(t)\, r^2 v_\phi \, dv_\phi + \left(m(t)\, r v_\phi^2 - \frac{K}{r^2}\right) dr$$
$$= dE_\mathcal{L} + \frac{\partial \mathcal{L}}{\partial t}\, dt,$$

which leads to

$$f_r = v_r, \quad f_\phi = v_\phi, \quad v_r f_r + r^2 v_\phi f_\phi = v_r^2 + r^2 v_\phi^2,$$

$$m(t)\, g_r = m(t)(2r v_\phi f_\phi - r v_\phi^2) - \dot{m}(t)\, v_r + \frac{K}{r^2} \quad g_\phi = -\frac{2v_\phi f_r}{r} - \frac{\dot{m}(t)}{m(t)}v_\phi,$$

and then the Euler–Lagrange vector field is

$$\Gamma_\mathcal{L} = \frac{\partial}{\partial t} + v_r \frac{\partial}{\partial r} + v_\phi \frac{\partial}{\partial \phi} + \left(r v_\phi^2 - \frac{\dot{m}(t)}{m(t)}v_r + \frac{K}{m(t)\, r^2}\right)\frac{\partial}{\partial v_r} - \left(\frac{\dot{m}(t)}{m(t)}v_\phi + \frac{2v_\phi v_r}{r}\right)\frac{\partial}{\partial v_\phi}.$$

Then, its integral curves $(r(t), \phi(t), v_r(t), v_\phi(t))$ are the solutions to

$$\frac{dr}{dt} = v_r, \quad \frac{d\phi}{dt} = v_\phi,$$

$$m(t)\frac{dv_r}{dt} = m(t)\,rv_\phi^2 - \dot{m}(t)\,v_r + \frac{K}{r^2}, \quad \frac{dv_\phi}{dt} = \frac{\dot{m}(t)}{m(t)}v_\phi - \frac{2}{r}v_\phi v_r,$$

that is,

$$m(t)\frac{d^2r}{dt^2} = m(t)\,r\left(\frac{d\phi}{dt}\right)^2 - \dot{m}(t)\frac{dr}{dt} + \frac{K}{r^2}, \quad \frac{d^2\phi}{dt^2} = -\frac{\dot{m}(t)}{m(t)}\frac{d\phi}{dt} - \frac{2}{r}\frac{d\phi}{dt}\frac{dr}{dt};$$

or equivalently

$$\frac{d}{dt}\left(m(t)\frac{dr}{dt}\right) = m(t)\,r\left(\frac{d\phi}{dt}\right)^2 + \frac{K}{r^2}, \quad \frac{d}{dt}\left(m(t)r^2\frac{d\phi}{dt}\right) = 0,$$

which are the Euler–Lagrange equations for this system.

There is an infinitesimal Lagrangian exact Noether symmetry, which is the vector field $Y = \frac{\partial}{\partial \phi}$, because $L(Y)dt = 0$ trivially, and a calculation similar to the one done in the example of Section 3.7.2 leads to $L(Y)\theta_\mathcal{L} = 0$ and $L(Y)E_\mathcal{L} = 0$. Therefore, the associated conserved quantity is

$$f_Y = i\left(\frac{\partial}{\partial \phi}\right)\theta_\mathcal{L} = m(t)r^2v_\phi,$$

which is the angular momentum, as the last Euler–Lagrange equation shows.

4.7.2.2 *Hamiltonian formalism*

The cosymplectic Hamiltonian formalism takes place on $\mathbb{R} \times T^*Q \simeq \mathbb{R}^5$, with local coordinates $(t, r, \phi, p_r, p_\phi)$. First, the Legendre transformation is

$$\mathcal{FL}^*t = t, \quad \mathcal{FL}^*r = r, \quad \mathcal{FL}^*\phi = \phi,$$

$$\mathcal{FL}^*p_r = m(t)\,v_r, \quad \mathcal{FL}^*p_\phi = m(t)\,r^2v_\phi,$$

which is a diffeomorphism because the Lagrangian is hyperregular. The canonical time-dependent Hamiltonian function is

$$h = \frac{p_r^2}{2m(t)} + \frac{p_\phi^2}{2m(t)\,r^2} + \frac{K}{r}.$$

As $\omega = dr \wedge dp_r + d\phi \wedge dp_\phi$, the Reeb vector field is $R = \frac{\partial}{\partial t}$. Then, the evolution vector field obtained from Eq. (4.13) is

$$\mathcal{E}_h = \frac{\partial}{\partial t} + \frac{p_r}{m(t)}\frac{\partial}{\partial r} + \frac{p_\phi}{m(t)\,r^2}\frac{\partial}{\partial \phi} + \left(\frac{p_\phi^2}{m(t)\,r^3} + \frac{K}{r^2}\right)\frac{\partial}{\partial p_r},$$

and its integral curves $(r(t), \phi(t), p_r(t), p_\phi(t))$ are the solutions to

$$m(t)\frac{dr}{dt} = p_r, \; m(t)\, r^2 \frac{d\phi}{dt} = p_\phi, \; \frac{dp_r}{dt} = \frac{p_\phi^2}{m(t)\, r^3} + \frac{K}{r^2}, \; \frac{dp_\phi}{dt} = 0,$$

which are the Hamilton equations for this system.

The Hamilton equations have the same aspect as in the autonomous case (see Section 3.7.2), but now $m = m(t)$. Combining these equations and using the Legendre map, we obtain the Euler–Lagrange equations for this nonautonomous system, and $\mathcal{FL}_*\Gamma_\mathcal{L} = \mathcal{E}_\mathrm{h}$.

Once again, a Hamiltonian exact Noether symmetry is given by the vector field $Y = \dfrac{\partial}{\partial\phi}$, since $\mathrm{L}(Y)\mathrm{d}t = 0$, $\mathrm{L}(Y)\theta = 0$ and $\mathrm{L}(Y)\mathrm{h} = 0$. Then, as the last Hamilton equation shows, the associated conserved quantity is again the angular momentum $f_Y = i\left(\dfrac{\partial}{\partial\phi}\right)\theta = p_\phi$.

4.7.2.3 Unified Lagrangian–Hamiltonian formalism

The extended unified bundle is $\mathcal{M} = \mathbb{R} \times TQ \times_Q T^*Q \simeq \mathbb{R}^7$, with natural coordinates $(t, r, \phi, v_r, v_\phi, p_r, p_\phi)$. The Reeb vector field is $R = \dfrac{\partial}{\partial t}$, the canonical forms are

$$\Theta_\mathcal{M} = p_r\, \mathrm{d}r + p_\phi\, \mathrm{d}\phi, \quad \Omega_\mathcal{M} = -\mathrm{d}\Theta_\mathcal{M} = \mathrm{d}r \wedge \mathrm{d}p_r + \mathrm{d}\phi \wedge \mathrm{d}p_\phi,$$

and the Hamiltonian function is

$$H = p_r v_r + p_\phi v_\phi - \frac{1}{2}m(t)(v_r^2 + r^2 v_\phi^2) + \frac{K}{r}.$$

For $X_H = f\dfrac{\partial}{\partial t} + f_r\dfrac{\partial}{\partial r} + f_\phi\dfrac{\partial}{\partial\phi} + F_r\dfrac{\partial}{\partial v_r} + F_\phi\dfrac{\partial}{\partial v_\phi} + G_r\dfrac{\partial}{\partial p_r} + G_\phi\dfrac{\partial}{\partial p_\phi}$, Eq. (4.15) gives $g = 1$ and

$$\begin{aligned}
i(X_H)\Omega_\mathcal{M} &= f_r\, \mathrm{d}p_r + f_\phi\, \mathrm{d}p_\phi - G_r\, \mathrm{d}r - G_\phi\, \mathrm{d}\phi \\
&= -\left(\frac{K}{r^2} + m(t)\, r v_\phi^2\right)\mathrm{d}r + (p_r - m(t)\, v_r)\, \mathrm{d}v_r \\
&\quad + (p_\phi - m(t)\, r^2 v_\phi)\, \mathrm{d}v_\phi + v_r\, \mathrm{d}p_r + v_\phi\, \mathrm{d}p_\phi \\
&= \mathrm{d}H - R(H)\mathrm{d}t,
\end{aligned}$$

which leads to

$$f_r = v_r, \quad f_\phi = v_\phi, \quad G_r = \frac{K}{r^2} + m(t)\, r v_\phi^2,$$
$$G_\phi = 0, \quad p_r = m(t)\, v_r, \quad p_\phi = m(t)\, r^2 v_\phi.$$

The last two equations are constraints defining the submanifold $\mathcal{M}_0 \hookrightarrow \mathcal{M}$, which give the Legendre map. The evolution vector field is

$$X_H|_{\mathcal{M}_0} = \frac{\partial}{\partial t} + v_r \frac{\partial}{\partial r} + v_\phi \frac{\partial}{\partial \phi} + F_r \frac{\partial}{\partial v_r} + F_\phi \frac{\partial}{\partial v_\phi} + \left(\frac{K}{r^2} + m(t)\, r v_\phi^2 \right) \frac{\partial}{\partial p_r},$$

and the tangency condition leads to the following equations (on $\mathcal{M}_0$)

$$X_H(p_r - m(t) v_r) = \frac{K}{r^2} + m(t)\, r v_\phi^2 - F_r m(t) - \dot{m}(t)\, v_r = 0$$

$$\Longleftrightarrow F_r = \frac{K}{m(t)\, r^2} + r v_\phi^2 - \frac{\dot{m}(t)}{m(t)} v_r,$$

$$X_H(p_\phi - m(t)\, r^2 v_\phi) = -m(t)\, F_\phi r^2 - m(t)\, 2 f_r r v_\phi - \dot{m}(t)\, r^2 v_\phi = 0$$

$$\Longleftrightarrow F_\phi = -\frac{2 v_r v_\phi}{r} - \frac{\dot{m}(t)}{m(t)} r^2 v_\phi.$$

Therefore,

$$X_H|_{\mathcal{M}_0} = \frac{\partial}{\partial t} + v_r \frac{\partial}{\partial r} + v_\phi \frac{\partial}{\partial \phi} + \left(r v_\phi^2 + \frac{K}{m(t)\, r^2} - \frac{\dot{m}(t)}{m(t)} v_r \right) \frac{\partial}{\partial v_r}$$
$$- \left(\frac{2 v_r v_\phi}{r} + \frac{\dot{m}(t)}{m(t)} r^2 v_\phi \right) \frac{\partial}{\partial v_\phi} + \left(\frac{K}{r^2} + m(t)\, r v_\phi^2 \right) \frac{\partial}{\partial p_r},$$

and its integral curves $(t, r(t), \phi(t), v_r(t), v_\phi(t), p_r(t), p_\phi(t))$ are the solutions to

$$\frac{dr}{dt} = v_r, \quad \frac{d\phi}{dt} = v_\phi,$$

$$\frac{dv_r}{dt} = \frac{K}{m(t)\, r^2} + r v_\phi^2 - \frac{\dot{m}(t)}{m(t)} v_r, \quad \frac{dv_\phi}{dt} = -\frac{2 v_r v_\phi}{r} - \frac{\dot{m}(t)}{m(t)} r^2 v_\phi,$$

$$\frac{dp_r}{dt} = \frac{K}{r^2} + m(t)\, r v_\phi^2, \quad \frac{dp_\phi}{dt} = 0.$$

The first four equations are equivalent to

$$m(t) \frac{d^2 r}{dt^2} = m(t)\, r \left(\frac{d\phi}{dt} \right)^2 + \frac{K}{r^2} - \dot{m}(t) \frac{dr}{dt}, \quad \frac{d^2 \phi}{dt^2} = -\frac{2}{r} \frac{d\phi}{dt} \frac{dr}{dt} - \frac{\dot{m}(t)}{m(t)} r^2 \frac{d\phi}{dt},$$

which are the Euler–Lagrange equation of the system. Furthermore, using the constraints $p_r = m(t)\, v_r$ and $p_\phi = m(t)\, r^2 v_\phi$ (i.e., the Legendre map), the first, second, fifth, and sixth equations are

$$\frac{dr}{dt} = \frac{p_r}{m(t)}, \quad \frac{d\phi}{dt} = \frac{p_\phi}{m(t)\, r^2}, \quad \frac{dp_r}{dt} = \frac{p_\phi^2}{m(t)\, r^3} + \frac{K}{r^2}, \quad \frac{dp_\phi}{dt} = 0,$$

which are the Hamilton equations for the system.

4.8   Exercises

Exercise 4.1
Prove Proposition 4.4 and the local expression (4.9).

Exercise 4.2
Prove Proposition 4.7.

Exercise 4.3
Prove Theorem 4.2.

Exercise 4.4
Prove that if $Y_1, Y_2 \in \mathfrak{X}(\mathbb{R} \times M)$ are infinitesimal cosymplectic Noether symmetries, then $[Y_1, Y_2]$ is also an infinitesimal cosymplectic Noether symmetry.

Exercise 4.5
Using natural coordinates on $\mathbb{R} \times T^*Q$ and $\mathbb{R} \times TQ$, prove the assertions of Remark 4.8.
(*Hint: See [1].*)

Exercise 4.6 (Quadratic nonautonomous regular Lagrangians).
In the manifold $Q \times \mathbb{R}$, with coordinates (q^i, t), $i = 1, \ldots, n$, consider the family of Lagrangian functions,

$$\mathcal{L} = g_{ij}(q^k, t)\, v^i\, v^j - g(q^k, t),$$

where the matrix $(g_{ij}(q^k, t))$ is regular everywhere (this is a time-dependent Riemannian metric). Study their cosymplectic Lagrangian, Hamiltonian, and unified formalisms.

Exercise 4.7
Study the cosymplectic Lagrangian, Hamiltonian, and unified formalisms of a pendulum in a vertical plane of $\mathbb{R}^3$, with mass m and length l, whose suspension point:

(1) moves with oscillations $A \cos \omega t$ ($A > 0$) in a horizontal straight line,
(2) moves with vertical oscillations $A \cos \omega t$ in the vertical line,
(3) moves on a circle with radius A and constant angular velocity ω in a horizontal plane.

Exercise 4.8
Repeat the same analysis as in the previous exercise for the double pendulum with masses m_1 and m_2 and lengths l_1 and l_2.

Exercise 4.9

Repeat the same study as in the previous exercises for a particle with mass m, which is moving without friction on the surface defined by the equation $z = k(x^2 + y^2)e^{-t}, k > 0$, subjected to the force of gravity $F = (0, 0, -mg)$. Compare the dynamics, when $t = 1$, with those obtained in Exercise 3.19 in Chapter 3.

Study the dynamics of the system when $t \longrightarrow \infty$ and interpret the result so obtained.

Exercise 4.10

Study the systems described in Exercises 4.6–4.9 using the contact formulation, both the Lagrangian and the Hamiltonian formalisms.

Exercise 4.11

Study the systems described in Exercises 4.6–4.9, using the extended symplectic formulation, both the Lagrangian and the Hamiltonian formalisms.

Exercise 4.12

Develop the unified Lagrangian–Hamiltonian formalism for regular nonautonomous dynamical systems in the contact formulation.

Exercise 4.13

Study the systems of Exercises 4.6–4.9 using the unified Lagrangian–Hamiltonian formalism in the contact formulation.

Exercise 4.14

Develop the unified Lagrangian–Hamiltonian formalism for regular nonautonomous dynamical systems in the extended symplectic formulation.

Exercise 4.15

Study the systems of Exercises 4.6–4.9 using the unified Lagrangian–Hamiltonian formalism in the extended symplectic formulation.

Exercise 4.16 (Time-dependent Hamiltonian mechanics and Reeb dynamics).

Let Q be an n-dimensional manifold and $h \colon \mathbb{R} \times T^*Q \longrightarrow \mathbb{R}$ be a nonautonomous Hamiltonian function. We can use the Hamiltonian function h and the natural forms in $\mathbb{R}$ and T^*Q to obtain another cosymplectic structure on the manifold $\mathbb{R} \times T^*Q$ as follows: let t be the global coordinate on $\mathbb{R}$ and $\mathrm{d}t \in \Omega^1(\mathbb{R})$ be the associated closed 1-form, and let $\Theta \in \Omega^1(T^*Q)$ be the Liouville canonical 1-form and $\Omega = -\mathrm{d}\Theta$ be the canonical symplectic form on T^*Q (we denote by the same symbol the pullback of $\mathrm{d}t$ and Ω to the product $\mathbb{R} \times T^*Q$). Then, define $\eta = \mathrm{d}t$ and $\Omega_{\mathrm{h}} = \Omega + \mathrm{d}h \wedge \mathrm{d}t$:

(1) Prove that $\Omega_{\mathrm{h}}^n \wedge \eta \neq 0$, so $(\mathbb{R} \times T^*Q, \eta, \Omega_{\mathrm{h}})$ is a cosymplectic manifold.

In a natural coordinate system (t, q^i, p_i) in $\mathbb{R} \times \mathrm{T}^*Q$, we have

$$\Omega_{\mathrm{h}} = \mathrm{d}q^i \wedge \mathrm{d}p_i + \mathrm{dh} \wedge \mathrm{d}t.$$

(2) Prove that the associated Reeb vector field for this structure is, locally,

$$\mathcal{R} = \frac{\partial}{\partial t} + \frac{\partial h}{\partial p_i} \frac{\partial}{\partial q^i} - \frac{\partial h}{\partial q^i} \frac{\partial}{\partial p_i},$$

which coincides with the evolution vector field for h given in (4.7).

(**Comment:** *We have noted that every evolution vector field of a time-dependent Hamiltonian can be understood as a Reeb vector field changing the corresponding cosymplectic structure. In cosymplectic dynamics, the Reeb vector field is a well-studied dynamical vector field whose properties have been widely analysed (see, for instance, [59, 359]).*)

Chapter 5

Riemannian Mechanics: Newtonian Dynamical Systems

In this chapter, we analyse the geometry of *Newtonian dynamical systems*, which are classical mechanical systems on a (semi)Riemannian manifold; they are also called *purely mechanical systems*. This is the geometrisation of the classical *Newtonian mechanics* or *(semi)Riemannian mechanics*. As we will see, a special type of these systems, the conservative ones, are just a particular kind of the Lagrangian systems studied in the previous chapter. Some basic references are $[1, 11, 46, 75, 91, 179, 291]$.

As in the previous chapters, we start by reviewing the basic notions and properties on connections in manifolds and Riemannian geometry, which are needed for the development of this chapter. Next, we introduce the general features of Newtonian dynamical systems and their properties: we state the classical Newton equations of dynamics geometrically, study conservation laws, and describe some particular types of these kinds of systems. This geometric framework is also used to study the case of systems with holonomic and nonholonomic constraints and their corresponding variational principles. Finally, nonautonomous Newtonian systems are also briefly analysed.

5.1 Connections in Manifolds: Riemannian Manifolds

(For a detailed account of these subjects and the proof of the results, see, for example, $[97, 179, 199, 238, 241, 242, 328]$.)

5.1.1 *Connections and covariant derivatives*

Definition 5.1. A *connection* on a manifold M is a map

$$\nabla \colon \mathfrak{X}(M) \times \mathfrak{X}(M) \longrightarrow \mathfrak{X}(M), \quad (X, Y) \mapsto \nabla_X Y$$

satisfying the following properties:

(1) It is $\mathbb{R}$-linear in Y.
(2) It is $C^\infty(M)$-linear in X.
(3) If $f \in C^\infty(M)$, then $\nabla_X(fY) = (L(X)f)Y + f\nabla_X Y$.

Then $\nabla_X Y$ is said to be the **covariant derivative** of Y with respect to X.

Connections and covariant derivatives satisfy the following properties:

- For every $p \in M$, the result of $(\nabla_X Y)(p)$ depends only on the value of X at p and on the value of Y on a neighbourhood of p. Thus, we can calculate $(\nabla_X Y)(p)$ in a local coordinate system around p, and $\nabla_u Y$ is well defined for $u \in T_p M$.

- (**Localisation of a connection**). If $U \subset M$ is an open set, then there exists a connection ∇^U on U defined as:

$$\nabla^U_X|_U Y|_U := (\nabla_X Y)|_U.$$

- Let (U, x^i) be a local chart; then the vector fields $\dfrac{\partial}{\partial x^i}$ define a basis of $\mathfrak{X}(U)$ and

$$\nabla_{\frac{\partial}{\partial x^i}} \frac{\partial}{\partial x^j} = \Gamma^k_{ij} \frac{\partial}{\partial x^k},$$

where $\Gamma^k_{ij} \in C^\infty(U)$ are the *Christoffel symbols* of the connection ∇ for the local basis $\left\{ \dfrac{\partial}{\partial x^i} \right\}$, or in the local chart (U, x^i). Then, if $X = X^i \dfrac{\partial}{\partial x^i}$ and $Y = Y^j \dfrac{\partial}{\partial x^j}$,

$$(\nabla_X Y)|_U = (L(X)Y^j)\frac{\partial}{\partial x^j} + X^i Y^j \Gamma^k_{ij} \frac{\partial}{\partial x^k}.$$

Observe that this expression shows that at $p \in M$, the covariant derivative depends only on the values of Y along a curve $\gamma \colon (-\epsilon, \epsilon) \subset \mathbb{R} \longrightarrow M$ such that $\gamma'(0) = X_p$.

This also holds if we have a local basis (X_i) for vector fields on an open set $U \subset M$, not necessarily local coordinate vector fields.

- If M is a parallelisable manifold, and $X_1, \ldots, X_n$ is a basis of $\mathfrak{X}(M)$, then any family of differentiable functions $\Gamma^k_{ij} \in C^\infty(M)$ defines a connection on M. That is, the Christoffel symbols determine the connection and they allow calculating the covariant derivative in the open set where they are defined.

  In particular, if $M = \mathbb{R}^n$, the standard connection is defined taking the Christoffel symbols equal to zero in the natural coordinate system.

- Let $(x^1, \ldots, x^n)$ and $(\tilde{x}^1, \ldots, \tilde{x}^n)$ be two coordinate systems on the same open set, and Γ_{ij}^k and $\tilde{\Gamma}_{\alpha\beta}^\gamma$ be the corresponding Christoffel symbols, then

$$\Gamma_{ij}^k = \tilde{\Gamma}_{\alpha\beta}^\gamma \frac{\partial \tilde{x}^\alpha}{\partial x^i} \frac{\partial \tilde{x}^\beta}{\partial x^j} \frac{\partial x^k}{\partial \tilde{x}^\gamma} + \frac{\partial^2 \tilde{x}^\delta}{\partial x^i \partial x^j} \frac{\partial x^k}{\partial \tilde{x}^\delta} \, ;$$

hence, they are not the components of any tensor field. Thus, it is illogical to say that a connection is null: the Christoffel symbols can vanish in a coordinate system and be different from zero in another one.

- The set of connections on a manifold is not a vector space but an affine space.

- Let ∇^α be a family of connections on M and f_α be a family of differentiable functions such that their support is a locally finite family of subsets of M. Then $\nabla = f_\alpha \nabla^\alpha$ defined as

$$\nabla_X Y := f_\alpha \nabla_X^\alpha Y$$

is a connection if, and only if, $\sum_\alpha f_\alpha = 1$.

- Using the above item, we can state that any paracompact manifold admits a connection, and conversely, if there exists a connection on M, then M is paracompact. (Remember that to say that M is paracompact is equivalent to stating that every open covering has a subordinate partition of unity, or that M is metrizable, or that every connected component has a numerable basis of open sets).

5.1.2 *Covariant derivative of tensor fields*

Definition 5.2. Let (M, ∇) be a differentiable manifold with a connection, and let $\mathcal{T}(M)$ be the algebra of tensor fields on M. For $X \in \mathfrak{X}(M)$, the **covariant derivative** by X defined by the connection is the unique map $\nabla_X \colon \mathcal{T}(M) \longrightarrow \mathcal{T}(M)$ verifying the following properties:

(1) It is the Lie derivative $\mathrm{L}(X)$ on $C^\infty(M)$.
(2) It reduces to the well-known ∇_X on $\mathfrak{X}(M)$.
(3) It is $\mathbb{R}$-linear.
(4) It respects the type of tensor field, that is, it maps $\mathcal{T}_h^k(M)$ onto $\mathcal{T}_h^k(M)$.
(5) (**Leibniz rule**). $\nabla_X(S \otimes T) = (\nabla_X S) \otimes T + S \otimes (\nabla_X T)$.
(6) It commutes with the internal contractions.

That is, ∇_X is a derivation of the $\mathbb{R}$-algebra $\mathcal{T}(M)$, with zero degree, which commutes with the internal contractions.

Observe that, as a consequence of the definition, for $f \in C^\infty(M)$ and $S, T \in \mathcal{T}(M)$, we have

$$\nabla_X(fS \otimes T) = (\mathrm{L}(X)f)(S \otimes T) + f(\nabla_X(S \otimes T)) = \nabla_X(S \otimes fT).$$

The covariant derivative has the following properties:

- For $p \in M$, the value of $(\nabla_X T)(p)$ only depends on $X(p)$ and the value of the tensor field T on an open set containing p. This property has the same consequences as in the case of derivation of vector fields.
- The above properties defining ∇_X allow us to calculate $\nabla_X T$ for any tensor field T provided that we know how ∇_X acts on functions and vector fields. For example, if $X, Y \in \mathcal{T}^1(M)$ and $\alpha \in \Omega^1(M) = \mathcal{T}_1(M)$, then,

$$\langle \nabla_X \alpha, Y \rangle = \nabla_X \langle \alpha, Y \rangle - \langle \alpha, \nabla_X Y \rangle.$$

  In particular, we have $\nabla_{\frac{\partial}{\partial x^i}} dx^j = -\Gamma_{ik}^j dx^k$.
- In a local coordinate system (U, x^i), if

$$X = X^j \frac{\partial}{\partial x^i}, \quad T = T_{j_1,\ldots,j_s}^{i_1,\ldots,i_r} dx^{j_1} \otimes \ldots \otimes dx^{j_s} \otimes \frac{\partial}{\partial x^{i_1}} \otimes \ldots \frac{\partial}{\partial x^{i_r}},$$

then

$$\nabla_X T = S_{j_1,\ldots,j_s}^{i_1,\ldots,i_r} dx^{j_1} \otimes \ldots \otimes dx^{j_s} \otimes \frac{\partial}{\partial x^{i_1}} \otimes \ldots \frac{\partial}{\partial x^{i_r}};$$

where

$$S_{j_1,\ldots,j_s}^{i_1,\ldots,i_r} = X^k \left(\frac{\partial T_{j_1,\ldots,j_s}^{i_1,\ldots,i_r}}{\partial x^k} + \sum_{n=1}^{r} T_{j_1,\ldots,j_s}^{i_1,\ldots,l,\ldots,i_r} \Gamma_{kl}^{i_n} - \sum_{n=1}^{s} T_{j_1,\ldots,l,\ldots,j_s}^{i_1,\ldots,i_r} \Gamma_{k_n}^{l} \right).$$

Definition 5.3. Let $T \in \mathcal{T}_s^r(M)$, $X_i, Y \in \mathfrak{X}(M)$, and $\beta^j \in \Omega^1(M)$. The tensor field $\nabla T \in \mathcal{T}_{s+1}^r(M)$ defined by

$$(\nabla T)(X_1, \ldots, X_s, Y, \beta^1, \ldots, \beta^r) := (\nabla_Y T)(X_1, \ldots, X_s, \beta^1, \ldots, \beta^r)$$

is called the **covariant differential** of T.

With the same notation for T as in the above item, we have

$$\nabla T = S_{j_1,\ldots,j_s;k}^{i_1,\ldots,i_r} dx^{j_1} \otimes \ldots \otimes dx^{j_s} \otimes dx^k \otimes \frac{\partial}{\partial x^{i_1}} \otimes \ldots \otimes \frac{\partial}{\partial x^{i_r}},$$

where

$$S^{i_1,\ldots,i_r}_{j_1,\ldots,j_s;k} = \frac{\partial T^{i_1,\ldots,i_r}_{j_1,\ldots,j_s}}{\partial x^k} + \sum_{n=1}^{r} T^{i_1,\ldots,l,\ldots,i_r}_{j_1,\ldots,j_s} \Gamma^{i_n}_{kl} - \sum_{n=1}^{s} T^{i_1,\ldots,i_r}_{j_1,\ldots,l,\ldots,j_s} \Gamma^{l}_{k_n}.$$

And for functions $f \in C^\infty(M)$, we have $\nabla f = \mathrm{d}f$.

For a vector field $X = X^i \frac{\partial}{\partial x^i}$, we have

$$\nabla X = \left(\frac{\partial X^i}{\partial x^k} X^i + X^l \Gamma^i_{kl}\right) \mathrm{d}x^k \otimes \frac{\partial}{\partial x^i} \in \mathcal{T}^1_1(U).$$

Observe that $(\nabla X)(Y) = \nabla_Y X$.

5.1.3 *Covariant derivative along curves*

Remember that a *vector field along a map*, $F: N \longrightarrow M$, is another map $Z: N \longrightarrow TM$ such that $Z(q) \in \mathrm{T}_{F(q)}M$, for every $q \in N$, that is, $\tau_Q \circ Z = F$. The set of vector fields along a map F is a $C^\infty(N)$-modulus denoted by $\mathfrak{X}(F)$.

If $\gamma: I \subset \mathbb{R} \longrightarrow M$ is a curve, then we have the set $\mathfrak{X}(\gamma)$ as a $C^\infty(I)$-modulus. One distinguished element of this set is the velocity vector $\gamma' \in \mathfrak{X}(\gamma)$. If $X \in \mathfrak{X}(M)$, then $X \circ \gamma \in \mathfrak{X}(\gamma)$, but not every element of $\mathfrak{X}(\gamma)$ is of this kind.

If $Y = Y^j \frac{\partial}{\partial x^i}$ is a vector field, then, by the above properties of the covariant derivative, $\nabla_{\gamma'(t)} Y$ is well defined as

$$\nabla_{\gamma'(t)} Y := (D(Y^j \circ \gamma)(t) \frac{\partial}{\partial x^i}|_{\gamma(t)} + \Gamma^k_{ij}(\gamma(t)) D\gamma^i(t)(Y^j(\gamma(t)) \frac{\partial}{\partial x^k}|_{\gamma(t)},$$

where $D = \frac{d}{dt}$ on the functions $f : I \subset \mathbb{R} \longrightarrow \mathbb{R}$, and γ^i are the components of γ. Then:

Definition 5.4. Given a curve $\gamma: I \subset \mathbb{R} \longrightarrow M$, the **covariant derivative** along γ is the unique map $\nabla_t: \mathfrak{X}(\gamma) \longrightarrow \mathfrak{X}(\gamma)$ such that

(1) It is $\mathbb{R}$-linear.
(2) If $f \in C^\infty(I)$ and $\mathbf{w} \in \mathfrak{X}(\gamma)$, then $\nabla_t(f\mathbf{w}) = (Df)\mathbf{w} + f\nabla_t\mathbf{w}$.
(3) If $X \in \mathfrak{X}(M)$, then $\nabla_t(X \circ \gamma)(t) = \nabla_{\gamma'(t)}X$.

The vector field along γ given by $\nabla_t\mathbf{w}$ is called the **covariant derivative** of $\mathbf{w}$ along γ.

Notation: Other notations for $\nabla_t\mathbf{w}$ are $\nabla^\gamma_t\mathbf{w} \equiv \nabla_{\gamma'(t)}\mathbf{w} \equiv \nabla_{\frac{d}{dt}}\mathbf{w}$.

If (U, x^i) is a local chart with $\gamma(I) \subset U$, then a basis for $\mathfrak{X}(\gamma)$ is given by $\dfrac{\partial}{\partial x^i} \circ \gamma$, and for $\mathbf{w} \in \mathfrak{X}(\gamma)$, then $\mathbf{w} = w^i \left(\dfrac{\partial}{\partial x^i} \circ \gamma \right)$, with $w^i \in C^\infty(I)$. Writing $\gamma' = \dot{q}^i \dfrac{\partial}{\partial x^i} \circ \gamma$, we have

$$\nabla_t \mathbf{w} = \left(\dot{w}^i + (\Gamma^k_{ij} \circ \gamma) \dot{q}^i w^j \right) \frac{\partial}{\partial x^k} \circ \gamma.$$

Similarly, we can define the concept of tensor fields along a map or along a curve and the covariant derivative of tensor fields along a curve with the same rules as above. If we know the action of ∇_t on functions and vector fields, we can obtain the covariant derivative of any kind of tensor field along the curve γ. In particular, if $\boldsymbol{\beta} \in \Omega^1(\gamma)$ and $\mathbf{w} \in \mathfrak{X}(\gamma)$, then

$$D\langle \boldsymbol{\beta}, \mathbf{w} \rangle = \langle \nabla_t \boldsymbol{\beta}, \mathbf{w} \rangle + \langle \boldsymbol{\beta}, \nabla_t \mathbf{w} \rangle,$$

which, for $\boldsymbol{\beta} = b_j dx^j \circ \gamma$, gives the following expression in a local chart

$$\nabla_t \boldsymbol{\beta} = \left(\dot{b}_j - (\Gamma^k_{ij} \circ \gamma) \dot{q}^i b_k \right) dx^j \circ \gamma.$$

5.1.4 *Parallel transport along a curve: Geodesic curves*

Definition 5.5. Let (M, ∇) be a manifold with a connection and $\gamma \colon I \subset \mathbb{R} \longrightarrow M$ a curve. We say that $\mathbf{w} \in \mathfrak{X}(\gamma)$ is **parallel** along γ if $\nabla_t \mathbf{w} = 0$.

The following properties hold:

- (**Existence of parallel vector fields**). Given (M, ∇) and $\gamma \colon I \longrightarrow M$, if $t_o \in I$ and $\mathbf{w}_o \in T_{\gamma(t_o)}M$, then there exists a unique $\mathbf{w} \in \mathfrak{X}(\gamma)$ parallel and such that $\mathbf{w}(t_o) = \mathbf{w}_o$.

  This result comes from the linearity of the differential equation defining parallel vector fields along a curve $\gamma \colon I \longrightarrow M$. We say that $\mathbf{w}$ is the **parallel transport** of $\mathbf{w}_o$ along γ, with respect to the connection ∇.

- Observe that the map $T_{\gamma(t_o)}M \longrightarrow \mathfrak{X}(\gamma)$ given by $\mathbf{w}_o \mapsto \mathbf{w}$ is linear and bijective.

- If $t \in I$, then the vector $\mathbf{w}(t)$ is called the *parallel transport* of $\mathbf{w}_o$ from $\gamma(t_o)$ to $\gamma(t)$. Thus, for $t_o, t \in I$, we have a **parallel transport operator** along γ, denoted by $\tau_{t_o,t} \colon T_{\gamma(t_o)}M \longrightarrow T_{\gamma(t)}M$, which is a linear isomorphism.

- Let $(\mathbf{e_i})$ be a basis of $T_{\gamma(t_o)}M$ and $\mathbf{w}_i$ be the corresponding parallel transported elements of $\mathfrak{X}(\gamma)$. Then:

  (1) $(\mathbf{w}_i(t))$ is a basis of $T_{\gamma(t)}M$ for every $t \in I$.

(2) $(\mathbf{w}_i)$ is a basis of $\mathfrak{X}(\gamma)$ as $C^\infty(I)$-module.

- For $t_o, t \in I$, consider $\tau_{t_o,t} \colon T_{\gamma(t_o)}M \longrightarrow T_{\gamma(t)}M$, which is the parallel transport operator along γ. Then, it is easy to show that

$$(\nabla_t \mathbf{w})(t_o) = \lim_{t \longrightarrow t_o} \frac{\tau_{t_o,t}^{-1}(\mathbf{w}(t)) - \mathbf{w}(t_o)}{t - t_o},$$

that is, the parallel transport determines the covariant derivative.

Definition 5.6. A curve $\gamma \colon I \longrightarrow M$ is a **geodesic** of the connection ∇ if γ', as a vector field along γ, is parallel along γ, that is, $\nabla_t \gamma' = 0$.

- If $\gamma = (q^i(t))$ in local coordinates, then the curve γ is a geodesic if, and only if, it satisfies the following second-order differential equation

$$\ddot{q}^k + (\Gamma_{ij}^k \circ \gamma)\dot{q}^i \dot{q}^j = 0.$$

As a consequence, given $u_p \in T_p M$, there is a unique geodesic with initial condition u_p.

- **(Reparameterisation of geodesics).** Let $\gamma \colon I_1 \subset \mathbb{R} \longrightarrow M$ be a geodesic and let $\varphi \colon I_2 \subset \mathbb{R} \longrightarrow I_1$ be a diffeomorphism between open intervals on $\mathbb{R}$. Then, the reparameterised curve $\gamma \circ \varphi$ is a geodesic if, and only if, φ is an affine map.

- A vector field $X \in \mathfrak{X}(M)$ is said to be *parallel* if it is parallel along any curve on M. This is equivalent to each of the following expressions:

(1) $\nabla_u X = 0$, for every tangent vector u.
(2) $\nabla_Y X = 0$, for every vector field Y.
(3) $\nabla X = 0$.

- The above notions can be extended to tensor fields to define *tensor fields parallel along a curve* or *parallel tensor fields*.

5.1.5   *Torsion and curvature of a connection*

Definition 5.7. Let (M, ∇) be a manifold endowed with a connection. The map $T \colon \mathfrak{X}(M) \times \mathfrak{X}(M) \longrightarrow \mathfrak{X}(M)$ defined by

$$T(X, Y) := \nabla_X Y - \nabla_Y X - [X, Y]$$

is $C^\infty(M)$-bilinear and defines a tensor field $T \in \mathcal{T}_2^1(M)$, which is called the **torsion tensor** of the connection.

A connection ∇ is said to be **symmetric** if its torsion tensor is zero.

The properties of the torsion tensor are as follows:

- The torsion tensor T is antisymmetric.
- Let (E_i) be a local reference frame of vector fields, (E^i) its dual reference, Γ^k_{ij} the corresponding Christoffel symbols, and $[E_i, E_j] = c^k_{ij} E_k$, then
$$T = T^k_{ij} \, E^i \otimes E^j \otimes E_k, \quad T^k_{ij} = \Gamma^k_{ij} - \Gamma^k_{ji} - c^k_{ij}.$$
Hence, ∇ is symmetric if, in any system of coordinates, the Christoffel symbols Γ^k_{ij} are symmetric in (i, j).
- It is easy to show that:

  (1) Two connections on M are the same if, and only if, they have the same geodesics and the same torsion tensor.
  (2) Given a connection on M, there exists another, unique and with the same geodesics and null torsion.

Definition 5.8. The map $R \colon \mathfrak{X}(M) \otimes \mathfrak{X}(M) \otimes \mathfrak{X}(M) \longrightarrow \mathfrak{X}(M)$ defined by

$$R(X, Y)Z := \nabla_X \nabla_Y Z - \nabla_Y \nabla_X Z - \nabla_{[X, Y]} Z,$$

is $C^\infty(M)$-trilinear and defines a tensor field $R \in \mathcal{T}^1_3(M)$, which is called the **curvature tensor** of the connection ∇.

A connection is said to be **flat** if its curvature tensor is zero.

The properties of the curvature tensor are as follows:

- The curvature tensor R is antisymmetric in (X, Y).
- With the same notations as above, in a local reference frame (E_i), we have
$$R = R^l_{ijk} \, E^i \otimes E^j \otimes E^k \otimes E_l,$$
with
$$R^l_{ijk} = \Gamma^l_{jk,i} - \Gamma^l_{ik,j} + \Gamma^m_{jk} \Gamma^l_{im} - \Gamma^m_{ik} \Gamma^l_{jm} - c^m_{ij} \Gamma^l_{mk}.$$
- When the torsion of ∇ is zero, then
$$R(X, Y)Z + R(Z, X)Y + R(Y, Z)X = 0.$$
This is called the *Bianchi identity*.
- It can be proven that for a manifold with a connection, the following conditions are equivalent:

  (1) The connection is flat.
  (2) On a neighbourhood of any point, there exists a local chart with coordinate vector fields that are parallel.
  (3) The parallel transport along a curve between two points of the manifold does not depend on the chosen curve.

5.1.6 *Riemannian manifolds*

Definition 5.9. Let M be a differentiable manifold. A **Riemannian metric** on M is a two-covariant tensor field $\mathbf{g} \in \mathcal{T}_2(M)$, which is symmetric and positive-defined, that is

(1) the map $\mathbf{g}\colon \mathfrak{X}(M) \times \mathfrak{X}(M) \longrightarrow C^\infty(M)$ is $C^\infty(M)$-bilinear and symmetric;
(2) for every $p \in M$ and $u_p \in T_pM$, we have $\mathbf{g}(u_p, u_p) \geq 0$, and it is zero if, and only if, $u_p = 0$.

A **Riemannian manifold** $(M, \mathbf{g})$ is a differentiable manifold endowed with a Riemannian metric.

Changing the positive-defined condition by nondegeneracy of $\mathbf{g}$, we say that we have a **pseudo** or **semi-Riemannian manifold** and $\mathbf{g}$ is a **pseudo** or **semi-Riemannian metric**. In the case that the signature of a semi-Riemannian metric $\mathbf{g}$ is $(1, m-1)$ or $(m-1, 1)$, we say that $(M, \mathbf{g})$ is a **Lorentzian manifold** and $\mathbf{g}$ is a **Lorentz metric**.

In the following section, we will suppose that $(M, \mathbf{g})$ is a Riemannian manifold unless otherwise indicated.

We have the following properties:

- Take note of the following notations and terminologies:
  (1) If $u_p, v_p \in T_pM$, then $(u_p \mid v_p) := \mathbf{g}(u_p, v_p)$ is the **scalar product** of u_p and v_p.
  (2) For $u_p \in T_pM$, the **norm** of u_p is $\parallel u_p \parallel := \mathbf{g}(u_p, u_p)^{1/2}$.
  (3) For $X, Y \in \mathfrak{X}(M)$, the scalar product of X and Y is $(X \mid Y) := \mathbf{g}(X, Y) : M \longrightarrow \mathbb{R}$.
  (4) For $u_p, v_p \in T_pM$, we say that they are **orthogonal** if $\mathbf{g}(u_p, v_p) = 0$.
  (5) Two vector fields $X, Y \in \mathfrak{X}(M)$ are orthogonal if $\mathbf{g}(X, Y) = 0$.
  (6) If $u_p, v_p \in T_pM$ and $\mathbf{g}$ is a Riemannian metric, then the *angle* θ between u_p and v_p: θ is defined by $\cos\theta := \dfrac{\mathbf{g}(u_p, v_p)}{\parallel u_p \parallel \parallel v_p \parallel}$.

- If (U, x^i) is a local chart, the local expression of $\mathbf{g}$ is $\mathbf{g}|_U = g_{ij}\, \mathrm{d}x^i \otimes \mathrm{d}x^j$, where

$$g_{ij} = \mathbf{g}\left(\frac{\partial}{\partial x^i}, \frac{\partial}{\partial x^j} \right).$$

The matrix (g_{ij}) is symmetric and positive-defined.

- In general, if (E_i) is a basis of vector fields, then $\mathbf{g} = g_{ij}\, E^i \otimes E^j$, where $g_{ij} = \mathbf{g}(E_i, E_j)$. Remember that (E^j) is the dual basis corresponding to (E_i).

- (**The standard metric on** $\mathbb{R}^m$). The standard Riemannian metric on $\mathbb{R}^m$ is given by $\mathbf{g} = \delta_{ij}\mathrm{d}x^i \otimes \mathrm{d}x^j$ in the canonical coordinate system.

- (**The Minkowski space**). On the vector space $\mathbb{R}^m$, we introduce the Minkowski metric given by $\mathbf{g} = \eta_{ij}\, \mathrm{d}x^i \otimes \mathrm{d}x^j$, where the diagonal of (η_{ij}) is $(-1, 1, \ldots, 1)$ and all the other elements are zero. This is the **Minkowski metric**, which is a Lorentzian metric, and therefore, $(\mathbb{R}^m, \mathbf{g})$ is a Lorentzian manifold.

- Let $j\colon N \hookrightarrow M$, be an immersed submanifold of a Riemannian manifold $(M, \mathbf{g})$. Then $(N, j^*\mathbf{g})$ is a Riemannian manifold. Observe that this cannot be true in the pseudo Riemannian case.

- Let $(M_1, \mathbf{g}_1)$ and $(M_2, \mathbf{g}_2)$ be Riemannian manifolds. An **isometry** between M_1 and M_2 is a diffeomorphism $F\colon M_1 \longrightarrow M_2$ such that $F^*\mathbf{g}_2 = \mathbf{g}_1$. If F is defined only on an open set, then it is called a local isometry.

- An **isometry** of $(M, \mathbf{g})$ is a diffeomorphism on M, such that it leaves the metric tensor $\mathbf{g}$ invariant.

  An **infinitesimal isometry** of $(M, \mathbf{g})$ is a vector field on M such that the diffeomorphisms defined by its flux F^t are isometries of $(M, \mathbf{g})$. The infinitesimal isometries are called **Killing vector fields**. $X \in \mathfrak{X}(M)$ is a Killing vector field if, and only if, $\mathrm{L}(X)\mathbf{g} = 0$.

- In a paracompact differentiable manifold, there exists a Riemannian metric. This is a consequence of the existence of partitions of unity.
  For the existence of a Lorentzian metric on a paracompact manifold M, a vector field without critical points must exist. This holds for connected compact manifolds with null Euler–Poincaré characteristic.

On a Riemannian manifold $(M, \mathbf{g})$, we have the following natural constructions:

- For every point $p \in M$, the metric $\mathbf{g}$ defines an isomorphism $\hat{\mathbf{g}}_p\colon \mathrm{T}_p M \longrightarrow \mathrm{T}_p^* M$, such that $\langle \hat{\mathbf{g}}_p(u_p), v_p \rangle := \mathbf{g}(u_p, v_p)$. This family of isomorphisms collects into a global one $\hat{\mathbf{g}}\colon \mathrm{T}M \longrightarrow \mathrm{T}^* M$, which is an isomorphism of vector bundles.

  This isomorphism can be extended to the corresponding $\mathrm{C}^\infty(M)$-modules of sections, that is, we have the map $\mathbf{g}^\flat\colon \mathfrak{X}(M) \longrightarrow \Omega^1(M)$ defined by $\mathbf{g}^\flat(X) = X^\flat := \hat{\mathbf{g}} \circ X$, which is an isomorphism with inverse

$\mathbf{g}^\sharp : \Omega^1(M) \longrightarrow \mathfrak{X}(M)$ such that $\alpha \mapsto \alpha^\sharp = \hat{\mathbf{g}}^{-1} \circ \alpha$.

- In a local chart (U, x^i), the isomorphism $\hat{\mathbf{g}}$ and its inverse are given, respectively, by

$$(x^i, v^i) \mapsto (x^i, v^i g_{ij}(x)), \quad (x^i, \alpha_i) \mapsto (x^i, \alpha_i g^{ij}(x)),$$

where (g^{ij}) is the inverse matrix of (g_{ij}).

- The above isomorphisms are usually called *musical isomorphisms*, and they are written as $v^\flat$ for the linear form with coordinates $v_j = v^i g_{ij}$, and $\alpha^\sharp$ for the tangent vector with coordinates $\alpha^j = \alpha_i g^{ij}$.

  These isomorphisms can be extended to any kind of tensor fields, and we can lift and lower indexes using the metric tensor.

- Let $f \in C^\infty(M)$; the **gradient** of f is the vector field $\operatorname{grad} f := \hat{\mathbf{g}}^{-1} \circ df$. Then we have $(\operatorname{grad} f | X) = \langle df, X \rangle$. In local coordinates, $\operatorname{grad} f = g^{ij} \dfrac{\partial f}{\partial x^i} \dfrac{\partial}{\partial x^j}$.

- In a Riemannian manifold, there exists a canonical measure $V_{\mathbf{g}}$ called **Riemannian volume**. First, in the open set of a local chart (U, x^i), the integral of a function $f : U \longrightarrow \mathbb{R}$ is given by

$$\int_U f \, dV_{\mathbf{g}} = \int_{\varphi(U)} \hat{f}(x^1, \ldots, x^m) \sqrt{|\det(g_{ij})|} \, dx^1 \ldots dx^m,$$

where $\hat{f}(x^1, \ldots, x^m)$ is the local expression of f.

- If M is an oriented manifold, then this measure comes from a canonical volume form Ω_g usually called *canonical Riemannian volume form*. In a local chart, this volume form is given by

$$\Omega_{\mathbf{g}} = \sqrt{|\det(g_{ij})|} \, dx^1 \ldots dx^m.$$

5.1.7 The Levi–Civita connection

Definition 5.10. Let $(M, \mathbf{g})$ be a Riemannian manifold. A connection ∇ on M is a **Riemannian connection** if the metric tensor field $\mathbf{g}$ is parallel, that is, $\nabla \mathbf{g} = 0$.

This is equivalent to stating that $\nabla_X \mathbf{g} = 0$ for every $X \in \mathfrak{X}(M)$.

Another interesting way to state this property is the following: for every $X, Y, Z \in \mathfrak{X}(M)$,

$$L(Z)(\mathbf{g}X, Y) = \mathbf{g}(\nabla_Z X, Y) + \mathbf{g}(X, \nabla_Z Y).$$

Using the covariant derivative along a curve, this condition is equivalent to stating that for every curve $\gamma : I \subset \mathbb{R} \longrightarrow M$ and $X, Y \in \mathfrak{X}(\gamma)$, the following equation holds:

$$D(X \mid Y) = (\nabla_{\dot{\gamma}(t)} X \mid Y) + (X \mid \nabla_{\dot{\gamma}(t)} Y).$$

Remember that for a function $f : I \subset \mathbb{R} \longrightarrow \mathbb{R}$, we have $Df = \dfrac{df}{dt}$.

In particular, observe that if X, Y are parallel along γ, then $(X \mid Y)$ is constant, and we get that the parallel transport with a Riemannian connection is an isometry. Furthermore, if γ is a geodesic line, then $(\gamma' \mid \gamma')$ is constant.

Finally, the *fundamental theorem of Riemannian geometry* states that if $(M, \mathbf{g})$ is a Riemannian manifold, there exists one, and only one, connection on M that is Riemannian and symmetric.

Definition 5.11. The above connection, called the **Levi–Civita connection** associated with the metric, is defined by the *Koszul formula*:

$$2(\nabla_X Y \mid Z) = \mathrm{L}(X)(Y \mid Z) + \mathrm{L}(Y)(Z \mid X) - \mathrm{L}(Z)(X \mid Y)$$
$$+ ([X, Y] \mid Z) + ([Z, X] \mid Y) - ([Y, Z] \mid X).$$

In the following section, unless indicated, when we use a connection ∇ in a Riemannian manifold, we will assume that it is the Levi–Civita connection.

In a local chart (U, x^i) in M, the Christoffel symbols of the Levi–Civita connection in the corresponding local basis for the vector fields are given by:

$$\Gamma_{ij}^k = \frac{1}{2} g^{kl} \left(\frac{\partial g_{jl}}{\partial x^i} + \frac{\partial g_{il}}{\partial x^j} - \frac{\partial g_{ij}}{\partial x^l} \right).$$

They are also written as $\Gamma_{ij}^k = g^{kl}[ij, l]$. In this case, $[ij, l]$ are called *Christoffel symbols of the first class.*

5.1.8 *Submanifolds of a Riemannian manifold*

Let $j \colon M \hookrightarrow \widetilde{M}$ be an embedded submanifold of a Riemannian manifold $(\widetilde{M}, \widetilde{g})$ with the Levi–Civita connection $\widetilde{\nabla}$. We know that $\mathbf{g} = j^* \widetilde{\mathbf{g}}$ is a Riemannian metric on M. Let ∇ be its Levi–Civita connection.

- For every $p \in M$, the embedding j allows us to identify $\mathrm{T}_p M$ as a subspace of $\mathrm{T}_{j(p)} \widetilde{M}$; thus, we have a splitting

$$\mathrm{T}_{j(p)} \widetilde{M} = \mathrm{T}_p M \oplus (\mathrm{T}_p M)^\perp.$$

Then, every tangent vector $u \in \mathrm{T}_{j(p)} \widetilde{M}$ can be decomposed as a sum $u = u^\top + u^\perp$, the tangent part to M and the orthogonal one. This splitting can be extended to vector fields on the manifold $\widetilde{M}$.

- Let $X, Y \in \mathfrak{X}(M)$. On a neighbourhood of every $p \in M$, we can extend the vector fields X, Y to vector fields defined on an open set of $\widetilde{M}$. Hence, we can calculate $(\widetilde{\nabla}_X Y)(p) \in \mathrm{T}_{j(p)}\widetilde{M}$, as this value is independent of the extensions. With this construction, we have a vector field along the embedding j, that is, $\widetilde{\nabla}_X Y \colon M \longrightarrow \mathrm{T}\widetilde{M}$. This vector field can be split into,

$$\widetilde{\nabla}_X Y = (\widetilde{\nabla}_X Y)^\top + (\widetilde{\nabla}_X Y)^\perp .$$

- It is easy to prove that for $X, Y \in \mathfrak{X}(M)$, the map $(X, Y) \mapsto (\widetilde{\nabla}_X Y)^\top$ defines a connection on $(M, \mathbf{g})$. This connection is the Levi–Civita connection.
- Now we consider the other component of the above splitting: The expression $\Pi(X, Y) = (\widetilde{\nabla}_X Y)^\perp$ is $C^\infty(M)$-bilinear and symmetric in $X, Y \in \mathfrak{X}(M)$. Symmetry holds because $[X, Y] \in \mathfrak{X}(M)$, for $X, Y \in \mathfrak{X}(M)$, so $[X, Y]^\perp = 0$.
- The above property of bilinearity implies that the map $\Pi_p \colon \mathrm{T}_p M \times \mathrm{T}_p M \longrightarrow (\mathrm{T}_p M)^\perp$, for every $p \in M$, is well defined and gives us a vector valued bilinear symmetric form, which takes values on $(\mathrm{T}_p M)^\perp$. It is called the *second fundamental form* of M.

  The decomposition $\widetilde{\nabla}_X Y = \nabla_X Y + \Pi(X, Y)$ is called the *Gauss formula*.

  Furthermore, if $N \in \mathfrak{X}(\widetilde{M})$ is orthogonal to M, then we have the *Weingarten equation*: $(\widetilde{\nabla}_X N \mid Y) = -(N \mid \Pi(X, Y))$.
- **Nash embedding theorem**: Every Riemannian manifold with a countable basis of open sets is isometric to a submanifold of some $\mathbb{R}^n$.

It is interesting to analyse the following particular cases of submanifolds of Riemannian manifolds:

5.1.8.1 *Hypersurfaces*

Let $(\widetilde{M}, \mathbf{g})$ be a Riemannian manifold and $M \subset \widetilde{M}$ be a hypersurface of $\widetilde{M}$, and suppose that both manifolds are oriented.

If $(X_1, \ldots, X_m)$ is an oriented positive basis of vector fields on M, then, on the points of M, we can take a unique vector field N, unitary and orthogonal to M, such that $(X_1, \ldots, X_m, N)$ is an oriented positive basis for $\widetilde{M}$ on the points of M. We say that N is the positive-oriented normal vector field to M. In this situation, the second fundamental form can be understood as $\mathbb{R}$-valued, and we write $h(X, Y) = (\Pi(X, Y) \mid N)$. We have $\Pi(X, Y) = h(X, Y)N$.

With this construction, we have a two-covariant tensor field on M. With this quadratic form, using the metric $\mathbf{g}$ at every point $p \in M$, we can define a symmetric endomorphism S on $\mathrm{T}_p M$, called the *Weingarten map*, which is defined by

$$(S(X) \mid Y) = h(X, Y) := ((\widetilde{\nabla}_X Y)^\perp \mid N).$$

The eigenvalues of S are called *principal curvatures* and the product of all of them is called the *Gauss curvature*. The eigenvectors are called *principal directions*.

5.1.8.2   *Curves*

Let $(M, \mathbf{g})$ be a Riemannian manifold and $C \subset M$ a connected orientable regular curve, that is, a submanifold of dimension one.

For every $p \in C$, there exists a neighbourhood and a local parameterisation of C with a curve $c \colon I \subset \mathbb{R} \longrightarrow M$ such that $\parallel c'(t) \parallel = 1$, the *arc parameter*, and with the orientation of C. If $\mathbf{t} = c' \in \mathfrak{X}(c)$ is the tangent vector field, for every $s \in I$, we have $\mathbf{t}(s)$ as a positive basis of $\mathrm{T}_{c(s)}C$.

Suppose that $(\mathbf{t}, \nabla_{\mathbf{t}(s)}\mathbf{t}, \ldots, \nabla_{\mathbf{t}(s)}^{m-1}\mathbf{t})$ are linearly independent in some $s_o \in I$. If we apply the Gram–Schmidt orthogonalisation procedure, we get an orthonormal basis $(\mathbf{f}_1(s) = \mathbf{t}, \ldots, \mathbf{f}_m(s))$ of $\mathrm{T}_{c(s)}C$ for s on a neighbourhood of s_o. This is the *Frenet reference*. The covariant derivatives of these vector fields are linear combinations of themselves, and the coefficients are the curvatures of C. These linear combinations are the *Frenet–Serret formulas*.

5.1.9   *Curvature and distance on a Riemannian manifold*

Definition 5.12. Let $(M, \mathbf{g})$ be a Riemannian manifold and R be the curvature tensor of the Levi–Civita connection ∇ of $(M, \mathbf{g})$. The **Riemann curvature tensor** is the tensor field $\mathrm{Rie} \in \mathcal{T}_4(M)$ obtained by lowering the contravariant index of R to the last place:

$$\mathrm{Rie}(X, Y, Z, W) = (R(X, Y)Z \mid W).$$

The **Ricci curvature tensor** is the tensor field $\mathrm{Ric} \in \mathcal{T}_2(M)$ obtained after the contraction of the first covariant index with the contravariant index of R, that is, $\mathrm{Ric} = c_1^1(R)$.

The **scalar curvature** is the function $S \in \mathcal{C}^\infty(M)$ obtained by taking the trace, with respect to **g**, of the Ricci tensor:

$$S = \mathrm{tr}_{\mathbf{g}}(\mathrm{Ric}) = \mathrm{tr}(\mathrm{Ric}^\sharp).$$

In a local reference for vector fields, we have

- $\mathrm{Rie} = R_{ijkl}\, E^i \otimes E^j \otimes E^k \otimes E^l$, where $R_{ijkl} = R^s_{ijk} g_{sl}$.
- $\mathrm{Ric} = R_{ij}\, E^i \otimes E^j$ with $R_{jk} = R^i_{ijk}$.
- $S = R^k_k = g^{kj} R_{jk}$.

Remark 5.1. There are other alternative definitions of these concepts that consist of making the contraction with respect to other different indices, and then the only difference is a change in the sign (and the same happens with the scalar curvature). For instance, in General Relativity, to constructing the Riemann curvature tensor is common when making the contraction of the contravariant index of R with the second covariant index, $R_{rkij} = g_{rl} R^l_{kij}$, so

$$g_{rl} R^l_{kij} = R_{rkij} = R_{ijrk} = -R_{ijkr} = -R^l_{ijk}\, g_{lr}.$$

In the same way, for the Ricci tensor, the contraction is made between the second covariant index with the contravariant index of R, $R_{ij} = R^k_{ikj}$, so

$$R^k_{ikj} = g^{kl} R_{likj} = -g^{kl} R_{kjil} = -R_{kjil} g^{lk} = -R^k_{kji}.$$

Remark 5.2. Let $(M, \mathbf{g})$ be a Riemannian manifold.

- The Riemannian manifold is **flat** if it is locally isometric to the Euclidean space. Therefore, (M, g) is flat if, and only if, its curvature tensor vanishes, that is, if its Levi–Civita connection is flat.
- Let $\gamma\colon I \longrightarrow M$ be a $\mathcal{C}^1$-piecewise curve. The **length** of γ is

$$\ell(\gamma) = \int_I \| \gamma' \| \, .$$

Observe that this length is invariant by reparameterisations, that is, if $\varphi\colon J \longrightarrow I$ is a diffeomorphism between open real intervals, then $\ell(\gamma \circ \varphi) = \ell(\gamma)$.

- Supposing that the manifold M is connected, any two points can be connected by a piecewise curve. Therefore, if $p, q \in M$, the *distance* between p and q is defined as:

$$d(p, q) := \inf\{\ell(\gamma) \mid \gamma \text{ is a } \mathcal{C}^1\text{-piecewise curve from } p \text{ to } q\}.$$

The function d is a distance defining the topology of M and is called **Riemannian distance**.

- Let M be a connected differentiable manifold; suppose it is Hausdorff but not necessarily paracompact. Then the following conditions are equivalent:

  (1) M admits a Riemannian metric.

  (2) M is metrizable as topological space.

  (3) The topology of M has a countable basis of open sets.

  (4) M is paracompact.

- If $(M, \mathbf{g})$ is a connected Riemannian manifold, we say that it is **geodesically complete** if the domain of definition of all the geodesic curves on M is $\mathbb{R}$.

- (**Hopf–Rinow Theorem**): Let $(M, \mathbf{g})$ be a connected Riemannian manifold and d its Riemannian distance. Then, the following conditions are equivalent:

  (1) For a subset of M, if it is bounded by d and closed, then it is compact.

  (2) As a metric space, M is complete.

  (3) $(M, \mathbf{g})$ is geodesically complete.

5.2 Newtonian Dynamical Systems

Next, in this section, we use the above geometric structures to state the description of dynamical systems on a (semi)Riemannian manifolds (see, for instance, [1, 11, 46, 75, 91, 179, 291] as general references).

5.2.1 *Newton dynamical equations: Kinetic energy*

From a geometric approach, mechanical systems in Newtonian physics have a common background which, as in the above cases, is collected in the following postulates:

Postulate 5.1 (First postulate of Newtonian mechanics). The **configuration space** Q of a dynamical system with n degrees of freedom is an n-dimensional differentiable manifold, which is endowed with a *Riemannian metric*.

The **state space**, or **phase space**, of coordinates–velocities is the tangent bundle TQ, and the **phase space** of coordinates–momenta is the cotangent bundle T^*Q of the manifold Q.

This formulation can be extended to describe relativistic systems, and in this case, the metric is semi-Riemannian.

Postulate 5.2 (Second postulate of Newtonian mechanics). The **observables** or physical magnitudes of the dynamical system are functions of $C^\infty(TQ)$ and $C^\infty(T^*Q)$ respectively.

The result of the measure of an observable is the value of its representing function at a point of the phase space.

Postulate 5.3 (Third postulate of Newtonian mechanics). The dynamics of the system is given by a 1-form on Q, the **work form** or **force form**, or by a vector field on Q, the **force field** or simply the **force**.

With these ideas in mind, we have:

Definition 5.13. A **Newtonian mechanical system** is a triple $(Q, \mathbf{g}, \omega)$, where

(1) Q is a differentiable manifold (dim $Q = n$).
(2) $\mathbf{g}$ is a Riemannian metric on Q. Therefore, $(Q, \mathbf{g})$ is a Riemannian manifold.
(3) ω is a differential 1-form on Q, called the **work form**.

As $\mathbf{g}$ is a Riemannian metric, the work form $\omega \in \Omega^1(Q)$ is associated with a unique vector field $F \in \mathfrak{X}(Q)$ such that $i(F)\mathbf{g} = \omega$. We call F the **force field** of the system. In this case, we denote the system as $(Q, \mathbf{g}, F)$).

For a Newtonian system $(Q, \mathbf{g}, F)$, let ∇ be the *Levi–Civita connection* associated with $\mathbf{g}$.

Postulate 5.4 (Fourth postulate of Newtonian mechanics). The dynamical trajectories of the Newtonian dynamical system $(Q, \mathbf{g}, F)$ are the curves $\gamma\colon [a,b] \subset \mathbb{R} \longrightarrow Q$ solution to the equation

$$\nabla_{\dot\gamma}\dot\gamma = \mathrm{F} \circ \gamma \qquad (5.1)$$

which is called the **Newton equation** of the system.

Remark 5.3 (Inertia law). Note that if $\mathrm{F} = 0$, then the dynamical trajectories are the geodesic curves of the metric **g**. This corresponds to the first Newton law.

If (U, x^i) is a local chart on Q, and $\{\Gamma_{ij}^k\}$ are the corresponding Christoffel symbols of the connection ∇, then the dynamical equation is locally given by

$$\ddot\gamma^k + \Gamma_{ij}^k \dot\gamma^i \dot\gamma^j = \mathrm{F}^k \circ \gamma.$$

Furthermore, if $\omega = \omega_i dx^i$ and $\mathrm{F} = \mathrm{F}^i \dfrac{\partial}{\partial x^i}$, in this chart, then we have

$$\omega_i = g_{ij}\mathrm{F}^j, \qquad \mathrm{F}^i = g^{ij}\omega_j,$$

where g^{ij} are the components of the inverse matrix of **g** in this local chart (the "inverse metrics").

This formalism has a dual counterpart as follows:

Definition 5.14. Let $(Q, \mathbf{g})$ be a Riemannian manifold. The musical diffeomorphism associated with the Riemannian metric **g**, defined by

$$\begin{aligned}
\theta : TQ &\longrightarrow \quad T^*Q \\
(q, v) &\longmapsto (q, i(v)g)
\end{aligned}$$

gives us the commutative diagram

Hence, θ can be understood as a differential 1-form on Q along τ_Q (denoted by $\theta \in \Omega^1(Q, \tau_Q)$), called the **linear momentum 1-form** associated with the metric.

Using θ, the Newton equation (5.1) can be written as:

> **Proposition 5.1 (Dual form of the dynamical equations).** *Given a Newtonian dynamical system* $(Q, \mathbf{g}, \omega)$, *a curve* $\gamma \colon I \subset \mathbb{R} \longrightarrow Q$ *is a solution to the dynamical equation if, and only if, it satisfies*
>
> $$\nabla_{\dot{\gamma}}(\theta \circ \dot{\gamma}) = \omega \circ \gamma. \tag{5.2}$$

Proof. If $\gamma \colon [a, b] \subset \mathbb{R} \longrightarrow Q$ is a curve on Q, then $\theta \circ \dot{\gamma} \in \Omega^1(Q, \gamma)$, and we have

$$\nabla_{\dot{\gamma}}(\theta \circ \dot{\gamma}) = \nabla_{\dot{\gamma}}(i(\dot{\gamma})\mathbf{g}) = i(\nabla_{\dot{\gamma}}\dot{\gamma})\mathbf{g} + i(\dot{\gamma})\nabla_{\dot{\gamma}}\mathbf{g} = i(\nabla_{\dot{\gamma}}\dot{\gamma})\mathbf{g} = \theta \circ \nabla_{\dot{\gamma}}\dot{\gamma}.$$

Furthermore, if γ is a dynamical trajectory, that is $\nabla_{\dot{\gamma}}\dot{\gamma} = \mathrm{F} \circ \gamma$, then

$$\nabla_{\dot{\gamma}}(\theta \circ \dot{\gamma}) = \theta \circ \nabla_{\dot{\gamma}}\dot{\gamma} = \theta \circ \mathrm{F} \circ \gamma = i(\mathrm{F})\mathbf{g} \circ \gamma = \omega \circ \gamma,$$

which is Eq. (5.2). $\qquad \square$

Note that as $\theta \circ \dot{\gamma} \in \Omega^1(Q, \gamma)$, then, for every $X \in \mathfrak{X}(Q)$, we have

$$\nabla_{\dot{\gamma}}((\theta \circ \dot{\gamma})(X)) = (\nabla_{\dot{\gamma}}(\theta \circ \dot{\gamma}))(X) + (\theta \circ \dot{\gamma})(\nabla_{\dot{\gamma}}X).$$

Hence, we obtain the following:

> **Theorem 5.1 (Linear momentum conservation).** *If the work form (or the force field) of a Newtonian mechanical system* $(Q, \mathbf{g}, \omega)$ *is zero, then the linear momentum is invariant (i.e., constant) along the trajectories of the system. (Geometrically, for every trajectory of the system* γ, *the form* $\theta \circ \dot{\gamma} \in \Omega^1(Q, \gamma)$ *is parallel along the trajectory* γ.)

Proof. If $\omega = 0$ or $F = 0$, then the Newton equation is $\nabla_{\dot{\gamma}}(\theta \circ \dot{\gamma}) = 0$; hence, $\theta \circ \dot{\gamma}$ is parallel along the trajectory γ. $\qquad \square$

We have maintained the statement as in the classical form in physics: in $\mathbb{R}^3$, with a natural Cartesian system of coordinates, we have $\Gamma_{ij}^k = 0$, for every i, j, k, so the components of $\theta \circ \dot{\gamma}$ are constants.

Finally, with the Riemannian metric, we can associate the following function:

Definition 5.15. Given a Riemannian manifold $(Q, \mathbf{g})$, the associated **kinetic energy** is the function $K \in C^\infty(TQ)$ defined by

$$K : \begin{array}{ccc} TQ & \longrightarrow & \mathbb{R} \\ (q, v) & \mapsto & \frac{1}{2}\mathbf{g}(v, v) \end{array}.$$

Its local expression is

$$K(q^i, v^j) = \frac{1}{2}g_{ij}(q)v^i v^j.$$

5.2.2 *Euler–Lagrange equations*

Now, we transform the Newton equation into a new form, which is easier to state when we know the elements defining the system. First, we need a technical result:

Lemma 5.1. *Let $(Q, \mathbf{g})$ be a Riemannian manifold, $K \in C^\infty(TQ)$ its kinetic energy, ∇ the Levi–Civita connection of $\mathbf{g}$, and $\gamma \colon I \subset \mathbb{R} \longrightarrow Q$ a curve. If $(U, \varphi = (q^i))$ is a local chart of Q with $\gamma(t) \in U$, and $(\tau_Q^{-1}(U), q^i, v^i)$ is the natural chart on TQ, then*

$$\frac{d}{dt}\left(\frac{\partial K}{\partial v^j} \circ \dot\gamma\right) - \frac{\partial K}{\partial q^j} \circ \dot\gamma = \mathbf{g}\left(\nabla_{\dot\gamma}\dot\gamma, \frac{\partial}{\partial q^j}\right).$$

Proof. We compare the local expressions of both sides of the equation. As $K = \frac{1}{2}g_{ij}v^i v^j$, we have

$$\frac{\partial K}{\partial v^j} = g_{ij}v^i, \quad \frac{\partial K}{\partial q^j} = \frac{1}{2}\frac{\partial g_{ik}}{\partial q^j}v^i v^k,$$

thus, if $\gamma = (\gamma^1, \ldots, \gamma^n)$,

$$\frac{\partial K}{\partial v^j} \circ \dot\gamma = (g_{ij} \circ \gamma) \circ \dot\gamma^i, \quad \frac{\partial K}{\partial q^j}\dot\gamma = \frac{1}{2}\left(\frac{\partial g_{ik}}{\partial q^j} \circ \gamma\right)\dot\gamma^i\dot\gamma^k.$$

Hence, we obtain

$$\frac{d}{dt}\left(\frac{\partial K}{\partial v^j} \circ \dot\gamma\right) = \left(\frac{\partial g_{ij}}{\partial q^k} \circ \gamma\right)\dot\gamma^k\dot\gamma^i + (g_{ij} \circ \gamma)\ddot\gamma^i$$

and

$$\frac{d}{dt}\left(\frac{\partial K}{\partial v^j} \circ \dot\gamma\right) - \frac{\partial K}{\partial q^j} \circ \dot\gamma = \left(\frac{\partial g_{ij}}{\partial q^k} \circ \gamma\right)\dot\gamma^k\dot\gamma^i + (g_{ij}\circ\gamma)\ddot\gamma^i - \frac{1}{2}\left(\frac{\partial g_{ik}}{\partial q^j} \circ \gamma\right)\dot\gamma^i\dot\gamma^k.$$

$$\tag{5.3}$$

Furthermore,

$$g\left(\nabla_{\dot\gamma}\dot\gamma, \frac{\partial}{\partial q^j}\right) = g\left((\ddot\gamma^i + \Gamma^i_{kl}\dot\gamma^k\dot\gamma^l)\frac{\partial}{\partial q^i}, \frac{\partial}{\partial q^j}\right) = (g_{ij}\circ\gamma)\ddot\gamma^i + (g_{ij}\circ\gamma)\Gamma^i_{kl}\dot\gamma^k\dot\gamma^l.$$

However, we have

$$[kl, j] = g_{ij}\Gamma^i_{kl} = \frac{1}{2}\left(\frac{\partial g_{jk}}{\partial q^l} + \frac{\partial g_{jl}}{\partial q^k} - \frac{\partial g_{lk}}{\partial q^j}\right),$$

and by substitution in Eq. (5.3), we obtain the desired result. $\qquad\square$

Theorem 5.2 (Lagrange). *Let $(Q, \mathbf{g}, \omega)$ be a Newtonian mechanical system, and $\gamma \colon I \subset \mathbb{R} \longrightarrow Q$ a curve contained in the domain $U \subset Q$ of a chart $(U, \varphi = (q^i))$ of Q. Then, γ is a solution to the dynamical equation (5.1) if, and only if, it satisfies the equations*

$$\frac{d}{dt}\left(\frac{\partial K}{\partial v^j}\circ\dot\gamma\right) - \frac{\partial K}{\partial q^j}\circ\dot\gamma = (\omega\circ\gamma)\left(\frac{\partial}{\partial q^j}\right) = \omega_j\circ\gamma, \qquad (5.4)$$

*which are called **Euler–Lagrange equations of the second kind** of the system.*

Proof. If γ is a solution to the Newton equation (5.1), then

$$\mathbf{g}\left(\nabla_{\dot\gamma}\dot\gamma, \frac{\partial}{\partial q^j}\right) = \mathbf{g}\left(\mathrm{F}\circ\gamma, \frac{\partial}{\partial q^j}\right) = (\omega\circ\gamma)\left(\frac{\partial}{\partial q^j}\right).$$

Hence, by the previous lemma, the curve γ satisfies Eq. (5.4).

Conversely, if γ satisfies (5.4), then,

$$\mathbf{g}\left(\nabla_{\dot\gamma}\dot\gamma, \frac{\partial}{\partial q^j}\right) = (\omega\circ\gamma)\left(\frac{\partial}{\partial q^j}\right) = \mathbf{g}\left(F\circ\gamma, \frac{\partial}{\partial q^j}\right).$$

Therefore, $\nabla_{\dot\gamma}\dot\gamma = \mathrm{F}\circ\gamma$, and γ is a solution to the Newton equation. $\qquad\square$

Remark 5.4. To calculate the Newton equation in a local chart, we need to know the Christoffel symbols of the Levi–Civita connection ∇ associated with $\mathbf{g}$, but to write the Euler–Lagrange equations, we do not need the local expression of the connection. Moreover, observe that the way to get Euler–Lagrange equations does not depend on the local chart we use: we need only to know the local expression of K and ω, and calculate the suitable derivatives.

Another way to understand Euler–Lagrange equations is as follows: let (U, q^i) be a local chart of Q; the Newton equations of the system in this chart are written as

$$\ddot{\gamma}^i + \Gamma^i_{jk}\dot{\gamma}^j\dot{\gamma}^k = F^i \circ \gamma,$$

or as it is usually expressed,

$$\ddot{q}^i + \Gamma^i_{jk}\dot{q}^j\dot{q}^k = F^i.$$

This is a second-order system of differential equations. To transform it into a first-order system, we introduce new variables $v^i = \dot{q}^i$, and we obtain the first-order system on TQ

$$\dot{q}^i = v^i, \quad \dot{v}^i = F^i - \Gamma^i_{jk}v^jv^k.$$

The associated vector field is given by

$$X = v^i\frac{\partial}{\partial q^i} + (F^i - \Gamma^i_{jk}v^jv^k)\frac{\partial}{\partial v^i} = v^i\frac{\partial}{\partial q^i} - \Gamma^i_{jk}v^jv^k\frac{\partial}{\partial v^i} + F^v,$$

where F^v is the vertical lift of F from Q to TQ. Observe that we obtain the geodesic vector field plus the vertical lift of the force field.

As $X \in \mathfrak{X}(\tau_Q^{-1}(U))$, for every point $(q, v) \in \tau_Q^{-1}(U)$, there exists a unique solution, with (q_o, v_o) as initial condition. Consider now the functions we need to obtain the Euler–Lagrange equations and calculate the action of X on them. Using the well-known properties of Christoffel symbols, we have:

$$X\left(\frac{\partial K}{\partial v^k}\right) = X(g_{lk}v^l) = \frac{\partial g_{lk}}{\partial q^i}v^iv^l + g_{lk}(F^l - \Gamma^l_{ij}v^iv^j),$$

hence

$$X\left(\frac{\partial K}{\partial v^k}\right) - \frac{\partial K}{\partial q^k} = \frac{\partial g_{lk}}{\partial q^i}v^iv^l + g_{lk}F^l - g_{lk}\Gamma^l_{ij}v^iv^j - \frac{1}{2}\frac{\partial g_{ij}}{\partial q^k}v^iv^j = g_{lk}F^l.$$

So, we have

Proposition 5.2. *Let $(Q, \mathbf{g}, \omega)$ be a Newtonian mechanical system, (U, q^i) a local chart on Q, and $(\tau_Q^{-1}(U), q^i, v^i)$ the corresponding natural chart on TQ. Then, there exists a unique vector field $X \in \mathfrak{X}(\tau_Q^{-1}(U))$ which satisfies,*

(1) $L(X)q^k = v^k.$

(2) $L(X)\left(\dfrac{\partial K}{\partial v^k}\right) = \dfrac{\partial K}{\partial v^k} + g_{ik}F^i.$

> *Furthermore, its integral curves* $\sigma \colon I \subset \mathbb{R} \longrightarrow TQ$ *are canonical lifts of curves* $\gamma \colon I \subset \mathbb{R} \longrightarrow Q$ *to* TQ, *which are solution to the Newton equations of the system.*

Proof. The existence and uniqueness of the vector field X are consequences of being $\left(q^k, \dfrac{\partial K}{\partial v^k} \right)$ a local chart on TQ, because the Riemannian metric **g** is nondegenerate. The properties of X have been proven in the previous discussion.

Furthermore, the integral curves of X are the lifts to TQ of the solutions to the system of Euler–Lagrange equations and hence solutions to the Newton equations. $\qquad\qquad\square$

5.2.2.1 *Velocity-dependent forces*

It is typical that the mechanical forces, or the work forms, depend not only on the position coordinates but also on the velocities. This is the case of dissipative systems or electromagnetic Lorentz forces. Geometrically, this means that $\omega \in \Omega^1(Q, \tau_Q)$ and $\mathbf{F} \in \mathfrak{X}(Q, \tau_Q)$. In this case, the only changes we need to make in the above paragraphs are the following:

(1) The Newton equations change to
$$\nabla_{\dot\gamma}\dot\gamma = \mathbf{F} \circ \dot\gamma,$$
or in dual form,
$$\nabla_{\dot\gamma}(\theta \circ \dot\gamma) = \omega \circ \dot\gamma.$$
(2) The Euler–Lagrange equations are given as
$$\frac{d}{dt}\left(\frac{\partial K}{\partial v^j} \circ \dot\gamma \right) - \frac{\partial K}{\partial q^j} \circ \dot\gamma = (g_{ij}\mathbf{F}^i) \circ \dot\gamma = (\omega_j \circ \dot\gamma).$$

If the force field of the system depends on the velocities, the work form ω cannot be the exterior differential of a function defined on the configuration space Q.

As can be seen, the only changes consist of substituting γ by $\dot\gamma$ when we compose with ω or with $\mathbf{F}$ to consider the new domain of definition.

5.2.3 *Conservative systems: Mechanical Lagrangians and Euler–Lagrange equations*

In particular, we are interested in a special kind of Newtonian systems: those that are of conservative or mechanical Lagrangian type, called *simple mechanical systems*.

Definition 5.16. A Newtonian mechanical system (Q, g, ω) is **conservative** if the work form is exact, meaning there exists $V \in C^\infty(Q)$ such that $\omega = -dV$.

Then, the function V is called the **potential energy** of the system and the force vector field is $F = -\mathrm{grad}\, V$.

The negative sign in $\omega = -dV$ is a customary tradition in physics to identify the potential function of the field with the potential energy of the system. Then, for these systems, we define:

Definition 5.17. In a conservative Newtonian mechanical system (Q, g, ω), with $\omega = -dV$, the **total energy** or **mechanical energy** of the system is the function $E \in C^\infty(TQ)$ defined as

$$
\begin{aligned}
E : TQ &\longrightarrow \mathbb{R} \\
(q, v) &\mapsto K(q, v) + (\tau_Q^* V)(q, v)\,.
\end{aligned}
$$

(To simplify notation, we usually write $E = K + V$.)

As a direct consequence of the definition, we have:

Theorem 5.3 (Mechanical energy conservation). *Let (Q, g, ω) be a conservative Newtonian mechanical system. Then the mechanical energy E is invariant ("constant") along the trajectories of the system.*

Proof. If $\gamma \colon I \subset \mathbb{R} \longrightarrow Q$ is a solution to the Newton equations,

$$
\nabla_{\dot\gamma}\dot\gamma = F \circ \gamma, \quad i(F)g = \omega = -dV\,, \tag{5.5}
$$

then we have

$$
\begin{aligned}
\frac{d(E \circ \dot\gamma)}{dt} &= \nabla_{\dot\gamma}(E \circ \dot\gamma) = \nabla_{\dot\gamma}\left(\frac{1}{2}g(\dot\gamma, \dot\gamma) + V \circ \gamma\right) \\
&= g(\nabla_{\dot\gamma}\dot\gamma, \dot\gamma) + \nabla_{\dot\gamma}(V \circ \gamma) = g(\nabla_{\dot\gamma}\dot\gamma, \dot\gamma) + dV(\dot\gamma) \\
&= g(F, \dot\gamma) + dV(\dot\gamma) = \omega(\dot\gamma) + dV(\dot\gamma) = 0. \qquad \square
\end{aligned}
$$

Proposition 5.3. *Let (Q, g, ω) be a conservative Newtonian mechanical system with $\omega = -dV$ and let $\gamma \colon I \subset \mathbb{R} \longrightarrow Q$ be a curve with image on the domain $U \subset Q$ of a local chart $(U, \varphi = (q^i))$ of Q. Then the*

Euler–Lagrange equation (5.4) *becomes*

$$\frac{d}{dt}\left(\frac{\partial \mathcal{L}}{\partial v^j} \circ \dot{\gamma}\right) - \frac{\partial \mathcal{L}}{\partial q^j} \circ \dot{\gamma} = 0, \tag{5.6}$$

that is, they are the Euler–Lagrange equations for the Lagrangian function $\mathcal{L} = K - \tau_Q^* V$.

Proof. In fact, we have

$$\frac{d}{dt}\left(\frac{\partial K}{\partial v^j} \circ \dot{\gamma}\right) - \frac{\partial K}{\partial q^j} \circ \dot{\gamma} = (-dV \circ \gamma)\left(\frac{\partial}{\partial q^j}\right) = -\frac{\partial(\tau_Q^* V)}{\partial q^j} \circ \gamma,$$

and recalling that $\dfrac{\partial(\tau_Q^* V)}{\partial v^j} = 0$, we can write

$$\frac{d}{dt}\left(\frac{\partial(K - \tau_Q^* V)}{\partial v^j} \circ \dot{\gamma}\right) - \frac{\partial(K - \tau_Q^* V)}{\partial q^j} \circ \dot{\gamma} = 0. \qquad \square$$

Definition 5.18. The function $\mathcal{L} := K - \tau_Q^* V$ is said to be a **Lagrangian function of mechanical type** or a **mechanical Lagrangian function**, and Eq. (5.6) is the **Euler–Lagrange equation of the first class** of the conservative Newtonian system.

Going forward, we shorten the notation and write simply $\mathcal{L} = K - V$.

Remark 5.5.

(1) For conservative Newtonian mechanical systems, it is simple to prove that the vector field $X \in \mathfrak{X}(\tau_Q^{-1}(U))$ of Proposition 5.2 satisfies the condition

$$L(X)\left(\frac{\partial \mathcal{L}}{\partial v^k}\right) = \frac{\partial \mathcal{L}}{\partial v^k},$$

(instead of condition 2 in the aforementioned proposition).

(2) Observe that the Lagrangian functions of mechanical type,

$$\mathcal{L} = K - V \equiv \frac{1}{2}g_{ij}(q)v^i v^j - V(q),$$

are regular because $\Omega_{\mathcal{L}} \equiv g_{ij}(q)dq^i \wedge dv^j$ is a symplectic form.

5.2.4   *Coupled systems (systems in interaction)*

Next, we study the case of Newtonian mechanical systems that are made of several dynamical systems in interaction, that is, the *coupled systems*.

Let $(Q_1, \mathbf{g}_1, \omega_1)$, $(Q_2, \mathbf{g}_2, \omega_2)$ be two Newtonian mechanical systems and $F_1 \in \mathfrak{X}(Q_1)$, $F_2 \in \mathfrak{X}(Q_2)$ the corresponding force fields. The dynamical equations for both systems are

$$\nabla^1_{\dot\gamma_1} \dot\gamma_1 = F_1 \circ \gamma_1, \quad \nabla^2_{\dot\gamma_2} \dot\gamma_2 = F_2 \circ \gamma_2 \tag{5.7}$$

(with $\gamma_1 \colon I \subset \mathbb{R} \longrightarrow Q_1$, $\gamma_2 \colon I \subset \mathbb{R} \longrightarrow Q_2$). They constitute a non-coupled, separated, set of ordinary differential equations. Together, they are a system of non-coupled ordinary differential equations.

Consider the system $(Q, \mathbf{g}, \omega)$, where:

- $Q = Q_1 \times Q_2$; with $\pi_1 \colon Q \longrightarrow Q_1$, $\pi_2 \colon Q \longrightarrow Q_2$.
- $\mathbf{g} = \mathbf{g}_1 \oplus \mathbf{g}_2$.
- $\omega = \pi_1^* \omega_1 + \pi_2^* \omega_2$.

We can see that $\nabla = \nabla^1 \oplus \nabla^2$ is the Levi–Civita connection of the Riemannian metric $\mathbf{g}$, defined as follows: if $\gamma \colon I \subset \mathbb{R} \longrightarrow Q$ is given by $\gamma = (\gamma_1, \gamma_2)$, then

$$\nabla_{\dot\gamma} \dot\gamma = \nabla^1_{\dot\gamma_1} \dot\gamma_1 + \nabla^2_{\dot\gamma_2} \dot\gamma_2.$$

Moreover, from the expression of ω, we obtain the associated force field $F \in \mathfrak{X}(Q)$ of

$$F(q) = ((q_1, q_2), F_1(q_1), F_2(q_2)),$$

with $q \equiv (q_1, q_2) \in Q$. Therefore, the dynamical equation of the system is

$$\nabla_{\dot\gamma} \dot\gamma = F \circ \gamma,$$

which is equivalent to the non-coupled system (5.7). From the physical point of view, this situation is a model of two *joined mechanical systems* but *without interaction*. With this is mind, we have:

Definition 5.19. We say that N Newtonian systems (Q_μ, g_μ, F_μ), with $\mu = 1, \ldots, N$, are **coupled** (or **in interaction**) if $F_\mu \in \mathfrak{X}(Q_\mu, \pi_\mu)$, that is, we have the following commutative diagram for the force field acting

on the system

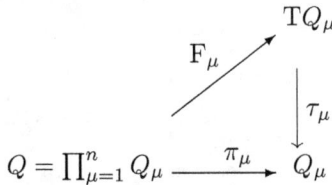

In this case, if $N = 2$, we have a system of differential equation as in (5.7), but now it is coupled, so we have

$$\nabla^1_{\dot{\gamma}_1} \dot{\gamma}_1 = F_1 \circ \gamma = F_1 \circ (\gamma_1, \gamma_2); \quad \nabla^2_{\dot{\gamma}_2} \dot{\gamma}_2 = F_2 \circ \gamma = F_1 \circ (\gamma_1, \gamma_2).$$

We want to formulate this system as a single Newtonian mechanical system; so we need the following result:

Lemma 5.2. *Given two manifolds* M_1, M_2, *let* $M = M_1 \times M_2$ *and* $\pi_i \colon M \longrightarrow M_i$ *($i = 1, 2$) be the natural projections. Then*

(1) $\mathfrak{X}(M)$ *is canonically isomorphic to* $\mathfrak{X}(M_1, \pi_1) \times \mathfrak{X}(M_2, \pi_2)$ *(as* $C^\infty(M)$-*módules).*

(2) $\Omega^1(M)$ *is canonically isomorphic to* $\Omega^1(M_1, \pi_1) \times \Omega^1(M_2, \pi_2)$ *(as* $C^\infty(M)$-*módules).*

Proof. Recall that if $q \equiv (q_1, q_2) \in M$, we have the isomorphism

$$\begin{aligned} \alpha_q : T_q M &\longrightarrow T_{q_1} M_1 \times T_{q_2} M_2 \\ v &\longmapsto (T_q \pi_1(v), T_q \pi_2(v)) \end{aligned}.$$

Hence, we obtain the sequence

$$\mathfrak{X}(M) \overset{\phi}{\longrightarrow} \mathfrak{X}(M_1, \pi_1) \times \mathfrak{X}(M_2, \pi_2) \overset{\psi}{\longrightarrow} \mathfrak{X}(M),$$

where the maps are defined as

$$\phi(X)(q) := \alpha_q(X(q)),$$
$$\psi(X_1, X_2)(q) := \alpha_q^{-1}(X_1(q), X_2(q)).$$

If $\rho_i \colon \mathfrak{X}(M_1, \pi_1) \times \mathfrak{X}(M_2, \pi_2) \longrightarrow \mathfrak{X}(M_i, \pi_i)$ ($i = 1, 2$) are the natural projections, we have:

(1) The map ϕ is well defined and $\phi(X)$ is differentiable: It is enough to observe that $\rho_i(\phi(X))$ is differentiable. To obtain $f \in C^\infty(M_i)$, we have

$$\begin{aligned} ((\rho_i(\phi(X)))f)(q) &= \rho_i(\alpha_p(X(q)))f = T_q \pi_i(X(q))f \\ &= X(q)(f \circ \pi_i) = (X(f) \circ \pi_i))(q), \end{aligned}$$

which depends differentially on $q \in M$.

(2) The map ϕ is injective because the maps α_q are isomorphisms.
(3) The map ϕ is a $C^\infty(M)$-modules isomorphism as can be seen directly.
(4) $\psi \circ \phi = \mathrm{Id}_{\mathfrak{X}(M)}$; in fact, if $X \in \mathfrak{X}(M)$ and $p \in M$, we have

$$((\psi \circ \phi)(X))(q) = \psi(\phi(X))(q) = \alpha_p^{-1}(\phi(X)(q))$$
$$= \alpha_q^{-1}(\alpha_p(X(q))) = X(q).$$

The statement (2) is a direct consequence of (1).   □

From this lemma, we obtain:

Theorem 5.4. *Two Newtonian mechanical systems in interaction,* $(Q_\mu, g_\mu, \mathrm{F}_\mu)$, $\mu = 1, 2$, *are equivalent to a unique Newtonian mechanical system* $(Q, \mathbf{g}, \mathrm{F})$, *where*

(1) $Q = Q_1 \times Q_2$, *with* $\pi_1 \colon Q \longrightarrow Q_1$, $\pi_2 \colon Q \longrightarrow Q_2$.
(2) $\mathbf{g}_1 \oplus \mathbf{g}_2$.
(3) $\omega = (\omega_1, \omega_2)$ *and* $\mathrm{F} = (\mathrm{F}_1, \mathrm{F}_2)$.

(We have identified $\mathfrak{X}(M)$ *with* $\mathfrak{X}(M_1, \pi_1) \times \mathfrak{X}(M_2, \pi_2)$, *and* $\Omega^1(Q)$ *with* $\Omega^1(Q_1, \pi_1) \times \Omega^1(Q_2, \pi_2)$, *in agreement with the above lemma.)*

Proof. The metric $\mathbf{g} = \mathbf{g}_1 \oplus \mathbf{g}_2$ is Riemannian on Q. Its Levi–Civita connection ∇ is $\nabla_1 \oplus \nabla_2$; then the dynamical equation associated to this Newtonian system can be split into the two components on Q_1 and Q_2, according to the dynamical equations of the two different systems in interaction Q_1, Q_2, so we obtain this result.

Observe that

$$i(\mathrm{F})\mathbf{g} = i(\mathrm{F}_1, \mathrm{F}_2)(\mathbf{g}_1 \oplus \mathbf{g}_2) = i(\mathrm{F}_1)\mathbf{g}_1 + i(\mathrm{F}_2)\mathbf{g}_2 = \omega_1 + \omega_2 \,,$$

and the dynamical equation $\nabla_{\dot\gamma}\dot\gamma = \mathrm{F} \circ \gamma$ splits into two different ones:

$$\nabla^1_{\dot\gamma_1}\dot\gamma_1 = \mathrm{F}_1 \circ \gamma = \mathrm{F}_1 \circ (\gamma_1, \gamma_2); \quad \nabla^2_{\dot\gamma_2}\dot\gamma_2 = \mathrm{F}_2 \circ \gamma = \mathrm{F}_1 \circ (\gamma_1, \gamma_2).$$

□

The meaning of this equivalence is the following: if $\gamma \colon I \subset \mathbb{R} \longrightarrow Q$ is a solution to the dynamical equation of $(Q, \mathbf{g}, \mathrm{F})$ and $\gamma = (\gamma_1, \gamma_2)$, then $\gamma_\mu \colon \subset \mathbb{R} \longrightarrow Q_\mu$, $(\mu = 1, 2)$ is a solution to the coupled dynamical equations of both systems $(Q_\mu, g_\mu, \mathrm{F}_\mu)$ and conversely.

Remark 5.6. When studying a physical complex system, we typically begin by considering the two component systems as independent, that is, without interaction, (Q_1, g_1, F_1), (Q_2, g_2, F_2), and in a second approach, we introduce the interaction $F^{int} = (F_1^{int}, F_2^{int})$, that is, the force fields depending on the positions of both systems. Then, we obtain the system $(Q_1 \times Q_2, g_1 \oplus g_2, F_1 + F_1^{int}, F_2 + F_2^{int})$.

Corollary 5.1. *A collection of n coupled Newtonian mechanical systems (Q_μ, g_μ, F_μ), $\mu = 1, \ldots n$, is "equivalent" to a unique Newtonian mechanical system (Q, g, F), where*

(1) $Q = Q_1 \times \ldots \times Q_N$; with $\pi_\mu : Q \longrightarrow Q_\mu$.
(2) $g = g_1 \oplus \ldots g_N$.
(3) $\omega = (\omega_1, \ldots, \omega_N)$ and $F = (F_1, \ldots, F_N)$.

(Observe that we have identified $\mathfrak{X}(M)$ with $\displaystyle\prod_{\mu=1}^{N} \mathfrak{X}(M_\mu, \pi_\mu)$, and $\Omega^1(Q)$ with $\displaystyle\prod_{\mu=1}^{N} \Omega^1(M_\mu, \pi_\mu)$ in agreement with the above lemma.)

The system (Q, g, F) is a Newtonian mechanical system composed of different systems with interaction.

Local coordinate expressions: To write the dynamical equation in a local chart, consider the curve $\gamma \equiv (\gamma_1, \ldots, \gamma_n)$ (where γ_μ is the component of γ on the manifold Q_μ). The dynamical equation is written as

$$\nabla^\mu_{\dot\gamma_\mu} \dot\gamma_\mu = F_\mu \circ \gamma.$$

To obtain the local expression of ∇, let (U_μ, q^i_μ) be local charts on Q_μ. We have (U, q^j_μ), with $U = U_1 \times \ldots \times U_N$ as a local chart on Q, and if $\{(\Gamma_\mu)^i_{jk}\}$ are the Christoffel symbols of ∇_μ, we have

$$\nabla_{\frac{\partial}{\partial q^j_\mu}} \frac{\partial}{\partial q^k_\mu} = (\Gamma_\mu)^i_{jk} \frac{\partial}{\partial q^i_\mu}, \quad \nabla_{\frac{\partial}{\partial q^j_\mu}} \frac{\partial}{\partial q^k_\nu} = 0 \quad (\mu \neq \nu),$$

because ∇ is the Levi–Civita connection. From these expressions, we have

$$\nabla_{\dot\gamma} \dot\gamma = \sum_{\mu=1}^{N} \left(\ddot\gamma^j_\mu \frac{\partial}{\partial q^j_\mu} + (\Gamma_\mu)^i_{jk} \dot\gamma^j_\mu \dot\gamma^k_\mu \frac{\partial}{\partial q^i_\mu} \right) = F \circ \gamma$$

and, for every $\mu = 1, \ldots, N$,

$$\ddot\gamma^i_\mu \frac{\partial}{\partial q^i_\mu} + (\Gamma_\mu)^i_{jk} \dot\gamma^j_\mu \dot\gamma^k_\mu \frac{\partial}{\partial q^i_\mu} = F^i_\mu \circ \gamma.$$

Hence, if $\dim Q_\mu = N_\mu$, for every $\mu = 1, \ldots, N$ and every $i = 1, \ldots, N_\mu$, we have

$$\ddot{\gamma}^i_\mu + (\Gamma_\mu)^i_{jk} \dot{\gamma}^j_\mu \dot{\gamma}^k_\mu = \mathrm{F}^i_\mu \circ \gamma.$$

To write the Euler–Lagrange equations of the system, we need to obtain the associated natural chart $(\tau_\mu^{-1}(U), q^j_\mu, v^j_\mu)$ in TQ, and we have

$$\frac{d}{dt}\left(\frac{\partial K}{\partial v^j_\mu} \circ \dot{\gamma}\right) - \frac{\partial K}{\partial q^j_\mu} \circ \dot{\gamma} = (g_\mu)_{jk} \mathrm{F}^k_\mu,$$

where the kinetic energy function is $K = \sum\limits_{\mu=1}^{N} K_\mu$, because $\mathbf{g} = \mathbf{g}_1 \oplus \ldots \oplus \mathbf{g}_N$. Hence, considering

$$\frac{d}{dt}\left(\frac{\partial K_\nu}{\partial v^j_\mu}\right) = 0, \quad \frac{\partial K_\nu}{\partial q^j_\mu} = 0 \quad ; \quad \text{if } \mu \neq \nu,$$

we write

$$\frac{d}{dt}\left(\frac{\partial K_\mu}{\partial v^j_\mu} \circ \dot{\gamma}\right) - \frac{\partial K_\mu}{\partial q^j_\mu} \circ \dot{\gamma} = (g_\mu)_{jk} \mathrm{F}^k_\mu.$$

Moreover, if the system is conservative, then $\omega = -dV$ (with $V \in C^\infty(Q)$), that is,

$$\omega = -\frac{\partial V}{\partial q^j_\mu} dq^j_\mu.$$

If we consider the Lagrangian function of the system $\mathcal{L} = K - V$, with local expression

$$\mathcal{L} = \frac{1}{2} \sum_{\mu=1}^{N} (g_\mu)_{jk} v^j_\mu v^k_\mu - V = \sum_{\mu=1}^{N} K_\mu - V,$$

we have

$$\frac{d}{dt}\left(\frac{\partial \mathcal{L}}{\partial v^j_\mu} \circ \dot{\gamma}\right) - \frac{\partial \mathcal{L}}{\partial q^j_\mu} \circ \dot{\gamma} = 0.$$

Finally, if the force fields depend on the velocities, the expression would be the same writing $\mathrm{F} \circ \dot{\gamma}$ instead of $\mathrm{F} \circ \gamma$ in every equation. In this case, $\mathrm{F}_\mu \in \mathfrak{X}(Q_\mu, \pi_\mu \circ \tau_\mu)$, where $\tau_\mu : TQ_\mu \longrightarrow Q_\mu$ are the natural projections.

5.3 Systems with Holonomic and Nonholonomic Constraints

Another relevant topic is the study of dynamical systems in classical mechanics with holonomic or nonholonomic constraints. Nonholonomic systems are those subjected to constraints depending on positions and velocities, and they have been thoroughly studied (see, for instance, [11, 34, 100, 127, 131, 159, 165, 248, 273, 282, 344]).

5.3.1 *Holonomic constraints: Holonomic d'Alembert's principle*

Let $(Q, \mathbf{g}, \omega)$ be a Newtonian mechanical system and $F \in \mathfrak{X}(Q)$ its force field. Let S be a submanifold of Q (usually called *submanifold of holonomic constraints*) and $j_S \colon S \hookrightarrow Q$ the natural embedding. The problem consists of describing the dynamics of the system when it is forced to evolve on the submanifold S. To force this behaviour, we must apply a new force field R, called *constraint force*, which obliges the system to remain on S. In general, this force depends not only on the position but also on the velocity; therefore, $R \in \mathfrak{X}(Q, \tau_Q)$, and moreover, it is a new unknown to find.

As a result we have a dynamical equation for curves $\gamma \colon I \subset \mathbb{R} \longrightarrow S$, which is

$$\nabla_{\dot{\gamma}} \dot{\gamma} = F \circ \gamma + R \circ \dot{\gamma}. \tag{5.8}$$

To solve this problem, we introduce the following:

> **Assumption 5.1 (d'Alembert's principle).** The constraint force R is orthogonal to the submanifold S, that is, for every $q \in S$ and for every $u, v \in T_q S$, we have $\mathbf{g}(u, R(q, v)) = 0$.

The question is how to solve the problem and obtain some information about the constraint force. Let $\mathbf{g}_S := j_S^* \mathbf{g}$. It is clear that $(S, \mathbf{g}_S)$ is a Riemannian manifold. Let ∇^S be the Levi–Civita connection associated with $\mathbf{g}_S$. We have the following natural orthogonal splitting:

$$T_q Q = T_q S \oplus (T_q S)^{\perp} \quad (\forall q \in S \subset Q),$$

and the projections

$$\pi_S(q) \colon T_q Q \to T_q S, \quad \pi_S^{\perp}(q) \colon T_q Q \to (T_q S)^{\perp},$$

and as they are defined at every point, we have

$$\pi_S \colon TQ|_S \to TS, \quad \pi_S^\perp \colon TQ|_S \to TS^\perp.$$

Then d'Alembert's principle reduces to $\pi_S \circ R = 0$. Hence, we have:

Proposition 5.4. $\nabla^S = \pi_S \circ \nabla$.

Proof. It is a direct computation to prove that on the vector fields tangent to S, the map $\pi_S \circ \nabla$ is a connection on S and it is symmetrical. Furthermore, for every $X, Y, Z \in \mathfrak{X}(S)$, we have

$$(\pi_S \circ \nabla_Z) g(X, Y) = g((\pi_S \circ \nabla_Z) X, Y) + g(X, (\pi_S \circ \nabla_Z) Y),$$

and hence $\pi_S \circ \nabla$ is the Levi–Civita connection of g_S. $\square$

Now, if we take the dynamical equation (5.8) and we split it into the tangent and orthogonal components to S, we obtain

$$\pi_S(\nabla_{\dot\gamma}\dot\gamma) = \pi_S \circ F \circ \gamma + \pi_S \circ R \circ \dot\gamma = \pi_S \circ F \circ \gamma, \tag{5.9}$$

$$\pi_S^\perp(\nabla_{\dot\gamma}\dot\gamma) = \pi_S^\perp \circ F \circ \gamma + \pi_S^\perp \circ R \circ \dot\gamma = \pi_S^\perp \circ F \circ \gamma + R \circ \dot\gamma. \tag{5.10}$$

Denoting $F^S := \pi_S \circ F \in \mathfrak{X}(S)$, the projection of F on S, then Eq. (5.9) is

$$\nabla_{\dot\gamma}^S \dot\gamma = F^S \circ \gamma, \tag{5.11}$$

and this is the dynamical equation of the Newtonian mechanical system (S, g_S, ω_S), where $\omega_S = i(F^S) g_S$.

The solutions to Eq. (5.11) are curves $\gamma \colon I \subset \mathbb{R} \longrightarrow S$ such that, if we introduce them into Eq. (5.10), it allows us to calculate the constraint force R for that trajectory, obtaining

$$\nabla_{\dot\gamma}\dot\gamma - \nabla_{\dot\gamma}^S \dot\gamma = F \circ \gamma - F^S \circ \gamma + R \circ \dot\gamma.$$

Then $R \circ \dot\gamma \in \mathfrak{X}(Q, \dot\gamma)$. Observe that we can calculate the constraint force only for each trajectory of the system, but not as a vector field depending on the velocities.

We also have:

Proposition 5.5. $\omega_S = j_S^* \omega$.

Proof. If $q \in S$ and $v \in T_q S$, we have

$$\omega_S(v) = (i(F^S) g_S)(v) = g_S(F^S, v) = g(F^S, v) = g(F, v),$$

$$(j_S^* \omega)(v) = (j_S^* i(F) g)(v) = g(F, v),$$

and the result follows. $\square$

We have obtained that the dynamical system, describing the motion of the systems $(Q, \mathbf{g}, \omega)$ constrained to move on the submanifold S, is the Newtonian mechanical system $(S, j_S^* \mathbf{g}, j_S^* \omega)$.

D'Alembert's principle tells us that the constraint force $R \in \mathfrak{X}(Q, \tau_Q)$, the force that obliges the system to move on the submanifold S, is orthogonal to S, but we can set this principle in a dual way:

Proposition 5.6 (Dual d'Alembert's principle). *Let* $\rho = i(R)\mathbf{g}$. *Then* $j_S^* \rho = 0$.

Proof. Let $q \in S$ and $u, v \in T_q S$, then

$$(j_S^* \rho_{(q,v)})(u) = \mathbf{g}_S(R(q, v), u)) = 0. \qquad \square$$

Consider now the linear momentum form of the constrained system, $\theta_S \colon TS \longrightarrow T^*S$. The equation of motion using θ_S is

$$\nabla_{\dot{\gamma}}^S (\theta_S \circ \dot{\gamma}) = \omega_S \circ \gamma,$$

and, as $\omega_S = j_S^* \omega$, we have

$$\nabla_{\dot{\gamma}}^S (\theta_S \circ \dot{\gamma}) = j_S^* \nabla_{\dot{\gamma}}(\theta \circ \dot{\gamma}).$$

Next, we discuss some particular cases:

5.3.1.1 *Systems with one constraint*

Let $S = \{q \in Q \; ; \; \varphi(q) = 0\}$, with $\varphi \in C^\infty(Q)$ and suppose that $d\varphi(q) \neq 0$, for every $q \in S$. This implies that S is a submanifold of Q. Let $X \in \mathfrak{X}(Q)$ such that $i(X)\mathbf{g} = d\varphi$, then X is orthogonal to S. Hence,

$$\pi_S^\perp(F) = \frac{\mathbf{g}(F, X)}{\mathbf{g}(X, X)} X = \frac{d\varphi(F)}{\|d\varphi\|^2} X,$$

and we have

$$F^S = \pi_S(F) = F - \frac{d\varphi(F)}{\|d\varphi\|^2} X.$$

As a consequence,

$$\omega_S = \omega - \frac{d\varphi(F)}{\|d\varphi\|^2} d\varphi,$$

and this allows us to find the trajectories of the system as solutions to the differential equation

$$\nabla_{\dot{\gamma}}^S \dot{\gamma} = F \circ \gamma - \frac{d\varphi(F)}{\|d\varphi\|^2} X.$$

We have the constraint force along the solution γ given by the following equation:

$$\nabla_{\dot{\gamma}}\dot{\gamma} - \nabla_{\dot{\gamma}}^S \dot{\gamma} = \frac{\mathrm{d}\varphi(\mathrm{F})}{\|\mathrm{d}\varphi\|^2} \circ \gamma - \mathrm{R} \circ \dot{\gamma},$$

where the only unknown is $\mathrm{R} \circ \dot{\gamma}$.

5.3.1.2 *Systems with several constraints*

Consider now $S = \{q \in Q \; ; \; \varphi_1(q) = 0, \ldots, \varphi_h(q) = 0\}$, with $\varphi_1 \ldots, \varphi_h \in C^{\infty}(Q)$, such that $\mathrm{d}\varphi_1(q), \ldots \mathrm{d}\varphi_h(q)$ are linearly independent at every point $q \in S$ (we assume that S is not empty). Let $Z_1, \ldots, Z_{n-h} \in \mathfrak{X}(Q)$ such that:

(1) $i(Z_i)\mathrm{d}\varphi_j = 0.$
(2) $\mathbf{g}(Z_i, Z_j) = 0; \; i \neq j.$

To obtain these vector fields Z_i, it is enough to consider vector fields $X_1, \ldots, X_{n-h} \in \mathfrak{X}(Q)$, satisfying the first condition (a linear equation), and applying the well-known *Gram–Schmidt method*. In this situation, we have

$$\pi_S(\mathrm{F}) = \sum_{i=1}^{m} \frac{\mathbf{g}(\mathrm{F}, Z_i)}{\mathbf{g}(Z_i, Z_i)} Z_i;$$

hence, as in the previous case, we obtain the dynamical equation and the expression of the constraint force along every trajectory.

5.3.2 **Euler–Lagrange equations**

In the above paragraph, we have studied the dynamics of a Newtonian mechanical system $(Q, \mathbf{g}, \omega)$ constrained to move on the submanifold $j_S : S \hookrightarrow Q$. We have proven that its dynamics are given by the Newtonian mechanical system $(S, \mathbf{g}_S, \omega_S)$. To write the corresponding Euler–Lagrange equation of this last system, take a local chart (U, q^i) on S and the corresponding natural lift $(\tau_Q^{-1}(U), q^i, v^i)$ to TS. Then we have

$$\frac{d}{dt}\left(\frac{\partial K_S}{\partial v^k} \circ \dot{\gamma}\right) - \frac{\partial K_S}{\partial q^k} \circ \dot{\gamma} = (\mathbf{g}_S)_{ik}(\mathrm{F}^S)^i, \qquad (5.12)$$

where $K_S \in C^{\infty}(TS)$ is the *kinetic energy* of the system, defined by

$$\begin{aligned} K_S : TS &\longrightarrow \quad \mathbb{R} \\ (q, v) &\longmapsto \tfrac{1}{2}\mathbf{g}_S(v, v) \end{aligned}.$$

Then we have:

Proposition 5.7. $K_S = (\mathrm{T}j_S)^*K$.

Proof. If $q \in S$ and $v \in T_qS$, then

$$K_S(q, v) = \frac{1}{2}(\mathbf{g}_S)_{ik}(q)v^iv^k = \frac{1}{2}g_{ik}(q)v^iv^k,$$

and the result follows. □

Thus, Eq. (5.12) is

$$\frac{d}{dt}\left(\frac{\partial(\mathrm{T}j_S)^*K}{\partial v^k} \circ \dot{\gamma}\right) - \frac{\partial(\mathrm{T}j_S)^*K}{\partial q^k} \circ \dot{\gamma} = (\omega_S)_k \circ \gamma = (j_S^*\omega)_k \circ \gamma.$$

If the dynamical system is conservative, that is, $\omega = -\mathrm{d}V$, then

$$\omega_S = j_S^*\omega = -j_S^*\mathrm{d}V = -\mathrm{d}j_S^*V,$$

and the above equation takes the expression

$$\frac{d}{dt}\left(\frac{\partial(\mathrm{T}j_S)^*K}{\partial v_S^j} \circ \dot{\gamma}\right) - \frac{\partial(\mathrm{T}j_S)^*K}{\partial q^j} \circ \dot{\gamma} = -\frac{\partial j_S^*V}{\partial q^k} \circ \gamma,$$

or equivalently,

$$\frac{d}{dt}\left(\frac{\partial\mathcal{L}_S}{\partial v^j} \circ \dot{\gamma}\right) - \frac{\partial\mathcal{L}_S}{\partial q^j} \circ \dot{\gamma} = 0,$$

where $\mathcal{L}_S := (\mathrm{T}j_S)^*\mathcal{L}$. Observe that the constraint force is not in these equations. This was one of the innovations developed by Lagrange.

If (W, x^i) is a local chart on Q, and (U, q^i) is another on S, both adapted to the map $j\colon W \hookrightarrow U$, then we have the local expression $x^i = f^i(q)$, and hence $\dot{x}^i = \dfrac{\partial f^i}{\partial q^j}\dot{q}^j$. This shows that it is enough to know the Lagrangian function $\mathcal{L}$ of the unconstrained system to introduce the expression of $x^i, \dot{x}^i$ using the above local expressions and by direct derivation, to obtain the Euler–Lagrange equations of the constrained system.

5.3.3 *Some examples of Newtonian mechanical systems*

In the following examples, the question is to identify the elements defining the different systems, that is, the *configuration manifold*, the *Riemannian metric*, and the *force field* or the *work form*.

5.3.3.1 *Unconstrained particle in* $\mathbb{R}^3$

Consider a particle with mass m moving on an open set $Q \subset \mathbb{R}^3$ and subjected to a force defined by a vector field $F \in \mathfrak{X}(Q)$.

The *geometric metric* $\mathbf{g}$ is the original metric on $\mathbb{R}^3$. Consider, instead, the Riemannian metric given by $\widetilde{\mathbf{g}} := m\mathbf{g}$. Observe that the corresponding Levi–Civita connections ∇ and $\widetilde{\nabla}$ for both metrics $\mathbf{g}$ y $\widetilde{\mathbf{g}}$ are the same: it is enough to calculate their Christoffel symbols in any local chart, or recall the calculus of $\nabla_X Y$ for the Levi–Civita connection. Let $\widetilde{F} := \dfrac{F}{m} \in \mathfrak{X}(Q)$, and consider the Newtonian mechanical system $(Q, \widetilde{\mathbf{g}}, \widetilde{F})$. The Newton equation is given as

$$\nabla_{\dot\gamma}\dot\gamma = \widetilde{F}\circ\gamma = \frac{F}{m}\circ\gamma,$$

which, as it is usual in mechanics, can be written as

$$m\nabla_{\dot\gamma}\dot\gamma = F\circ\gamma.$$

Observe that the work form is given by

$$\omega = i(\widetilde{F})\widetilde{\mathbf{g}} = i(F)\mathbf{g},$$

and the linear momentum form is

$$\widetilde{\theta}\circ\dot\gamma = i(\dot\gamma)\widetilde{\mathbf{g}},$$

whose local expression is

$$\widetilde{\theta}\circ\dot\gamma = mg_{ij}\dot\gamma^i dq^j \circ\gamma.$$

5.3.3.2 *Particle constrained to a surface of* $\mathbb{R}^3$

Consider now a particle with mass m on an open set $Q \subset \mathbb{R}^3$. We have seen above that the Newtonian system describing this situation is given by $(Q, \widetilde{\mathbf{g}}, \widetilde{F})$. If $j_S \colon S \hookrightarrow Q$ is a regular surface and the particle is constrained to move on it, then the associated Newtonian system is $(S, \widetilde{\mathbf{g}}_S, \widetilde{F}^S)$, where $\widetilde{\mathbf{g}}_S = j_S^* \widetilde{\mathbf{g}}$ and $\widetilde{F}^S = \pi_S \circ \widetilde{F}$. Hence, the dynamical equation is

$$\nabla_{\dot\gamma}^S \dot\gamma = \widetilde{F}^S \circ\gamma.$$

If we know a solution γ, then the constraint force along γ is given by the equation

$$\nabla_{\dot\gamma}\dot\gamma - \nabla_{\dot\gamma}^S\dot\gamma = \widetilde{F}\circ\gamma - \widetilde{F}^S\circ\gamma + \widetilde{R}\circ\dot\gamma,$$

from where we can obtain $\widetilde{R}\circ\dot\gamma$. Recall that $R = m\widetilde{R}$.

We can write the dual formulation using the work from

$$\widetilde{\omega} = i(F)\mathbf{g} = i(\widetilde{F})\widetilde{\mathbf{g}}.$$

with $\widetilde{\omega}_S = j_S^*\widetilde{\omega}$. Hence, we have the dynamical equation being

$$\nabla_{\dot\gamma}^S(\widetilde{\theta}_S\circ\dot\gamma) = \widetilde{\omega}_S\circ\gamma,$$

with $\widetilde{\theta}_S \circ \dot\gamma = i(\dot\gamma)\widetilde{\mathbf{g}}_S$, and $\widetilde{\theta}_S = (Tj_S)^*\widetilde{\theta}$.

5.3.3.3 *Systems of particles in $\mathbb{R}^3$*

Consider now a system made of N particles, denoted by $P_1, \ldots, P_N$, with masses $m_1, \ldots, m_N$. We suppose that the particle P_μ moves in Q_μ, open set of $\mathbb{R}^3$, equipped with the usual metric g_μ and the associated dynamical metric $\widetilde{g}_\mu := m_\mu g_\mu$.

If the force field acting on every particle is $F_\mu \in \mathfrak{X}(Q_\mu)$, that is, F_μ depends only on the position of the corresponding particle; then we have N uncoupled mechanical systems, and their dynamical equation can be solved separately.

However, if $\pi_\mu \colon \prod_{\nu=1}^N Q_\nu \longrightarrow Q_\mu$, and the forces on the system are a family of vector fields $F_\mu \in \mathfrak{X}(Q_\mu, \pi_\mu)$, that is, the force F_μ on the particle P_μ depends on the position of the other particles, then the N particles are interact, and as we have shown in Section 5.2.4, the N systems $(Q_\mu, \widetilde{g}_\mu, \widetilde{F}_\mu)$ are equivalent to a unique Newtonian mechanical system $(Q, \widetilde{g}, \widetilde{F})$ with

(1) $Q = \displaystyle\prod_{\mu=1}^N Q_\mu$,

(2) $\widetilde{g} = \oplus_{\mu=1}^N \widetilde{g}_\mu$,

(3) $\omega = (\omega_1, \ldots, \omega_N)$ and $\widetilde{F} = (\widetilde{F}_1, \ldots, \widetilde{F}_N)$,

and the dynamical equation is corresponds to this last system.

5.3.3.4 *Systems of particles on a submanifold*

In this case, we have a system of N particles $P_1, \ldots, P_N$, with masses $m_1, \ldots, m_N$, moving on open sets $Q_1, \ldots, Q_N$ in $\mathbb{R}^3$. We suppose that each one of these sets is equipped with the corresponding metric g_μ. If F is the force field acting on the system, then the system is given by $(Q, \widetilde{g}, \widetilde{F})$, as we have seen in the above paragraph. Observe that it is the same situation if the particles interact or not.

If the dynamics is constrained to a submanifold $j_S \colon S \hookrightarrow Q$, then there exists a constraint force R, and the dynamical equation is

$$\nabla_{\dot\gamma}\dot\gamma = \widetilde{F} \circ \gamma + \widetilde{R} \circ \dot\gamma,$$

for curves $\gamma \colon I \subset \mathbb{R} \longrightarrow S$, where $\widetilde{R} = (\widetilde{R}_1, \ldots, \widetilde{R}_N)$ and $\widetilde{R}_\mu = m_\mu \widetilde{R}_\mu$, $\mu = 1, \ldots, N$.

To solve this equation, assuming *d'Alembert's principle*, that is, that $\widetilde{R}$ is $\widetilde{g}$-orthogonal to S, we need to decompose the equation projecting onto S and on its orthogonal complement at every point, as we did in the general study (see Section 5.3). Then we solve the S component and use the orthogonal one to compute the constraint force along every solution.

5.3.4 Nonholonomic constraints: Nonholonomic d'Alembert's principle

Let $(Q, \mathbf{g}, \omega)$ be a Newtonian mechanical system and $\mathrm{F} \in \mathfrak{X}(Q)$ be the force field. Let C be a submanifold of TQ, such that $\tau_Q(C) = Q$. In this situation, C is called the *submanifold of nonholonomic constraints*, and $j_C \colon C \hookrightarrow TQ$ is the natural embedding. We want to describe the dynamics of the system when it is constrained to evolve on the submanifold C. The system is given by $(Q, \mathbf{g}, F, C)$.

To solve this problem, we suppose that there exists a *constraint force* R usually depending on the velocities, that is $\mathrm{R} \in \mathfrak{X}(Q, \tau_Q)$, which forces the system to move on C and it is unknown. Then the Newton dynamical equation is given, in this case, for curves $\gamma \colon I \subset \mathbb{R} \longrightarrow Q$ satisfying:

(1) $\dot{\gamma}(t) \in C$, $t \in I$.
(2) $\nabla_{\dot{\gamma}} \dot{\gamma} = \mathrm{F} \circ \gamma + \mathrm{R} \circ \dot{\gamma}$.

Thus, we need to state the conditions allowing us to find the trajectories of the system and calculate R [1].

Let $(q, v) \in C$. The condition assumed on C, $\tau_Q(C) = Q$, tells us that the dimension of the subspace of $V_{(q,v)}(TQ)$, which is tangent to C, does not depend on the point (q, v). Let

$$T^V_{(q,v)}C = V_{(q,v)}(TQ) \cap T_{(q,v)}C = \{w \in V_{(q,v)}(TQ) \; ; \; w \in T_{(q,v)}C\}$$

be the vertical subspace tangent to C. This is a vector subbundle of TQ and we can write $T^V C = V(TQ)|_C \cap TC$ as vector bundles on the manifold C. Consider the vertical lift from the point $q \in Q$ to (q, v) given by

$$\lambda_q^{(q,v)} \colon T_q Q \longrightarrow V_{(q,v)}(TQ),$$

defined as

$$\lambda_q^{(q,v)}(u_q) : \phi \mapsto \lim_{t \to 0} \frac{\phi(q, v + tu) - \phi(q, v)}{t},$$

that is, the directional derivative of ϕ along u_q at the point $(q, v) \in TQ$. As $\lambda_q^{(q,v)}$ is an isomorphism from $T_q Q$ to $T_{(q,v)}(TQ)$, let $(T^V_{(q,v)}C)_q$ be the inverse image of $T^V_{(q,v)}C \subset T_{(q,v)}(TQ)$ by $\lambda_q^{(q,v)}$. Then, $T_q Q = (T^V_{(q,v)}C)_q \oplus (T^V_{(q,v)}C)_q^\perp$, being an orthogonal decomposition with respect to g.

Then, we state:

---

[1]In [326], *Arnold Sommerfeld* says that this force R is a "geometric force", versus F that is an "applied force".

Assumption 5.2 (Nonholonomic d'Alembert's principle). The constraint force $R \in \mathfrak{X}(Q, \tau_Q)$ satisfies

$$R(q,v) \in (T^V_{(q,v)}C)^\perp_q,$$

that is, $g(R(q,v), w) = 0$, for every $w \in (T^V_{(q,v)}C)_q$.

In the classical physics literature, the elements in $(T^V_{(q,v)}C)_q$ are called *virtual velocities*.

Remark 5.7. If there are no constraints, that is, $C = TQ$, then for every $(q,v) \in C$ we have $T^V_{(q,v)}C = V_{(q,v)}(TQ)$; hence $(T^V_{(q,v)}C)_q = T_qQ$ and $(T^V_{(q,v)}C)^\perp_q = \{0\}$, that is, $R(q,v) = 0$, and there is no constraint force.

This principle allows us to obtain the dynamical equations of the trajectories of the system and the constraint force along every trajectory, as we will see in the sequel.

To obtain this, we need to characterise the subspace $(T^V_{(q,v)}C)_q$ in relation with the constraints, that is the functions vanishing on the submanifold C. First, let $\phi \in C^\infty(TQ)$, and consider the 1-form $d^V\phi \in \Omega^1(Q, \tau_Q)$ defined by

$$(d^V\phi(q,v))(u) = d\phi(\lambda^{(q,v)}_q(u)), \quad (q,v) \in TQ, \quad u \in T_qQ;$$

whose expression in a local natural chart (q^i, v^i) of TQ can be calculated directly applying it to $\dfrac{\partial}{\partial q^i}$, and we obtain $d^V\phi = \dfrac{\partial\phi}{\partial v^i}dq^i$. We have the following result:

Proposition 5.8. *Let $(q,v) \in C$.*

(1) If $w \in T_qQ$, then $w \in (T^V_{(q,v)}C)_q$ if, and only if, $(d^V\phi(q,v))(w) = 0$, for every $\phi \in C^\infty(TQ)$ such that $j^*_C\phi = 0$.

(2) Let $((T^V_{(q,v)}C)_q)^\circ = \{\alpha \in T^*_q; \alpha(w) = 0, \ w \in (T^V_{(q,v)}C)_q\} \subset T^*_qQ$ be the annihilator of $(T^V_{(q,v)}C)_q$, then

$$((T^V_{(q,v)}C)_q)^\circ = \{d^V\phi(q,v); \forall \phi \in C^\infty(TQ), j^*_C\phi = 0\}.$$

(3) If $w \in T_qQ$, then $w \in (T^V_{(q,v)}C)_q$ if, and only if, $i(w)g \in ((T^V_{(q,v)}C)_q)^\circ$.

Proof. (1) Let $w \in T_q Q$, then we have

$$w \in (T^V_{(q,v)}C)_q \iff \lambda_q^{(q,v)}(w) \in T^V_{(q,v)}C \iff$$

$$\lambda_q^{(q,v)}(w)(\phi) = 0, \forall \phi \in C^\infty(TQ), \text{such that } j^*_C \phi = 0 \iff$$

$$d\phi(\lambda_q^{(q,v)}(w)) = 0, \forall \phi \in C^\infty(TQ), \text{such that } j^*_C \phi = 0 \iff$$

$$(d^V \phi(q,v))(w) = 0, \forall \phi \in C^\infty(TQ), \text{such that } j^*_C \phi = 0.$$

(2) It is a consequence of the previous item.

(3) It is a direct consequence of the definitions. $\qquad\qquad\square$

Corollary 5.2. *Let* $R \in \mathfrak{X}(Q, \tau_Q)$ *and* $(q,v) \in C$. *Therefore* $R(q,v) \in (T^V_{(q,v)}C)^\perp_p$ *if, and only if,* $i(R(q,v))g \in ((T^V_{(q,v)}C)_q)^o$.

Usually, the submanifold C is given by the vanishing of a finite family of constraint functions defined on TQ. We want to characterise $((T^V_{(q,v)}C)_q)^o$ using these constraints. Then, suppose that the submanifold C is defined by the vanishing of r functions $\{\phi^i\}$, with $r < n = \dim Q$, satisfying the condition rank $\left(\dfrac{\partial \phi^1, \ldots, \phi^r}{\partial v^1, \ldots, v^n}\right) = r$. Then, $\dim C = 2n - r$. We have:

Proposition 5.9. *Let* $(q,v) \in C$, *then*

(1) $\dim T^V_{(q,v)}C = r$.

(2) $(T^V_{(q,v)}C)_q = \{w \in T_q Q; (d^V \phi^i(q,v))(w) = 0, i = 1, \ldots, r\}$.

(3) *If* $\alpha \in T^*_q Q$ *satisfies* $\alpha|_{(T^V_{(q,v)}C)_q} = 0$, *then* α *is a linear combination of the elements* $d^V \phi^1(q,v), \ldots, d^V \phi^r(q,v)$, *that is, the vector space* $((T^V_{(q,v)}C)_q)^o$ *is generated by* $\{d^V \phi^1(q,v), \ldots, d^V \phi^r(q,v)\}$.

Proof. Let (q^i, v^i) be a natural coordinate system on TQ.

(1) The assumed condition rank $\left(\dfrac{\partial \phi^1, \ldots, \phi^r}{\partial v^1, \ldots, v^n}\right) = r$ implies that up to a change of order in the coordinates $q^1, \ldots, q^n$, we can suppose that

$$\det \left(\frac{\partial \phi^1, \ldots, \phi^r}{\partial v^1, \ldots, v^r}\right) \neq 0.$$

Then $(q^1, \ldots, q^n, \phi^1, \ldots, \phi^r, v^{r+1}, \ldots, v^n)$ is a local coordinate system of TQ by the inverse function theorem. The vector space $V_{q,v}(TQ)$ is generated by

$$\left\{\frac{\partial}{\partial\phi^1},\ldots,\frac{\partial}{\partial\phi^r},\frac{\partial}{\partial v^{r+1}},\ldots,\frac{\partial}{\partial v^n}\right\}_{(q,v)},$$

and the subspace $T^V_{(q,v)}C \subset V_{q,v}(TQ)$ is generated by

$$\left\{\frac{\partial}{\partial\phi^1},\ldots,\frac{\partial}{\partial\phi^r}\right\}_{(q,v)}.$$

(2) The inclusion part is proven in the first item of Proposition 5.8, and the equality follows from a dimensional analysis.

(3) The previous items imply that $\{d^V\phi^1(q,v),\ldots,d^V\phi^r(q,v)\}$ is a basis of $((T^V_{(q,v)}C)_q)^\circ$. $\qquad\square$

And as a straightforward consequence we obtain:

Proposition 5.10. *For every* $(q,v) \in C$, *the form* $\eta \in \Omega^1(Q,\tau_Q)$ *satisfies the condition* $j^*_C\eta|_{(T^V_{(q,v)}C)_q} = 0$ *if, and only if, there exists* $f_1,\ldots,f_r \in C^\infty(TQ)$ *such that* $\eta = f_i d^V\phi^i$.

In the case that the constraints only locally define the submanifold C, then the above results are valid only in the corresponding open set.

The last proposition allows us to state the *Dual d'Alembert's nonholonomic principle*, which states that "the work form $i(R)g$ corresponding to the constraint force R annihilates the virtual velocities of the system."

As a final result, we have:

Corollary 5.3. *If* R *is the nonholonomic constraint force, then there exists* $f^1,\ldots,f^r \in C^\infty(TQ)$ *such that*

$$i(R)g = f_i d^V\phi^i = f_i\frac{\partial\phi^i}{\partial v^j}dq^j,$$

and as a consequence,

$$R = f_i\frac{\partial\phi^i}{\partial v^j}g^{jk}\frac{\partial}{\partial q^k}.$$

Definition 5.20. The functions $f^1,\ldots,f^r$ are called **Lagrange multipliers** of the nonholonomic system.

Particular case: The submanifold $C \subset \mathrm{T}Q$ is a linear subbundle of $\mathrm{T}Q$.

(1) In this case, C is defined by the vanishing of a family of differential forms, that is, we have $\omega^i \in \Omega^1(Q)$, $i = 1, \ldots, r$, linearly independent at every point of Q, and

$$C = \{(q, v) \in \mathrm{T}Q; \; \omega_q^i(v) = 0, i = 1, \ldots, r\}.$$

In local coordinates, if $\omega^i = a_j^i(q)\mathrm{d}q^j$, then $\phi^i = a_j^i(q)v^j$, that is, the constraints are linear in the velocities, and the expression of the constraint force is

$$R = f_i a_j^i g^{jk} \frac{\partial}{\partial q^k}.$$

(2) Alternatively, we can suppose that the subbundle C is given as a regular distribution $\mathcal{D}$, the distribution annihilated by $\{\omega^i, i = 1, \ldots, r\}$. If $(q, v) \in \mathcal{D}$, by linearity, we have $\mathrm{T}_{(q,v)}^V C = \lambda_q^{(q,v)}(\mathcal{D}_q)$, hence $(\mathrm{T}_{(q,v)}^V C)_q = \mathcal{D}_q$ and $(\mathrm{T}_{(q,v)}^V C)_q^\perp = \mathcal{D}_q^\perp$, then the constraint force R is orthogonal to $\mathcal{D}$.

(3) If the distribution $\mathcal{D}$ is integrable and $(q, v) \in \mathcal{D}$ is the initial condition of the dynamical equation for the solution γ, then the image of γ is contained in the integral submanifold of $\mathcal{D}$ passing through the point $q \in Q$, because $\dot{\gamma}(t) \in \mathcal{D}_{\gamma(t)}$ for every t. The constraint force R, orthogonal to $\mathcal{D}$, forces the system to move on the integral submanifolds of the constraint distribution $\mathcal{D}$.

Remark 5.8.

(1) We can understand the solution as follows: if we have a constraint $\phi : \mathrm{T}Q \longrightarrow \mathbb{R}$, there is an associated 1-form, $\mathrm{d}^V\phi$, which gives a "constraint force" R^ϕ defined by $i(R^\phi)\mathbf{g} = \mathrm{d}^V\phi$. If we have r independent constraints $\{\phi^i\}$, we have the corresponding constraint forces, R^{ϕ^i}, and the subbundle generated by them, $\{R^{\phi^i}\}$, and then the resulting constraint force R is on this subbundle.

(2) For these systems, d'Alembert's principle states that if the system moves "along the vertical fibres" of C, the work realised by the constraint force along the trajectory is null. As we have said, these vertical velocities are called "virtual" because it is not possible to move the system with these velocities on the constraint manifold C.

5.3.5 Dynamical equations

Following the above results, if C is locally defined by the annihilation of r functions $\{\phi^i\}$, and they are independent constraints, then the dynamical equations are

$$\nabla_{\dot\gamma}\dot\gamma = \mathrm{F}\circ\gamma + f_j\,{}_i(\mathrm{d}^V\phi^j)g^{-1}\circ\dot\gamma,$$

or, in dual form,

$$\nabla_{\dot\gamma}(\theta\circ\dot\gamma) = \omega\circ\gamma + f_j\mathrm{d}^V\phi^j\circ\dot\gamma.$$

These equations together with the constraints defining the submanifold C, $\phi^1 = 0,\dots,\phi^r = 0$, are a system of $n+r$ equations with $n+r$ unknowns: the components of the trajectory γ and the multipliers f_i. Observe that some of them are differential equations and the remaining ones are the constraint functions.

The corresponding Euler–Lagrange equations are

$$\frac{d}{dt}\left(\frac{\partial K}{\partial v^j}\circ\dot\gamma\right) - \frac{\partial K}{\partial q^j}\circ\dot\gamma = \omega_j\circ\gamma + f_k\frac{\partial\phi^k}{\partial v^j}\circ\dot\gamma, \quad (j=1,\dots,n),$$

because $\omega = \omega_k\mathrm{d}q^k$, with $\omega_k = g_{kj}\mathrm{F}^j$.

If the system is conservative, then $\omega = -\mathrm{d}V$, where $V \in \mathrm{C}^\infty(Q)$ is the potential function. In this case, we can introduce the Lagrangian function $\mathcal{L} := K - \tau_Q^* V$, and we have

$$\frac{d}{dt}\left(\frac{\partial\mathcal{L}}{\partial v^j}\circ\dot\gamma\right) - \frac{\partial\mathcal{L}}{\partial q^j}\circ\dot\gamma = f_k\frac{\partial\phi^k}{\partial v^j}\circ\dot\gamma.$$

In any case, these equations, together with the constraints, defining C, are also a system of $n + r$ equations with $n + r$ unknowns.

If C is a vector subbundle, then $\phi^i = a_j^i(q)v^j$ and the Euler–Lagrange equations are

$$\frac{d}{dt}\left(\frac{\partial K}{\partial v^j}\circ\dot\gamma\right) - \frac{\partial K}{\partial q^j}\circ\dot\gamma = \omega_j\circ\gamma + f_k a_j^k\circ\dot\gamma, \quad (j=1,\dots,n).$$

In the case of conservative systems,

$$\frac{d}{dt}\left(\frac{\partial\mathcal{L}}{\partial v^j}\circ\dot\gamma\right) - \frac{\partial\mathcal{L}}{\partial q^j}\circ\dot\gamma = f_k a_j^k\circ\dot\gamma.$$

In summary, the fundamental data in the Lagrangian formulation of the autonomous dynamical systems are the following:

- A Newtonian mechanical system (independent of time) is a triple $(Q, \mathbf{g}, \omega)$, where $(Q, \mathbf{g})$ is a Riemannian manifold where the motion happens, which is called the *configuration space* of the system. Its points represent the *positions* of the particles of the system, and the form ω, or equivalently, the vector field F, are geometric elements carrying the dynamical information.
- The dynamical trajectories, which are the solutions to Newton equations, are curves on TQ, which are canonical lifts of curves on Q. The points of the manifold TQ are the possible initial conditions for these equations, and hence, they represent the possible *position* and *velocities* of the system and are the *physical states* of the system. Then, TQ is said to be the *state* or *phase space* (of *velocities*) of the system.
- In the particular case of conservative systems, the geometric element representing the dynamics is substituted by the *(mechanical) Lagrangian function* $\mathcal{L} \in C^\infty(TQ)$, and the dynamical equations are the *Euler–Lagrange equations (of the first kind)*.

Finally, if the degrees of freedom of the system have some kind of restriction, we have two different situations:

(1) If the dynamics takes place on some submanifold $j_S \colon S \hookrightarrow Q$, the situation is the same, but considering $(S, j_S^* \mathbf{g}, j_S^* \omega)$ as a Newtonian system. In this case, observe that the configuration space is S and the corresponding phase space (of velocities) is TS.
(2) If the configuration space is not restricted but the velocities are, then we have a nonholonomic system, and we need to introduce new unknowns, the Lagrange multipliers, to obtain the corresponding dynamical equations.

5.4   Nonautonomous Newtonian Systems

In some interesting cases, the force field acting on a Newtonian system depends not only on the positions and the velocities, but also on time. In the following paragraphs, we extend our geometric formulation to this situation.

5.4.1   *Mechanical systems with time-dependent forces*

The geometric model appropriate to this case is the following:

Definition 5.21. A **nonautonomous Newtonian mechanical system** is a triple $(\mathbb{R} \times Q, \mathbf{g}, \mathbf{F})$, where $(Q, \mathbf{g})$ is a Riemannian manifold and the force field is of the form $\mathbf{F} \in \mathfrak{X}(Q, \pi_2)$ (with $\pi_2 \colon \mathbb{R} \times Q \longrightarrow Q$), that is,

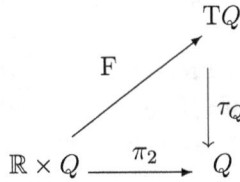

$$
\begin{array}{ccc}
 & & TQ \\
 & \overset{\mathbf{F}}{\nearrow} & \big\downarrow {\scriptstyle \tau_Q} \\
\mathbb{R} \times Q & \xrightarrow{\ \pi_2\ } & Q
\end{array}
$$

Moreover, if the force field depends on the velocities, then $\mathbf{F} \in \mathfrak{X}(Q, \tau_Q \circ \rho_2)$, with $\rho_2 \colon \mathbb{R} \times TQ \longrightarrow TQ$, that is,

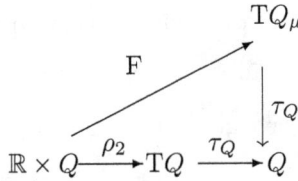

$$
\begin{array}{ccc}
 & & TQ_\mu \\
 & \overset{\mathbf{F}}{\nearrow} & \big\downarrow {\scriptstyle \tau_Q} \\
\mathbb{R} \times Q & \xrightarrow{\ \rho_2\ } TQ \xrightarrow{\ \tau_Q\ } & Q
\end{array}
$$

The Newton equations are written in the usual way:

- In the case where the force does not depend on the velocities, we have

$$
\nabla_{\dot{\gamma}} \dot{\gamma} = \mathbf{F} \circ \bar{\gamma},
$$

where $\bar{\gamma} = (t, \gamma) \colon I \subset \mathbb{R} \longrightarrow I \times Q$. We can also use the dual form by means of the corresponding work form $\omega \in \Omega^1(Q, \pi_2)$.

- If the force field depends on the velocities, then

$$
\nabla_{\dot{\gamma}} \dot{\gamma} = \mathbf{F} \circ \bar{\dot{\gamma}},
$$

where $\bar{\dot{\gamma}} = (t, \dot{\gamma}) \colon I \subset \mathbb{R} \longrightarrow I \times TQ$. Based on the above, we can use the corresponding work form $\omega \in \Omega^1(Q, \tau_Q \circ \rho_2)$ and obtain the equations in the dual form.

The Euler–Lagrange equations are the same as usual, but the second term depends on time $t \in \mathbb{R}$. In particular, if the work form depends on time, $\omega \in \Omega^1(Q, \pi_2)$, we say that the system is **conservative** if there exists $V \colon \mathbb{R} \times Q \longrightarrow \mathbb{R}$, such that $\omega = -dV_t$, where $V_t \colon Q \longrightarrow Q$ is defined by $V_t(p) := V(t, p)$, for every $p \in Q$ and $t \in \mathbb{R}$. In this situation, we can define the Lagrangian function $\mathcal{L} := K - V$, depending on time, and the Euler–Lagrange equation is as usual

$$
\frac{d}{dt}\left(\frac{\partial \mathcal{L}}{\partial v^i} \circ \bar{\dot{\gamma}} \right) - \frac{\partial \mathcal{L}}{\partial q^i} \circ \bar{\dot{\gamma}} = 0.
$$

Time-dependent Newtonian systems with holonomic and nonholonomic constraints can be studied in the same way, as we see in the next section.

5.4.2 *Time-dependent holonomic and nonholonomic constraints*

Definition 5.22. A **nonautonomous holonomic Newtonian mechanical system** is a triple $(Q, \mathbf{g}, \omega)$, where $(Q, \mathbf{g})$ is a Riemannian manifold, $\omega \in \Omega^1(Q))$, and we have an embedding $j_S \colon S \hookrightarrow \mathbb{R} \times Q$, where S is a submanifold of $\mathbb{R} \times Q$.

In this situation, it is necessary to assume that the constraint force depends not only on positions and velocities but also on time, that is, $\mathrm{R} \in \mathfrak{X}(Q, \tau_Q \circ \rho_2)$. The Newton equation is given as

$$\nabla_{\dot\gamma}\dot\gamma = \mathrm{F} \circ \bar\gamma + \mathrm{R} \circ \bar{\dot\gamma},$$

and d'Alembert's principle can be stated as follows:

Assumption 5.3 (Nonautonomous d'Alembert's principle). The constraint force $\mathrm{R} \in \mathfrak{X}(Q, \tau_Q \circ \rho_2)$ satisfies that, for every $t \in \mathbb{R}$, $p \in S$ and $u, v \in T_q S$,

$$\mathbf{g}(\mathrm{R}(t, (\mathrm{T}_q j_S)(u)), (\mathrm{T}_q j_S)(v)) = 0.$$

As in the autonomous case, this means that $\mathrm{R}(t, (\mathrm{T}_q j_S)(u)) \in (\mathrm{T}_{j_S(q)} S)^{\perp}$, and the results we obtain are the same as in Section 5.3.

Definition 5.23. A **nonautonomous nonholonomic Newtonian mechanical system** is a triple $(Q, \mathbf{g}, \omega)$, where $(Q, \mathbf{g})$ is a Riemannian manifold, $\omega \in \Omega^1(Q))$, and we have an embedding $j_C \colon C \hookrightarrow TQ$, such that $(\tau_Q \circ j_C)(C) = Q$, where C is a differentiable manifold.

Once again, we need to assume that the constraint force also depends on time, that is, $\mathrm{R} \in \mathfrak{X}(Q, \tau_Q \circ \rho_2)$, and this makes the dynamical trajectories of the system, $\gamma \colon I \subset \mathbb{R} \longrightarrow Q$, to satisfy $\dot\gamma(t) \in TC$, for every $t \in I$. Hence, the Newton equation is

$$\nabla_{\dot\gamma}\dot\gamma = \mathrm{F} \circ \bar\gamma + \mathrm{R} \circ \bar{\dot\gamma}.$$

To state the *nonholonomic d'Alembert's principle*, for every $t \in \mathbb{R}$ and $p \in C$, we need to define $(\mathrm{T}_{j_C(\mathrm{p})} C)_{\tau_Q(\mathrm{p})}$, as in the time-independent case (see Section 5.3.4). With this aim, let $\mathrm{p} \in C$ and $j_C(\mathrm{p}) = (q, v) \in j_C(C) \subset$

TQ, we take $T^V_{(q,v)}C$ as the vertical subbundle tangent to C and in the same way $(T^V_{(q,v)}C)_q$ using the vertical lift $\lambda^{(q,v)}_q$. Then:

Assumption 5.4 (Nonautonomous nonholonomic d'Alembert's principle). The constraint force $R \in \mathfrak{X}(Q, \tau_Q \circ \rho_2)$ satisfies

$$g(R(t, (q, v)), u) = 0, \quad u \in (T^V_{(q,v)}C)_q,$$

for every $p \in C$, $j_C(p) = (q, v) \in j_C(C)$, and every $t \in \mathbb{R}$; that is, $R(t, j_C(p)) \in (T_{j_C(p)}C)^{\perp}_{(\tau_Q \circ j_C)(p)}$.

The consequences of this principle on the dynamical equations and the calculus of the constraint force are similar to the autonomous case (see Section 5.3.4), with the only difference that now the Lagrange multipliers f_i are elements of $C^\infty(\mathbb{R} \times TQ)$.

Remark 5.9. Sometimes it is necessary to work with constraints depending on time, called **rheonomic** constraints instead of **scleronomic** constraints or not depending on time (see [318] for details). In this situation, we have a submanifold $B \subset \mathbb{R} \times TQ$ such that corresponding manifolds, B_t for any $t \in \mathbb{R}$, are diffeomorphic. Then d'Alembert's principle is stated at every t for the submanifold B_t.

5.5   Exercises

5.5.1   *Connections in manifolds: Riemannian manifolds*

Exercise 5.1

Let (x^i), $(\bar{x}^\alpha)$ be two local charts in a manifold M with a connection ∇. Let Γ^k_{ij}, $\Gamma^\gamma_{\alpha\beta}$ be the corresponding Christoffel symbols. Calculate the transformation of the Christoffel symbols from the expression of the change of coordinates.

Exercise 5.2

In the manifold $M = \mathbb{R}^2 - \{0\}$, we consider the vector fields X, Y such that for every $p \in M$, $X_p := \dfrac{\mathbf{p}}{\|\mathbf{p}\|}$, $Y_p := \varphi(X_p)$, where φ is the $\pi/2$ rotation in the positive sense. Calculate the torsion of the connection on M in which X and Y are parallel vector fields in any direction. Justify that this connection exists and is unique.

(**Hint:** *Use polar coordinates.*)

Exercise 5.3 (**Covariant differential**).

Let M be a manifold with a connection ∇. For a tensor field $T \in \mathcal{T}^r_s(M)$, the covariant differential of T, denoted $\nabla T \in \mathcal{T}^r_{s+1}(M)$, is defined as:

$$(\nabla T)(X_0, X_1, \ldots, X_s, \omega^1, \ldots, \omega^r) = (\nabla_{X_0} T)(X_1, \ldots, X_s, \omega^1, \ldots, \omega^r),$$

where $X_i \in \mathfrak{X}(M)$, $\omega^j \in \Omega^1(M)$. Prove that ∇T is a tensor field.

Exercise 5.4

Let M be a manifold with a connection ∇. Let X be a vector field on M, and σ an integral curve of X such that $\sigma(0) = p$. We denote by $\tau_t \colon \mathrm{T}_p(M) \longrightarrow \mathrm{T}_{\sigma(t)}(M)$ the parallel transport along σ. Prove that, if Y is another vector field, then

$$(\nabla_X Y)(p) = \lim_{t \to 0} \frac{\tau_t^{-1}(Y(\sigma(t))) - Y(p)}{t},$$

that is, the parallel transport determines the covariant derivative.

(**Hint:** *Use a basis made of parallel vector fields along σ.*)

Exercise 5.5 (**Curvature of a connection**).

Let ∇ be a connection on a manifold M. Given $X, Y, Z \in \mathfrak{X}(M)$, define the new vector field,

$$R(X, Y)Z := \nabla_X \nabla_Y Z - \nabla_Y \nabla_X Z - \nabla_{[X,Y]} Z.$$

(1) Prove that R is a one-contravariant and three-covariant tensor field.
(2) Prove that $R(X, Y)Z$ is skew symmetric in (X, Y).

(3) Assuming that the torsion of the connection vanishes, prove that

$$R(X,Y)Z + R(Z,X)Y + R(Y,Z)X = 0.$$

Exercise 5.6 (Reparameterisation of geodesics).
Let M be a manifold, ∇ a connection on M, and $\gamma: I \subset \mathbb{R} \longrightarrow M$ a geodesic curve.

(1) If $\varphi: J \subset \mathbb{R} \longrightarrow I \subset \mathbb{R}$ is a diffeomorphism, prove that the parameterised curve $\xi = \gamma \circ \varphi: J \subset \mathbb{R} \longrightarrow M$ satisfies the equation $\nabla_{\dot\xi}\dot\xi = f\,\dot\xi$, where $f: J \subset \mathbb{R} \longrightarrow \mathbb{R}$ is a function.
(2) Under which conditions is ξ a geodesic?
(3) Conversely, if $\xi: J \subset \mathbb{R} \longrightarrow M$ is a curve such that $\nabla_{\dot\xi}\dot\xi = f\,\dot\xi$, for some function f, prove that there exists a diffeomorphism $\varphi: J \longrightarrow I$ such that $\gamma = \xi \circ \varphi^{-1}$ is a geodesic.

Exercise 5.7 (Difference of connections).
Given a manifold M and two connections ∇, $\overline\nabla$ on M, we define $D, A, S: \mathfrak{X}(M) \times \mathfrak{X}(M) \longrightarrow \mathfrak{X}(M)$ by:

$$D(X, Y) := \overline\nabla_X Y - \nabla_X Y,$$
$$A(X, Y) := (1/2)\left(D(X, Y) - D(Y, X)\right),$$
$$S(X, Y) := (1/2)\left(D(X, Y) + D(Y, X)\right).$$

(1) Prove that D, A, y S are tensor fields with degrees $(2, 1)$, being A skew symmetric and S symmetric, and that $D = S + A$.
    (*Comment: We say that D is the tensor difference between ∇ and $\overline\nabla$.*)
(2) Prove that the following assertions are equivalent:

   (a) ∇ and $\overline\nabla$ have the same geodesic curves.
   (b) For every vector field X, $D(X,X) = 0$.
   (c) $S = 0$.
   (d) $D = A$.

(3) Prove that $\overline\nabla = \nabla$ if and only if they have the same geodesic curves and the same torsion tensor.
(4) Given a connection ∇ on M, prove that there exists only one connection with the same geodesic curves and with vanishing torsion.
    (*Hint: Take $\overline\nabla := \nabla - (1/2)\,\mathrm{T}$, where T is the torsion of ∇.*)

Exercise 5.8

Let M be a manifold, dim $M = m$, ∇ a connection on M, and T and R the corresponding torsion and curvature tensors, respectively. Let $U \subset M$ be an open set and $X_1, \ldots, X_m \in \mathfrak{X}(U)$ linearly independent vector fields in all the points of U. Let $\omega^1, \ldots, \omega^m \in \Omega^1(U)$ defined by

$$\omega^i(X_j) = \delta^i_j; \quad i, j = 1, \ldots, m.$$

Consider the differential forms $\theta^i_j \in \Omega^1(U)$ and $\tau^i, \Omega^i_j \in \Omega^2(U)$ defined as follows: for every $Y, Z \in \mathfrak{X}(U)$,

$$\theta^i_j(Y) := \omega^i(\nabla_Y X_j), \; \tau^i(Z,Y) := \omega^i(T(Z,Y)), \; \Omega^i_j(Z,Y) := \omega^i(R(Z,Y)X_j).$$

Prove that:

(1) $\theta^i_j, \tau^i, \Omega^i_j$ are well defined and of class C^∞.
(2) $\tau^i = 0 \iff \nabla$ is symmetric on U.
(3) $\Omega^i_j = 0 \iff R|_U = 0$.
(4) $\tau^i = d\omega^i + \theta^i_j \wedge \omega^j$.
(5) $\Omega^i_j = d\theta^i_j + \theta^i_k \wedge \theta^k_j$.

(*Comment: This problem is part of the proof of a known theorem of Cartan that states: Let $(M, \mathbf{g})$ be a Riemannian manifold and let ∇ be the Levi–Civita connection. Then, for every point of M, there exists a local chart $(U, \varphi = (x^i))$ such that $\mathbf{g}|_U = \sum_{i=1}^m dx^i \otimes dx^j$ if, and only if, $R = 0$.*)
(*Solution: For a complete proof, see [328], where seven different proofs are given under the name of "test case".*)

Exercise 5.9

Give the expression of the usual metric tensors of $\mathbb{R}^2$ and $\mathbb{R}^3$ in the following coordinate systems, indicating the open domain where they are defined:

(1) Polar coordinates (r, θ) in $\mathbb{R}^2$, given by $x = r\cos\theta$, $y = r\sin\theta$.
(2) Parabolic coordinates (u, v) in $\mathbb{R}^2$, given by $x = \dfrac{u^2 - v^2}{2}$, $y = uv$.
(3) Cylindrical coordinates (ρ, θ, z) in $\mathbb{R}^3$, given by $x = r\cos\theta$, $y = r\sin\theta$, $z = z$.
(4) Parabolic cylindrical coordinates (u, v, z) in $\mathbb{R}^3$, given by $x = \dfrac{u^2 - v^2}{2}$, $y = uv$, $z = z$.
(5) Spherical coordinates (r, θ, ϕ) in $\mathbb{R}^3$, given by $x = r\cos\phi\sin\theta$, $y = r\sin\phi\sin\theta$, $z = r\cos\theta$.

Calculate the Christopher symbols and the geodesic equation of the Levi–Civita connection of the ordinary metric on $\mathbb{R}^2$ and $\mathbb{R}^3$ in these coordinate systems.

Exercise 5.10
Consider the following differentiable manifold: $M = \{(x, y) \in \mathbb{R}^2; y > 0\}$ with the natural Cartesian coordinates and the *Lobachevski metric tensor* given by

$$\mathbf{g} = \frac{1}{y^2}(\mathrm{d}x^2 + \mathrm{d}y^2).$$

(1) Check that the usual metric on M is invariant by translations and rotations with center at the origin, but $\mathbf{g}$ is not.
(2) Prove that $\mathbf{g}$ is invariant by the transformations:

$$x + iy \mapsto \frac{a(x + iy) + b}{c(x + iy) + d},$$

with $ad - bc \neq 0$, but the ordinary metric is not.
(*Comment: The set of these transformations, with the composition as operation, is the* Möbius group.)
(3) Calculate the Christoffel symbols and the geodesic equation of the Levi–Civita connection of the Lobachevski metric. Check that they are the vertical straight lines and the semicircles with center in the x-axis.

Exercise 5.11
Let $(M, \mathbf{g})$ be a Riemannian manifold. Given $Y \in \mathfrak{X}(M)$, we construct the 1-form $\omega = i(Y)\mathbf{g} \in \Omega^1(M)$ (i.e., $\omega(X) := \mathbf{g}(X, Y)$, for every $X \in \mathfrak{X}(M)$). If T is a tensor field of contravariant degree, then the above operation can be generalised by increasing in one the covariant degree and decreasing in one the contravariant degree of T.
(*Comment: This contraction is called to lower the indexes with the metric.*)

(1) Let (U, φ), $\varphi = (x^1, \ldots, x^m)$, be a local coordinate chart in M. Write the expression of the above contraction for a vector field in these coordinates.
(2) As an example, let the vector field $Y = xy\frac{\partial}{\partial x} + (2y - z^2)\frac{\partial}{\partial y} + xz\frac{\partial}{\partial z}$, written in natural coordinates in $\mathbb{R}^3$. Give its expression in spherical coordinates and calculate its contraction with the tensor metric.
(3) Write the corresponding expression for a general tensor field T.

Exercise 5.12

Let $(M, \mathbf{g})$ be a Riemannian manifold. The contraction defined in the above problem is a bijective map, $\mathbf{g}^\sharp : \mathfrak{X}(M) \longrightarrow \Omega^1(M)$, because $\mathbf{g}$ is a Riemannian metric. Following this map, we can translate the scalar product to the differentiable 1-forms in the following way: If $\omega, \gamma \in \Omega^1(M)$

$$\mathbf{g}(\omega, \gamma) = \mathbf{g}((\mathbf{g}^\sharp)^{-1}(\omega), (\mathbf{g}^\sharp)^{-1}(\gamma)).$$

(1) Prove that $\mathbf{g}^\sharp$ is a 2-contravariant tensor field. Prove that the matrices of $\mathbf{g}$ and $\mathbf{g}^\sharp$ in a local chart are one inverse of the other.
    (**Comment:** We say that $\mathbf{g}^\sharp$ is the contravariant metric associated with $\mathbf{g}$.)
(2) With some examples, calculate the contravariant metrics of the above examples in $\mathbb{R}^2$ and $\mathbb{R}^3$.

Exercise 5.13

Let $F : M \longrightarrow N$ be a differentiable map with M and N manifolds. Let $\mathbf{g}$ be a Riemannian metric on N. Prove that $F^*\mathbf{g}$ is a 2-covariant tensor field on M. Which condition must F fulfil so that $F^*\mathbf{g}$ is a Riemannian metric?

Exercise 5.14

In $\mathbb{R}^2$ with the Cartesian chart, consider the connection ∇ with Christoffel symbols $\Gamma^1_{12} = \Gamma^1_{21} = 1$, and all the others $\Gamma^k_{ij} = 0$.

(1) Find the geodesic curve passing through the point $p = (2, 1)$, with tangent vector $\left(\dfrac{\partial}{\partial x} + \dfrac{\partial}{\partial y} \right)_p$.
(2) Find the parallel transport of $\mathbf{u} = (0, 1) \in T_p\mathbb{R}^2$ along the curve $\gamma(t) = (t, t)$, $t \in [0, 1]$.
(3) Given two arbitrary points in $\mathbb{R}^2$, is there any geodesic curve of piecewise C^1-class joining both points?
(4) Is this connection ∇ the Levi–Civita connection of any metric?

Exercise 5.15

Let $S \subset \mathbb{R}^3$ be the origin-centred sphere with radius R.

(1) Using the results of the above problems, calculate the Christoffel symbols of the induced connection on the sphere S in the local chart (θ, ϕ).
(2) Obtain the same result from the Riemannian metric induced on S by the ordinary metric of $\mathbb{R}^3$.

(3) Given the curve $\gamma(t) = (\theta_0, t)$ on S, following a parallel line of the sphere, calculate the expression $X(t) = f(t)\dfrac{\partial}{\partial \phi} + h(t)\dfrac{\partial}{\partial \theta}$ of a parallel vector field along γ.

Exercise 5.16

Let $(M, \mathbf{g})$ be a Riemannian manifold. Given a $\mathcal{C}^\infty$-function $\lambda \colon M \longrightarrow \mathbb{R}$, we define a new metric $\bar{\mathbf{g}} := e^{2\lambda}\mathbf{g}$, and we denote by ∇ and $\overline{\nabla}$ the corresponding Levi–Civita connections, respectively. Prove that if $X, Y \in \mathfrak{X}(M)$, then

$$\overline{\nabla}_X Y - \nabla_X Y = (X \cdot \lambda)\, Y + (Y \cdot \lambda)\, X - \mathbf{g}(X, Y)\, \mathrm{grad}\, \lambda.$$

Exercise 5.17

Let $(M, \mathbf{g})$ be a connected Riemannian manifold and ∇ its Levi–Civita connection. Remember that a *Killing vector field* or an *infinitesimal isometry* is a vector field $X \in \mathfrak{X}(M)$ such that $\mathrm{L}(X)\mathbf{g} = 0$. The set of such vector fields is denoted by $\mathbf{Isom}(M, \mathbf{g})$. For $X \in \mathfrak{X}(M)$, let $A_X := \mathrm{L}(X) - \nabla_X$.

(1) Prove that A_X is a derivation.
(2) Prove that X is parallel if, and only if, $A_X = 0$.
(3) Prove that $X \in \mathbf{Isom}(M)$ if, and only if, $\mathbf{g}(A_X Y, Z) + \mathbf{g}(Y, A_X Z) = 0$, for every $Y, Z \in \mathfrak{X}(M)$.
(4) If $X \in \mathbf{Isom}(M, \mathbf{g})$, prove that $A_X X = 0$ if, and only if, $\|X\|$ is a constant.
(5) If $X \in \mathbf{Isom}(M, \mathbf{g})$, prove that X is a geodesic if, and only if, $\|X\|$ is constant.

5.5.2 Newtonian mechanics

Exercise 5.18

Consider a simple pendulum with mass m_2 and length l whose suspension point has mass m_1 and moves freely in a horizontal line under the action of gravity.

(1) Describe the configuration space M as a Riemannian manifold and give an adequate coordinate system on M.
(2) Calculate the kinetic energy metric on the manifold M in local coordinates. Give the Christoffel symbols of the associated Levi–Civita connection.
(3) Give the expression of the force and obtain the dynamical Newton equations of the system.

Exercise 5.19
A pendulum is made with an oscillating rod (with negligible mass) of length l_1 and one mass m_1 on the free end of the rod. In the rod, there is another sliding mass m_2 attached to one spring with characteristic constant k. The spring is attached to the fixed oscillation point of the pendulum. The system is under the action of gravity.

(1) Give the configuration manifold and the kinetic energy metric on it.
(2) Give the force field of the system and the Newton equation.
(3) Give the Lagrange equations of the system.

Exercise 5.20 (Second Kepler law).
Consider the problem of a particle in a central force field. As is known, the angular momentum is conserved; hence, the movement is constrained to be in one plane. In this plane, the position vector of one trajectory $\mathbf{r}(t)$ sweeps a certain region of the plane. If $A(t)$ is the swept area from $t = 0$ to $t = t$, prove that $A'(t) = 0$.

Exercise 5.21
Consider a double pendulum on a horizontal plane. The rods of the pendulum have lengths l_1 and l_2. The masses at the ends are m_1 and m_2, and they have electric charges q_1 and q_2, respectively. A vertical and uniform magnetic field with value B acts on the system.

(1) Give the configuration manifold and kinetic energy on it.
(2) Give the force field and the Newton equation of the system.
(3) Obtain the equilibrium points of the system.
(4) Is the kinetic energy an invariant of the motion?

Exercise 5.22 (Dissipative forces).
Suppose we have a mechanical system given by its Newton equation $\nabla_{\dot\gamma}\dot\gamma = F \circ \gamma$ in a Riemannian manifold (Q, g), with a conservative force $F = -\mathrm{grad}\, V$, with $V \colon Q \longrightarrow \mathbb{R}$. Recall that the mechanical energy is given by $E = K + V$, with $K(v_q) = \frac{1}{2}g(v_q, v_q)$, and is a constant of the motion. The system is modified by adding a force depending on velocity, that is, a vector field, $R \colon TQ \longrightarrow TQ$, along the map τ_Q. Then the new dynamical equation is given by

$$\nabla_{\dot\gamma}\dot\gamma = F \circ \gamma + R \circ \dot\gamma.$$

Write $\alpha \colon TQ \longrightarrow \mathbb{R}$ for $\alpha(v_q) = g_q(R(v_q), v_q)$. We say that R is a *dissipative force* if $\alpha(v_q) \leq 0$, for every $v_q \in TQ$.

(1) Prove that, if γ is a solution to the Newton equation, then $D(E \circ \dot\gamma) = \alpha \circ \dot\gamma$.

(2) If R is a dissipative force, what can we say about the existence of periodical solutions?

(3) Let $R \colon TQ \longrightarrow TQ$ be an endomorphism of the vector bundle TQ such that for every $q \in Q$, R_q is a negative self-adjoint operator in the Euclidean space (T_qQ, g_q). Prove that R is a dissipative force.

(4) The function $\mathcal{F}(v_q) = -\alpha(v_q)/2$ is called *Rayleigh dissipation function*. Prove that at every $q \in Q$, $R(v_q) = -\text{grad}\,\mathcal{F}_q(v_q)$ (where the gradient is taken along the fibre in q).

(5) If $f(q) \geq 0$ is a function in Q, develop the above items for the case $R(v_q) = -f(q)v_q$.
    (*Comment:* We say that there is a friction, which is proportional to the velocity.)

Exercise 5.23 (Jacobi metric).

Consider a mechanical system defined on the Riemannian manifold $(Q, \mathbf{g})$ with potential energy function $V \colon Q \longrightarrow \mathbb{R}$. Recall that the energy is given by $E(v_q) = \frac{1}{2}\mathbf{g}(v_q, v_q) + V(q)$, and it is a constant of the motion. Suppose that $E_0 > V(q)$ for all $q \in Q$. We define the *Jacobi metric* as $\mathbf{g}_0 := (E_0 - V)\mathbf{g}$. Prove that the solutions γ to the Newton equation $\nabla_{\dot\gamma}\dot\gamma = -\text{grad}\,V \circ \gamma$ with fixed energy E_0 are, under a convenient reparameterisation, the geodesics of $\mathbf{g}_0$ (Jacobi Theorem).
(*Hint: Use Exercises 5.6 and 5.16.*)

Exercise 5.24

Let $(M, \mathbf{g})$ be a Riemannian manifold and ∇ the Levi–Civita connection. Given a vector field X, we define the 1-form in M
$$X^\flat = i(X)\mathbf{g}$$
and the function in TM
$$P_X(v_p) = \mathbf{g}(X(p), v_p) = \langle X^\flat(p), v_p \rangle.$$

(1) Calculate the covariant differential of $\nabla X^\flat \in T_2^0(M)$ (see *Exercise 5.3*), and prove that $(\nabla X^\flat)(Y, Z) = \mathbf{g}(\nabla_Y X, Z)$.

(2) Prove that $(\nabla X^\flat)(Y, Y) = \frac{1}{2}(L(X)\mathbf{g})(Y, Y)$.

(3) Prove that X is an infinitesimal isometry of $\mathbf{g}$ if, and only if, $\nabla X^\flat$ is a skew-symmetric tensor.

(4) Calculate the derivative $D(P_X \circ \dot\gamma)$ as a function of $\nabla X^\flat$, where $\gamma \colon I \longrightarrow M$ is a curve.
    (*Hint: Recall that D is the ordinary derivative with respect to t in $\mathbb{R}$.*)

(5) Prove that X is an infinitesimal isometry of g if, and only if, P_X is constant along the geodesics of g.

(6) Consider a mechanical system described by the Newton equation $\nabla_{\dot\gamma}\dot\gamma = F \circ \gamma$, where F is a force field in M. If X is an infinitesimal isometry of g, under which conditions can we say that P_X is a constant of motion?

(7) Suppose that, in a local coordinate system in Q, we have

$$X = \sum a^i \frac{\partial}{\partial x^i}, \quad X^b = \sum a_i dx^i, \quad g = g_{ij}dx^i \otimes dx^j.$$

Give the local expression of ∇X^b, using the property of being skew symmetric. Give also the local expression of P_X.

(8) Let $M = \mathbb{R}^3$ with the tensor metric $g = m(dx^2 + dy^2 + dz^2)$. Taking X as the infinitesimal generator of translations, $\dfrac{\partial}{\partial x}, \dfrac{\partial}{\partial y}, \dfrac{\partial}{\partial z}$, and rotations,

$-y\dfrac{\partial}{\partial x} + x\dfrac{\partial}{\partial y}, \; -x\dfrac{\partial}{\partial z} + z\dfrac{\partial}{\partial x}, \; -z\dfrac{\partial}{\partial y} + y\dfrac{\partial}{\partial z}$; determine the functions P_X.

Exercise 5.25 (Laplace–Runge–Lenz vector).

Consider the mechanical system in the Euclidean 3-space given by $\mu\ddot{\mathbf{r}} = -\operatorname{grad} V$, where $V(\mathbf{r}) = k/r$. If $\mathbf{p} = \mu\dot{\mathbf{r}}$ and $\mathbf{L} = \mathbf{r} \times \mathbf{p}$, the *Laplace–Runge–Lenz vector* is defined as

$$\mathbf{A} = \mathbf{p} \times \mathbf{L} + \mu k \frac{\mathbf{r}}{r}.$$

(1) Prove that $\mathbf{A}$ is a constant of motion.

   (**Hint:** *Recall that* $\mathbf{a} \times (\mathbf{b} \times \mathbf{c}) = (\mathbf{a} \cdot \mathbf{c})\mathbf{b} - (\mathbf{a} \cdot \mathbf{b})\mathbf{c}$.)

(2) Prove that $\mathbf{A} \cdot \mathbf{L} = 0$ and $\mathbf{A}^2 = \mu^2 k^2 + 2\mu E L^2$, where $L = \|\mathbf{L}\|$, and E is the total mechanical energy.

(3) If we have an attractive force and a bounded motion, that is an elliptical orbit, prove that the vector $\mathbf{A}$ has the direction of the pericenter of the orbit.

   (**Hint:** *Let θ be the angle between the Lenz vector and the position vector. Calculate $\mathbf{A} \cdot \mathbf{r}$ and try to see that r is minimal when $\theta = 0$.*)

Exercise 5.26

Let M be a differentiable manifold, $\dim M = m$, and $L : TM \longrightarrow \mathbb{R}$ be a differentiable function.

(1) For a Riemannian metric g in M, consider the function $K_g : TM \longrightarrow \mathbf{R}$ defined by $K_g(u_p) = g(u_p, u_p)$. Let $Z \in \mathfrak{X}(M)$ and ϕ_t^Z be the flux of Z. Taking into account that $(\phi_t^Z)^*g$ is another metric, prove that $(T\phi_t^Z)^*K_g = K_{(\phi_t^Z)^*g}$ and obtain the expression $K_{L(Z)g} = L(Z^C)K_g$.

(2) If $K = (1/2)K_{\mathbf{g}}$ is the kinetic energy defined by the metric $\mathbf{g}$; prove that X is a Killing vector field of $\mathbf{g}$ if, and only if, $L(X^C)K = 0$.

(3) If $V: M \longrightarrow \mathbb{R}$ a differentiable function and X a Killing vector field of $\mathbf{g}$, give the conditions for X to be a natural symmetry of the Lagrangian function $\mathcal{L} = K - V$.

(4) Take $M = \mathbb{R}^2$ and $\mathbf{g} = dx^2 + dy^2$. Find the Killing vector fields for $(M, \mathbf{g})$ and obtain the symmetries and the Noether invariants associated to the Lagrangian $\mathcal{L} = K - V$, where $V(x, y) = f(x^2 + y^2)$, with $V: \mathbb{R}^2 \longrightarrow \mathbb{R}$ and $f: \mathbb{R} \longrightarrow \mathbb{R}$.

(*Hint: Recall that if $X \in \mathfrak{X}(M)$, then $L(X^C)\Delta = 0$ and $L(X^C)J = 0$.*)

Exercise 5.27 (Geometric Hamilton–Jacobi Lagrangian equation).
Let $(M, \mathbf{g})$ be a Riemannian manifold with $\dim M = m$ and ∇ be the corresponding Levi-Civita connection. If $F \in \mathfrak{X}(M)$, we consider the Newtonian system defined in $(M, \mathbf{g})$ by F as the force field. Let $\alpha = i(F)\mathbf{g} \in \Omega^1(M)$ and $\beta = \tau_M^* \alpha \in \Omega^1(TM)$.

Prove that the integral curves of the vector field $\Gamma \in \mathfrak{X}(TM)$ which is the solution to the equation

$$i(\Gamma)\omega_K = dK - \beta,$$

(where K is the kinetic energy defined by $\mathbf{g}$ considered as a Lagrangian function) are the natural lift to TM of the curves, which are solutions to the Newtonian system $(M, \mathbf{g}, F)$.

(*Hint: If $X \in \mathfrak{X}(M)$ and $\sigma : I \longrightarrow M$ is an integral curve of X, prove the equation*

$$(\nabla_X X) \circ \sigma = \nabla_{\dot{\sigma}} \dot{\sigma}.$$

Recall that the Lagrangian dynamical vector field Γ_o corresponding to the Lagrangian function $\mathcal{L} = K$ is the geodesic second-order vector field.)

Exercise 5.28

(1) With the same hypotheses of the above problem, prove that the following conditions are equivalent for a vector field $X \in \mathfrak{X}(M)$:

(a) The integral curves of X are the solution to the Newton equation for the system $(M, \mathbf{g}, F)$.

(b) X satisfies the equation $\nabla_X X = F$.

(c) The vector field Γ is tangent to $X(M) \subset TM$, as a submanifold of TM, so $X(M)$ is an invariant submanifold by the flux of Γ.

If $X \in \mathfrak{X}(M)$ satisfies those conditions, characterise the integral curves of Γ coming from the integral curves of X.

(2) Suppose that F is a conservative force, that is $i(\mathbf{g})F = -dV$, with $V : M \longrightarrow \mathbf{R}$, and let $E = K + V \circ \tau_M$ be the energy of the system $(M, \mathbf{g}, F)$. Prove that:

(a) If $X \in \mathfrak{X}(M)$ satisfies that $\nabla_X X = F$, then $\mathrm{L}(X)(E \circ X) = 0$.

(b) If $X, Y, Z \in \mathfrak{X}(M)$, then

$$d(i(X)\mathbf{g})(Y, Z) = g(\nabla_Y X, Z) - \mathbf{g}((\nabla_Z X, Y).$$

(3) Let $X \in \mathfrak{X}(M)$ be a vector field satisfying the condition $d(i(X)\mathbf{g}) = 0$; prove that the following conditions are equivalent:

(a) $\nabla_X X = F$.   (b) $d(E \circ X) = 0$.

(*Comment*: *If a vector field X satisfies these conditions, we say that X is a Hamilton–Jacobi vector field of the mechanical system $(M, \mathbf{g}, F)$. Recall that we have $i(X)\mathbf{g} = dS$ for some local function S in M.*

If M is a connected manifold, condition $d(E \circ X) = 0$ implies that $E \circ X = \mathrm{cnt.}$, and this is the classical Hamilton–Jacobi equation. This equation allows us to obtain local vector fields X satisfying our conditions. These vector fields allow us to obtain families of solutions to the dynamical equations of the mechanical system.

The Hamilton–Jacobi equation is usually studied in a Hamiltonian setting. We have given here a Lagrangian approach. Most of the classical books on analytical mechanics include the study of the Hamilton–Jacobi equation and applications.)

Exercise 5.29

Let $(Q, \mathbf{g})$ be an n-dimensional Riemannian manifold, K the associated kinetic energy, and $V : Q \longrightarrow \mathbb{R}$. Let $\alpha \in \Omega^1(Q)$ and $\hat{\alpha} : TM \longrightarrow \mathbb{R}$ be the associated function.

(1) Consider the Lagrangian function $\hat{\mathcal{L}} = K - V + \hat{\alpha}$. Prove that $\hat{\mathcal{L}}$ is regular and that a point $(q, u) \in TQ$ is an equilibrium point of the Lagrangian system if, and only if, $u = 0$ and $dV(q) = 0$ (i.e., q is a critical point of the function V).

(2) Consider the particular case where $Q = \mathbb{R}^2$, $\mathbf{g}$ is the ordinary metric, and (x, y) are the Cartesian coordinates in $\mathbb{R}^2$. Let $V = \frac{1}{2}(a^2x^2 + b^2y^2)$ and $\alpha = ky\,dx$, where $a, b, k \in \mathbb{R} - \{0\}$. Give the Euler–Lagrange equation for the Lagrangians $\mathcal{L} = K - V$ and $\hat{\mathcal{L}} = \mathcal{L} + \hat{\alpha}$. Integrate the dynamical differential equations for $\mathcal{L}$ and prove that their solutions are oscillatory functions. Prove that the corresponding solutions for $\hat{\mathcal{L}}$ are oscillatory, but with different frequencies.

(3) Give the expressions of the Legendre maps for $\mathcal{L}$ and $\hat{\mathcal{L}}$ and obtain the corresponding Hamiltonian functions h and $\hat{h}$. Justify why $E_{\mathcal{L}} = E_{\hat{\mathcal{L}}}$, but $h \neq \hat{h}$.

Chapter 6

An Introduction to Contact Mechanics and Dissipative Dynamical Systems

In Chapter 4, we studied nonautonomous or time-dependent systems; one of the main characteristics of which is that they are not conservative, but dissipative, that is, unlike the conservative systems studied in Chapters 3 and 5, energy is not a conserved quantity. However, in mechanics, many other indistinctly autonomous or nonautonomous systems also manifest this characteristic [163, 304]. They are those in which forces of a nonconservative type appear, such as those described in Section 5.2.2.1 (see also Exercise 5.22 in Chapter 5).

In recent years, there has been a great interest in geometrically studying these types of systems by using techniques from *contact geometry* [3, 20, 43, 174]. In fact, *contact structure* is quite similar to cosymplectic structure and allows us to give a natural Hamiltonian description of mechanical systems with dissipation [38, 40, 41, 68, 189, 191, 235, 251, 354]. Their Lagrangian formalism [93, 117, 169, 191] and the unified Lagrangian–Hamiltonian formalism [114] have also been stated in many different situations and applications, such as damped oscillators, motion on viscous fluids, and motion with friction, in general. However, the Lagrangian formulation for these kinds of systems was introduced by G. Herglotz [213, 214], who used a generalisation of the Hamilton variational principle to obtain the *Herglotz–Euler–Lagrange equations*, which are the same equations obtained using contact geometry (see also [130, 175, 240, 251] for a modern geometric version of this variational approach). In all these descriptions, the Lagrangian and the Hamiltonian functions depend also on an additional variable that, as we will see, can be identified with the "action" of the system, and for this reason, physicists often refer to them as *action-dependent systems*.

Contact geometry has also been used to describe different types of physical systems in thermodynamics (for instance, in [38, 39, 321, 342]), quantum mechanics [93, 215, 224], circuit theory [93, 187], astrophysics [176], theoretical physics [224], control theory [121, 300], etc., and even to describe other mechanical systems than just dissipative ones [145]. Finally, Herglotz's variational methods and the contact structure itself have been generalised in different ways for the treatment of action-dependent field theories [115, 168, 170, 171, 353] (see also [9, 15, 37, 111, 162, 278, 307, 308, 335] for other contributions and less general approaches).

This whole body of doctrine, generically referred to as *contact mechanics*, is currently a topic of active research, both in its foundation, and in its extensions and applications.

In this chapter, we review the main foundations of contact geometry and its application to describe dissipative autonomous dynamical systems. The extension for the treatment of dissipative nonautonomous dynamical systems (*cocontact formalism*) has been done in [112, 141, 309], as well as to other types of systems and problems, such as higher order dissipative systems [113], nonholonomic systems [116, 122], or reduction theorems [190, 362].

In this chapter, we start with the basic concepts and properties of *contact manifolds* and the generic definition of *contact Hamiltonian systems*. Next, we develop the Lagrangian, Hamiltonian, and unified formalisms for these kinds of systems. We also study symmetries and the fundamental concept of *dissipated quantities* geometrically, defining, in particular, the notion of *contact Noether symmetry* for contact Hamiltonian and Lagrangian systems, and establishing the statements of the dissipation theorems. These are analogous to the conservation theorems of conservative systems and show how to associate dissipated and conserved quantities to these symmetries. Finally, the examples of the damped harmonic oscillator and the Kepler problem with dissipation are analysed in this context.

6.1   A Survey on Contact Geometry

(See, for instance, [3, 20, 39, 57, 93, 169, 224, 249] for more information.)

6.1.1   *Contact manifolds*

Definition 6.1. A *contact manifold* is a pair (M, η), where M is a $(2n + 1)$-dimensional manifold and $\eta \in \Omega^1(M)$ is a differential 1-form such that $\eta \wedge (\mathrm{d}\eta)^n$ is a volume form in M. Then, the form η is called a **contact form** (or a **contact structure**).

As a straightforward consequence of this definition, we have:

Theorem 6.1. *Given a contact manifold* (M, η), *the condition that* $\eta \wedge (\mathrm{d}\eta)^n$ *is a volume form is equivalent to the splitting*

$$TM = \ker \mathrm{d}\eta \oplus \ker \eta \equiv \mathcal{D}^{\mathrm{R}} \oplus \mathcal{D}^{\mathrm{C}}. \qquad (6.1)$$

Then, there exists a unique vector field $\mathcal{R} \in \mathfrak{X}(M)$ *such that*

$$i(\mathcal{R})\mathrm{d}\eta = 0, \quad i(\mathcal{R})\eta = 1, \qquad (6.2)$$

and hence it generates the distribution $\mathcal{D}^{\mathrm{R}}$.

Conversely, if we have two distributions $\mathcal{D}^{\mathrm{R}}$ *and* $\mathcal{D}^{\mathrm{C}}$ *of ranks 1 and 2n, respectively, such that (6.1) holds, they define a contact structure on* M.

Definition 6.2. The above vector field $\mathcal{R} \in \mathfrak{X}(M)$ is called the **Reeb vector field**, and $\mathcal{D}^{\mathrm{R}}$ and $\mathcal{D}^{\mathrm{C}}$ are known as the **Reeb** (or **horizontal distribution**), and the **contact** (or **vertical distribution** of (M, η)).

Remark 6.1. It is relevant to point out that although Definition 6.1 is the most assumed, there are other different terminologies in relation to the concept of *contact manifold*. For instance, it can be defined by demanding the existence of two distributions $\mathcal{D}^{\mathrm{R}}$ and $\mathcal{D}^{\mathrm{C}}$, satisfying the properties stated in Theorem 6.1, instead of using differential forms (see, for instance, [11,174,189,224]), and when these distributions are associated with a fixed form, it is called a **strict contact manifold**. Also, in [1], a more generic definition is given, by demanding the existence of a 2-form of maximal rank in an odd-dimensional manifold, and then calling **exact contact** the case in which this form is exact (which is the case of Definition 6.1).

Given a contact manifold (M, η), as a consequence of the splitting (6.1), there exists a vector bundle isomorphism

$$\flat\eta : TM \longrightarrow T^*M$$
$$(\mathrm{p}, X_{\mathrm{p}}) \mapsto \left(\mathrm{p}, i(X_{\mathrm{p}})\mathrm{d}\eta_{\mathrm{p}} + [i(X_{\mathrm{p}})\eta_{\mathrm{p}}]\eta_{\mathrm{p}}\right)$$

and its inverse $\sharp_\eta = \flat_\eta^{-1} \colon \mathrm{T}^*\mathrm{M} \longrightarrow \mathrm{TM}$. Their natural extensions are the $C^\infty(\mathrm{M})$-module isomorphisms, which are denoted with the same notation,

$$\flat_\eta \colon \mathfrak{X}(\mathrm{M}) \longrightarrow \Omega^1(\mathrm{M})$$
$$X \longmapsto i(X)\mathrm{d}\eta + (i(X)\eta)\eta \tag{6.3}$$

and its inverse $\sharp_\eta = \flat_\eta^{-1} \colon \Omega^1(\mathrm{M}) \longrightarrow \mathfrak{X}(\mathrm{M})$. In particular, for the Reeb vector field, we have $\flat_\eta(\mathcal{R}) = \eta$.

Remark 6.2. If in Definition 6.1 the form η is such that $\eta \wedge (\mathrm{d}\eta)^n$ is not a volume form, but the rank of the distribution $\ker \mathrm{d}\eta \cap \ker \eta$ is constant and $\dim \mathrm{M} - \mathrm{rank}(\ker \mathrm{d}\eta \cap \ker \eta)$ is odd, then we say that η is a **precontact form** on M and that (M, η) is a **precontact manifold** (moreover, $\dim \mathrm{M}$ could be arbitrary). Under these hypotheses, there exist Reeb vector fields defined by (6.2), but they are not uniquely defined, and the map $\flat_\eta$ is not an isomorphism [112]. (See also [190] for a more general definition of precontact structure using *contact distributions*.)

Proposition 6.1. *On a contact manifold* (M, η), *there are charts of coordinates* $(U; z^I; s)$, $I = 1, \ldots, 2n$, *such that*

$$\eta|_U = \mathrm{d}s - f_I(z^J)\,\mathrm{d}z^I, \quad \mathcal{R}|_U = \frac{\partial}{\partial s},$$

where f_I *are functions depending only on the* z^J. *(They are called* **adapted coordinates** *of the contact structure.)*

Proof. On M, we can take charts of coordinates $(U; z^I, s)$, $I = 1, \ldots, 2n$, such that they rectify the vector field $\mathcal{R}$, that is, $\mathcal{R} = \dfrac{\partial}{\partial s}$ on U. Then $\eta|_U = a\,\mathrm{d}s - f_I(z^J, s)\mathrm{d}z^I$, but the conditions defining $\mathcal{R}$ imply that $\dfrac{\partial f_I}{\partial s} = 0$ and $a = 1$, so the result follows. $\qquad\square$

Furthermore, one can prove the existence of Darboux-type coordinates:

Theorem 6.2 (Darboux Theorem for contact manifolds). *Let* (M, η) *be a contact manifold. Then, for every point* $p \in \mathrm{M}$, *there exists a chart of coordinates* $(U; x^i, y_i, s)$, $1 \leq i \leq n$, *such that*

$$\eta|_U = \mathrm{d}s - y_i\,\mathrm{d}x^i, \quad \mathcal{R}|_U = \frac{\partial}{\partial s}.$$

> These are the **Darboux** or **canonical coordinates** of the contact manifold (M, η).

Proof. Taking local charts of adapted coordinates $(U; z^I, s)$, if we do the quotient $U/\mathcal{D}^R$, the form $d\eta|_U = df_I(z^J) \wedge dz^I$ projects to the quotient and is a symplectic form on it. Then we can take symplectic Darboux coordinates on the quotient and pull them back to U, thus obtaining contact chart Darboux coordinates $(U; x^i, y_i, s)$ on M. $\qquad\square$

Relevant examples of contact manifolds are as follows:

Canonical model: The canonical model for contact manifolds is the manifold $T^*Q \times \mathbb{R}$. In fact, if $\Theta \in \Omega^1(T^*Q)$ is the canonical 1-form in T^*Q, s is the Cartesian coordinate of $\mathbb{R}$, and $\pi_1 \colon T^*Q \times \mathbb{R} \longrightarrow T^*Q$ is the canonical projection, then $\eta = ds - \pi_1^*\Theta = ds - p_i dq^i$ is a contact form in $T^*Q \times \mathbb{R}$, the Reeb vector field is $\mathcal{R} = \dfrac{\partial}{\partial s}$, and (s, q^i, p_i) are Darboux coordinates on $T^*Q \times \mathbb{R}$.

Contactification of a symplectic manifold: If (P, ω) is an exact symplectic manifold such that $\omega = -d\theta \in \Omega^2(P)$, consider the manifold $M = P \times \mathbb{R}$. If s is the Cartesian coordinate of $\mathbb{R}$, then the 1-form $\eta = ds - \theta \in \Omega^1(M)$ (where we have denoted θ the pull-back of θ to M) is a contact form, and (M, η) is a contact manifold, which is called the **contactified** of P. Observe that the canonical model, $T^*Q \times \mathbb{R}$, is the contactified of T^*Q endowed with its canonical symplectic structure.

6.1.2 Hamiltonian, gradient, and evolution vector fields on a contact manifold

As in the case of cosymplectic manifolds, for a contact manifold (M, η), the existence of the natural $C^\infty(M)$-modules isomorphism $\flat_\eta$, introduced in the above section, allows us to associate some characteristic vector fields to a function $f \in C^\infty(M)$:

Definition 6.3. Let (M, η) be a contact manifold and $f \in C^\infty(M)$.

The **Hamiltonian vector field** associated with f is the vector field $X_f \in \mathfrak{X}(M)$ defined by $\flat_\eta(X_f) := df - (\mathcal{R}(f) + f)\eta$.

The **gradient vector field** associated with f is the vector field $\mathbf{grad}\, \mathbf{f} \in \mathfrak{X}(M)$ defined by $\flat_\eta(\mathrm{grad}\, f) := df$.

The **evolution vector field** associated with f is the vector field $\varepsilon_f \in \mathfrak{X}(M)$ defined as $\varepsilon_f := f\mathcal{R} + X_f$ or equivalently, $\flat_\eta(\varepsilon_f) := df - \mathcal{R}(f)\eta$.

Lemma 6.1. $\operatorname{grad} f = X_f + (\mathcal{R}(f) + f)\,\mathcal{R}.$

Proof. Observe that

$$\flat\eta(X_f) = \mathrm{d}f - (\mathcal{R}(f) + f)\,\eta = \flat\eta(\operatorname{grad} f) - (\mathcal{R}(f) + f)\eta,$$

then, being $\flat\eta$ a diffeomorphism and considering that $\flat\eta(\mathcal{R}) = \eta$, the result follows. $\qquad\square$

Considering this, these vector fields can be equivalently characterised as follows:

Proposition 6.2. *The Hamiltonian vector field associated with f is determined by the equations:*

$$i(X_f)\eta = -f, \quad i(X_f)\mathrm{d}\eta = \mathrm{d}f - \mathcal{R}(f)\,\eta. \tag{6.4}$$

The gradient vector field associated with f is determined by the equations:

$$i(\operatorname{grad} f)\eta = \mathcal{R}(f), \quad i(\operatorname{grad} f)\mathrm{d}\eta = \mathrm{d}f - \mathcal{R}(f)\,\eta. \tag{6.5}$$

The evolution vector field associated with f is determined by the equations:

$$i(\varepsilon_f)\eta = 0, \quad i(\varepsilon_f)\mathrm{d}\eta = \mathrm{d}f - \mathcal{R}(f)\,\eta. \tag{6.6}$$

Proof. For every $f \in C^\infty(M)$, for the Hamiltonian vector field X_f, using the definitions of the isomorphism $\flat\eta$ and of X_f, first we have

$$i(X_f)\mathrm{d}\eta + (i(X_f)\eta)\eta = \flat\eta(X_f) = \mathrm{d}f - (\mathcal{R}(f) + f)\eta, \tag{6.7}$$

therefore, contracting both members with the Reeb vector field and using (6.2), we get

$$(i(X_f)\eta)\,i(\mathcal{R})\eta = i(\mathcal{R})\mathrm{d}f - (\mathcal{R}(f) + f)\,i(\mathcal{R})\eta = -f\,i(\mathcal{R})\eta$$
$$\Longleftrightarrow i(X_f)\eta = -f.$$

Now, going to (6.7), we obtain

$$i(X_f)\mathrm{d}\eta = \mathrm{d}f - \mathcal{R}(f)\,\eta.$$

Conversely, using (6.4) in the Definition (6.3) of $\flat\eta$, we get $\flat\eta(X_f) :=$ $\mathrm{d}f - (\mathcal{R}(f) + f)\eta.$

For the gradient vector field, from the definition of **grad** f and Lemma 6.1, and using the above results and (6.2), we have

$$i(\mathbf{grad}\, f)\mathrm{d}\eta = i(\mathrm{X}_f)\mathrm{d}\eta + (\mathcal{R}(f) + f)\, i(\mathcal{R})\mathrm{d}\eta = i(\mathrm{X}_f)\mathrm{d}\eta = \mathrm{d}f - \mathcal{R}(f)\,\eta,$$
$$i(\mathbf{grad}\, f)\eta = i(\mathrm{X}_f)\eta + (\mathcal{R}(f) + f)\, i(\mathcal{R})\eta = -f + (\mathcal{R}(f) + f) = \mathcal{R}(f).$$

Conversely, using (6.5) in Definition 6.3 of $\flat\eta$, we obtain $\flat\eta\,(\mathbf{grad}\, f) := \mathrm{d}f$.

Finally, for the evolution vector field ε_f, from the definition of ε_f and using the above results,

$$i(\varepsilon_f)\mathrm{d}\eta = i(\mathrm{X}_f)\mathrm{d}\eta + f\, i(\mathcal{R})\mathrm{d}\eta = i(\mathrm{X}_f)\mathrm{d}\eta = \mathrm{d}f - (\mathcal{R}(f))\,\eta,$$
$$i(\varepsilon_f)\eta = i(\mathrm{X}_f)\eta + f\, i(\mathcal{R})\eta = -f + f = 0.$$

Once again, using (6.6) in Definition 6.3 of $\flat\eta$, we obtain $\flat\eta(\varepsilon_f) := \mathrm{d}f - \mathcal{R}(f)\,\eta$. $\square$

As a logical consequence of Proposition 6.2, we obtain:

Proposition 6.3. *The equations for the integral curves $c \colon I \subset \mathbb{R} \longrightarrow \mathrm{M}$ of the Hamiltonian, the gradient, and the evolution vector fields associated with $f \in C^\infty(\mathrm{M})$ are, respectively,*

$$i(\widetilde{c})(\eta \circ c) = -f \circ c, \quad i(\widetilde{c})(\omega \circ c) = (\mathrm{d}f - R(f)\eta) \circ c, \qquad (6.8)$$
$$i(\widetilde{c})(\eta \circ c) = R(f) \circ c, \quad i(\widetilde{c})(\omega \circ c) = (\mathrm{d}f - R(f)\eta) \circ c,$$
$$i(\widetilde{c})(\eta \circ c) = 0, \quad i(\widetilde{c})(\omega \circ c)) = (\mathrm{d}f - R(f)\eta) \circ c,$$

Local expressions: In Darboux coordinates (x^i, y_i, s) on M, we have $\mathcal{R}(f) = \dfrac{\partial f}{\partial s}$, so

$$\mathrm{d}f - \mathcal{R}(f)\eta = \left(\frac{\partial f}{\partial x^i} + y_i \frac{\partial f}{\partial s} \right) \mathrm{d}x^i + \frac{\partial f}{\partial y_i}\, \mathrm{d}y_i,$$

and from (6.4), (6.5), and (6.6) (or using (6.4), Lemma 6.1 and the definition of ε_f), we obtain,

$$\mathrm{X}_f = \frac{\partial f}{\partial y_i} \frac{\partial}{\partial x^i} - \left(\frac{\partial f}{\partial x^i} + y_i \frac{\partial f}{\partial s} \right) \frac{\partial}{\partial y_i} + \left(y_i \frac{\partial f}{\partial y_i} - f \right) \frac{\partial}{\partial s}, \qquad (6.9)$$

$$\mathbf{grad}\, f = \frac{\partial f}{\partial y_i} \frac{\partial}{\partial x^i} - \left(\frac{\partial f}{\partial x^i} + y_i \frac{\partial f}{\partial s} \right) \frac{\partial}{\partial y_i} + \left(y_i \frac{\partial f}{\partial y_i} + \frac{\partial f}{\partial s} \right) \frac{\partial}{\partial s}, \qquad (6.10)$$

$$\varepsilon_f = \frac{\partial f}{\partial y_i} \frac{\partial}{\partial x^i} - \left(\frac{\partial f}{\partial x^i} + y_i \frac{\partial f}{\partial s} \right) \frac{\partial}{\partial y_i} + y_i \frac{\partial f}{\partial y_i} \frac{\partial}{\partial s}. \qquad (6.11)$$

Therefore, if $c(t) = (x^i(t), y_i(t), s(t))$ is an integral curve of any of these vector fields, $c(t)$ should satisfy, respectively, the following systems of differential equations:

$$\frac{dx^i}{dt} = \frac{\partial f}{\partial y_i}, \quad \frac{dy_i}{dt} = -\left(\frac{\partial f}{\partial x^i} + y_i \frac{\partial f}{\partial s}\right), \quad \frac{ds}{dt} = y_i \frac{\partial f}{\partial y_i} - f, \quad (6.12)$$

$$\frac{dx^i}{dt} = \frac{\partial f}{\partial y_i}, \quad \frac{dy_i}{dt} = -\left(\frac{\partial f}{\partial x^i} + y_i \frac{\partial f}{\partial s}\right), \quad \frac{ds}{dt} = y_i \frac{\partial f}{\partial y_i} + \frac{\partial f}{\partial s},$$

$$\frac{dx^i}{dt} = \frac{\partial f}{\partial y_i}, \quad \frac{dy_i}{dt} = -\left(\frac{\partial f}{\partial x^i} + y_i \frac{\partial f}{\partial s}\right), \quad \frac{ds}{dt} = y_i \frac{\partial f}{\partial y_i}.$$

6.2 Contact Hamiltonian Dynamical Systems

As we shall see in the following sections, the contact structures and their underlying tools constitute an ideal framework for the geometric description of dissipative dynamic systems (see, for instance, [40, 93, 117, 169, 178, 235] for more details).

6.2.1 *Contact Hamiltonian systems*

Following the same guidelines as in all previous chapters, the geometric study of dissipative (nonconservative or action-dependent) Hamiltonian dynamical systems, in general, is based on the following set of postulates:

Postulate 6.1 (First postulate of contact Hamiltonian mechanics). The phase space of a regular (respectively, singular) dissipative dynamical system is a differentiable manifold M endowed with a contact (respectively, precontact) structure $\eta \in \Omega^1(M)$.

Postulate 6.2 (Second postulate of contact Hamiltonian mechanics). The observables or physical magnitudes of a dissipative dynamical system are functions of $C^\infty(M)$.

Postulate 6.3 (Third postulate of contact Hamiltonian mechanics). The dynamics of a dissipative dynamical system is given by a function $h \in C^\infty(M)$ (or in general, a closed 1-form $\alpha \in Z^1(M)$, such that

$\alpha = $ dh, locally) which is called the **Hamiltonian function** (or the **Hamiltonian 1-form**) of the system. This function represents the energy of the system.

Postulate 6.4 (Fourth postulate of contact Hamiltonian mechanics). The dynamical trajectories of the dissipative system are the integral curves of the Hamiltonian vector field $X_h \in \mathfrak{X}(M)$ associated with h, that is, of the vector field solution to equations (6.4), that are now written as,

$$i(X_h)\eta = -h, \quad i(X_h)\mathrm{d}\eta = \mathrm{d}h - \mathcal{R}(h)\,\eta. \tag{6.13}$$

Thus, these trajectories are the solutions to Eq. (6.12), which read as,

$$\frac{dx^i}{dt} = \frac{\partial h}{\partial y_i}, \quad \frac{dy_i}{dt} = -\left(\frac{\partial h}{\partial x^i} + y_i\frac{\partial h}{\partial s}\right), \quad \frac{ds}{dt} = y_i\frac{\partial h}{\partial y_i} - h. \tag{6.14}$$

It is interesting to point out the different roles of the Hamiltonian and the evolution vector fields in cosymplectic and in contact mechanics: in cosymplectic mechanics, the dynamic is given by the evolution vector field, but in contact mechanics, it is given by the Hamiltonian vector field [1].

Definition 6.4. A *regular dissipative* or *contact Hamiltonian dynamical system* is a set (M, η, h), where (M, η) is a contact manifold and $h \in C^\infty(M)$ is the Hamiltonian function of the system. If (M, η) is a precontact manifold, then (M, η, h) is said to be a **singular dissipative** or **precontact Hamiltonian dynamical system**.

Equations (6.13) and (6.14) are the **(pre)contact Hamiltonian equations** for X_h and its integral curves, respectively.

(We write "(pre)contact" to refer interchangeably to both situations, contact and precontact, and write "contact" or "precontact" to distinguish each of them in particular. In any case, we will refer to this formalism as *contact mechanics*, and we will talk about *contact dynamical systems*, in general.)

Definition 6.5. Given a dissipative Hamiltonian dynamical system (M, η, h), the **Hamiltonian problem** posed by the system consists of

---

[1]Although the terminology could be changed to make these aspects coherent, it has not been done because this is the usual nomenclature in the bibliography.

finding the Hamiltonian vector field $X_h \in \mathfrak{X}(M)$ associated with h (if it exists).

6.2.2 *Properties of contact Hamiltonian systems*

Proposition 6.4. *If (M, η, h) is a regular dissipative Hamiltonian system, then there exists a unique Hamiltonian vector field $X_h \in \mathfrak{X}(M)$, that is, a unique vector field that is the solution to Eq. (6.13).*

Proof. It is a consequence of the existence of the isomorphism $\flat_\eta$ in the regular case. $\qquad\square$

Remark 6.3. As it happens with singular dynamical systems, if $(M, \eta; h)$ is a precontact Hamiltonian system, Eq. (6.13) is not necessarily compatible everywhere on M and the corresponding constraint algorithm must be implemented to find a *final constraint submanifold* $P_f \hookrightarrow M$ (if it exists) where there are contact Hamiltonian vector fields $X_h \in \mathfrak{X}(M)$, tangent to P_f, which are the solutions (not necessarily unique) to Eq. (6.13) on P_f.

Proposition 6.5. *Let (M, η, h) be a contact Hamiltonian system. Then, the contact Hamiltonian equations (6.13) can be equivalently written as:*

$$L(X_h)\eta = -\mathcal{R}(h)\,\eta, \quad i(X_h)\eta = -h, \qquad (6.15)$$

and its integral curves $c \colon I \subset \mathbb{R} \longrightarrow M$ are solutions to the equations

$$L(\widetilde{c})(\eta \circ c)) = -((\mathcal{R}(h))\,\eta) \circ c, \quad i(\widetilde{c})(\eta \circ c)) = -h \circ c.$$

Proof. As $i(X_h)\eta = -h$, from Eq. (6.13), we obtain,
$$L(X_h)\eta = i(X_h)d\eta + d\,i(X_h)\eta = dh - \mathcal{R}(h)\,\eta - dh = -\mathcal{R}(h)\,\eta.$$
Conversely, from (6.15),
$$-\mathcal{R}(h)\,\eta \;=\; L(X_h)\eta = i(X_h)d\eta + d\,i(X_h)\eta = i(X_h)d\eta - dh$$
$$\Longleftrightarrow i(X_h)d\eta = dh - \mathcal{R}(h)\,\eta.$$
From here, the equation for the integral curves is apparent. $\qquad\square$

The first result shows that unlike in symplectic Hamiltonian systems, the geometric structure of contact Hamiltonian systems is not conserved by the dynamics, and the same happens with the Hamiltonian function. Indeed,

---

Proposition 6.6 (Dissipation of energy). *Let* (M, η, h) *be a contact Hamiltonian system. If* $X_h \in C^\infty(M)$ *is a solution to Eq. (6.13), then,*

$$L(X_h)h = -(\mathcal{R}(h))\, h. \qquad (6.16)$$

Proof. As a consequence of the above proposition, we have

$$L(X_h)h = -L(X_h)(i(X_h)\eta = -i(X_h)\,L(X_h)\eta = i(X_h)\big((L(\mathcal{R})h)\,\eta\big)$$
$$= -(L(\mathcal{R})h)\, h. \qquad \square$$

As a final result, there is another way of writing Eq. (6.13) without using the Reeb vector field $\mathcal{R}$.

Proposition 6.7. *Let* (M, η, h) *be a contact Hamiltonian system. If* $U = \{p \in M; h(p) \neq 0\}$ *and* $\Omega = -h\,d\eta + dh \wedge \eta$ *on* U, *Eq. (6.13) can also be written as:*

$$i(X_h)\Omega = 0, \quad i(X_h)\eta = -h; \quad \text{(on } U\text{)}, \qquad (6.17)$$

and its integral curves $\mathbf{c}\colon I \subset \mathbb{R} \longrightarrow M$ *are solutions to*

$$i(\tilde{c})(\Omega \circ c) = 0, \quad i(\tilde{c})(\eta \circ c)) = -h \circ \mathbf{c}; \quad \text{(on } U\text{)}.$$

Proof. If X_h satisfies Eq. (6.17), then,

$$0 = i(X_h)\Omega = -h\, i(X_h)d\eta + \big(i(X_h)dh\big)\eta - dh\, i(X_h)\eta$$
$$= -h\, i(X_h)d\eta + \big(i(X_h)dh\big)\eta + h\, dh.$$

Hence,

$$h\, i(X_h)d\eta = \big(i(X_h)dh\big)\eta + h\, dh. \qquad (6.18)$$

Contracting with the Reeb vector field,

$$0 = h\, i(\mathcal{R})\, i(X_h)d\eta = (i(X_h)dh)\, i(\mathcal{R})\eta + h\, i(\mathcal{R})dh$$

from which $i(X_h)dh = -h\, i(\mathcal{R})dh$, and using this in (6.18), we get

$$h\, i(X_h)d\eta = h\big(dh - (i(\mathcal{R})dh)\eta\big) = h\big(dh - (L(\mathcal{R})h)\,\eta\big).$$

Hence, $i(X_h)d\eta = dh - (L(\mathcal{R})h)\,\eta.$

Conversely, if X_h satisfies Eq. (6.13), then, considering (6.16),

$$i(X_h)\Omega = i(X_h)(-hd\eta + dh \wedge \eta) = -h\, i(X_h)d\eta + (i(X_h)dh)\,\eta + hdh$$
$$= h(L(\mathcal{R})h)\,\eta + (L(X_h)h)\eta = \big(h(L(\mathcal{R})h) + (L(X_h)h)\big)\,\eta = 0.$$

From here, the equation for the integral curves is straightforward. $\square$

This form of the dynamical equations is especially interesting in the case of singular systems because they do not depend on the Reeb vector field, and as we pointed out, Reeb vector fields are not uniquely determined in a precontact manifold.

6.3 Contact Lagrangian Dynamical Systems

In this section, we review the Lagrangian formulation for contact systems [93, 117, 169].

6.3.1 *Contact Lagrangian systems*

If Q is an n-dimensional manifold, consider the product manifold $TQ \times \mathbb{R}$ equipped with adapted coordinates (q^i, v^i, s). Now, the canonical projections are denoted as

$$s: TQ \times \mathbb{R} \longrightarrow \mathbb{R}, \quad \tau_1: TQ \times \mathbb{R} \longrightarrow TQ, \quad \tau_0: TQ \times \mathbb{R} \longrightarrow Q \times \mathbb{R}.$$

As discussed in Section 4.3.1, the canonical geometric structures of the tangent bundle TQ, namely the canonical endomorphism and the Liouville vector field, are extended naturally to $TQ \times \mathbb{R}$. We use the same notation as in that section to denote them: $\mathcal{J} \in \mathcal{T}_1^1(TQ \times \mathbb{R})$ and $\Delta \in \mathfrak{X}(TQ \times \mathbb{R})$, respectively, and they have the same coordinate expressions, $\mathcal{J} = \dfrac{\partial}{\partial v^i} \otimes \mathrm{d}q^i$ and $\Delta = v^i \dfrac{\partial}{\partial v^i}$.

Similarly, the canonical lift of a curve $\mathbf{c}: \mathbb{R} \to Q \times \mathbb{R}$ to $TQ \times \mathbb{R}$, with $\mathbf{c} = (\mathbf{c}^i(t), s(t))$, is defined as in Section 4.3.1, and it is $\widehat{\mathbf{c}}(t) = \left(c^i(t), \dfrac{\mathrm{d}c^i}{\mathrm{d}t}(t), s(t)\right)$. Finally, the definitions of *holonomic curves* and SODE vector fields in $TQ \times \mathbb{R}$ are also as in Section 4.3.1.

Then, the foundations of the Lagrangian formalism of the contact formulation for autonomous dissipative systems are established through the reformulation of the generic postulates stated in Section 6.2.1 for contact Hamiltonian systems:

> **Postulate 6.5 (First postulate of contact Lagrangian mechanics).**
> The configuration space of a dissipative dynamical system with n degrees of freedom is $Q \times \mathbb{R}$, where Q is an n-dimensional differentiable manifold, and n are the degrees of freedom of the system. The phase space is the bundle $TQ \times \mathbb{R}$.

Postulate 6.6 (Second postulate of contact Lagrangian mechanics). The observables or physical magnitudes of a dissipative dynamical system are functions of $C^\infty(TQ \times \mathbb{R})$.

Postulate 6.7 (Third postulate of contact Lagrangian mechanics). There is a function $\mathcal{L} \in C^\infty(TQ \times \mathbb{R})$, called the **contact Lagrangian function** that carries the dynamical information of the system.

Remark 6.4. In many cases, the contact Lagrangian function is of the following type: $\mathcal{L} = L + \phi$, where $L = \tau_1^* L_o$ for a function $L_o \in C^\infty(TQ)$, and $\phi = \tau_0^* \phi_o$, for $\phi_o \in C^\infty(Q \times \mathbb{R})$, that is, in coordinates, $\mathcal{L}(q^i, v^i, s) = L(q^i, v^i) + \phi(q^i, s)$. In particular, the case where $\phi(q^i, s) = \gamma s$, with $\gamma \in \mathbb{R}$, appears very frequently in physical applications.

Starting from a Lagrangian function, the following magnitudes and structures are defined:

Definition 6.6. Let $\mathcal{L} \in C^\infty(TQ \times \mathbb{R})$ be a Lagrangian function. The **Lagrangian energy** associated with $\mathcal{L}$ is the function

$$E_\mathcal{L} := \Delta(\mathcal{L}) - \mathcal{L} \in C^\infty(TQ \times \mathbb{R}).$$

The **Cartan Lagrangian forms** associated with $\mathcal{L}$ are defined as

$$\theta_\mathcal{L} := \mathcal{J}(\mathrm{d}\mathcal{L}) \in \Omega^1(TQ \times \mathbb{R}), \quad \omega_\mathcal{L} := -\mathrm{d}\theta_\mathcal{L} \in \Omega^2(TQ \times \mathbb{R}). \quad (6.19)$$

The **(pre)contact Lagrangian form** is

$$\eta_L = \mathrm{d}s - \theta_\mathcal{L} \in \Omega^1(TQ \times \mathbb{R}),$$

and it satisfies $\mathrm{d}\eta_L = \omega_L$.

In natural coordinates (q^i, v^i, s) on $TQ \times \mathbb{R}$, we have

$$\eta_L = \mathrm{d}s - \frac{\partial L}{\partial v^i} \, \mathrm{d}q^i,$$

$$\mathrm{d}\eta_L = -\frac{\partial^2 L}{\partial s \partial v^i} \mathrm{d}s \wedge \mathrm{d}q^i - \frac{\partial^2 L}{\partial q^j \partial v^i} \mathrm{d}q^j \wedge \mathrm{d}q^i - \frac{\partial^2 L}{\partial v^j \partial v^i} \mathrm{d}v^j \wedge \mathrm{d}q^i.$$

Definition 6.7. Given a Lagrangian $\mathcal{L} \in C^\infty(TQ \times \mathbb{R})$, the **Legendre map** associated with $\mathcal{L}$ is the fibre derivative of $\mathcal{L}$ and considered as a function on the vector bundle $\pi_0 \colon TQ \times \mathbb{R} \longrightarrow Q \times \mathbb{R}$, that is, the map $\mathfrak{F}\mathcal{L} \colon TQ \times \mathbb{R} \longrightarrow T^*Q \times \mathbb{R}$ which is given by

$$\mathfrak{F}\mathcal{L}(q, v_q, s) = (\mathfrak{F}\mathcal{L}_s(q, v_q), s),$$

where $\mathcal{L}_s \colon TQ \longrightarrow \mathbb{R}$ denotes the restriction of $\mathcal{L}$ to each fibre of the bundle $TQ \times \mathbb{R} \longrightarrow \mathbb{R}$ (i.e., the Lagrangian $\mathcal{L}$ with s "frozen").

In natural coordinates, we have

$$q^i \circ \mathfrak{F}\mathcal{L} = q^i, \quad p_i \circ \mathfrak{F}\mathcal{L} = \frac{\partial \mathcal{L}}{\partial v^i}, \quad s \circ \mathfrak{F}\mathcal{L} = s.$$

Observe that if $\theta \in \Omega^1(T^*Q \times \mathbb{R})$ is the canonical form and $\omega = -d\theta$,

$$\theta_\mathcal{L} = \mathfrak{F}\mathcal{L}^*\theta, \quad \omega_\mathcal{L} = \mathfrak{F}\mathcal{L}^*\omega.$$

Proposition 6.8. *For a Lagrangian function $\mathcal{L}$, the following conditions are equivalent:*

(1) The Legendre map $\mathfrak{F}\mathcal{L}$ is a local diffeomorphism.
(2) The pair $(TQ \times \mathbb{R}, \boldsymbol{\eta}_\mathcal{L})$ is a contact manifold.
(3) The Hessian matrix $W_{ij} \equiv \left(\dfrac{\partial^2 \mathcal{L}}{\partial v^i \partial v^j} \right)$ is nondegenerate everywhere.

Proof. The proof can be done using natural coordinates. $\qquad\square$

Definition 6.8. A Lagrangian function $\mathcal{L}$ is said to be **regular** if the equivalent conditions in the above Proposition 6.8 hold. Otherwise, $\mathcal{L}$ is called a **singular** Lagrangian. In particular, $\mathcal{L}$ is said to be **hyperregular** if $\mathfrak{F}\mathcal{L}$ is a global diffeomorphism.

Remark 6.5. If $\mathcal{L}$ is a regular Lagrangian, then $(TQ \times \mathbb{R}, \boldsymbol{\eta}_\mathcal{L}, E_\mathcal{L})$ is a contact Hamiltonian system. When $\mathcal{L}$ is not regular, it can induce a precontact structure, but also a structure, which is neither contact nor precontact. For example, the Lagrangian $\mathcal{L} = \sum_{i=1}^{n} v^i s$ in $TQ \times \mathbb{R}$ gives a form $\boldsymbol{\eta}_\mathcal{L} \in \Omega^1(TQ \times \mathbb{R})$ for which condition (6.1) does not hold and that has no Reeb vector fields associated with it (see Remark 6.2).

Definition 6.9. If $\mathcal{L} \in C^\infty(TQ \times \mathbb{R})$ is a Lagrangian and $(TQ \times \mathbb{R}, \eta_{\mathcal{L}})$ is a contact or a precontact manifold, then the pair $(TQ \times \mathbb{R}, \mathcal{L})$ is said to be a *(pre)contact **Lagrangian dynamical system.***

Given a contact Lagrangian system $(TQ \times \mathbb{R}, \mathcal{L})$, the Reeb vector field $\mathcal{R}_{\mathcal{L}} \in \mathfrak{X}(TQ \times \mathbb{R})$ is uniquely determined by the conditions

$$i(\mathcal{R}_{\mathcal{L}}) \mathrm{d}\eta_{\mathcal{L}} = 0, \quad i(\mathcal{R}_{\mathcal{L}}) \eta_{\mathcal{L}} = 1 \tag{6.20}$$

and its local expression is

$$\mathcal{R}_{\mathcal{L}} = \frac{\partial}{\partial s} - W^{ji} \frac{\partial^2 \mathcal{L}}{\partial s \partial v^j} \frac{\partial}{\partial v^i}, \tag{6.21}$$

where (W^{ij}) is the inverse of the partial Hessian matrix, namely, $W^{ij} W_{jk} = \delta^i_k$. Observe that the Reeb vector field does not appear in the simplest form $\frac{\partial}{\partial s}$ since, in general, the natural coordinates in $TQ \times \mathbb{R}$ are neither adapted nor Darboux coordinates for $\eta_{\mathcal{L}}$. If $(TQ \times \mathbb{R}, \mathcal{L})$ is a precontact Lagrangian system, then Reeb vector fields are not uniquely defined.

Postulate 6.8 (Fourth postulate of contact Lagrangian mechanics). The dynamical trajectories of the system are the integral curves of a vector field $X_{\mathcal{L}} \in \mathfrak{X}(TQ \times \mathbb{R})$ satisfying the conditions:

(1) $X_{\mathcal{L}}$ is the Hamiltonian vector field associated with $E_{\mathcal{L}}$, that is, if $\mathcal{R}_{\mathcal{L}}$ is a Reeb vector field determined by Eq. (6.20), then $X_{\mathcal{L}}$ is a solution to the equations

$$i(X_{\mathcal{L}}) \eta_{\mathcal{L}} = -E_{\mathcal{L}}, \quad i(X_{\mathcal{L}}) \mathrm{d}\eta_{\mathcal{L}} = \mathrm{d}E_{\mathcal{L}} - (\mathcal{R}_{\mathcal{L}}(E_{\mathcal{L}})) \eta_{\mathcal{L}}, \tag{6.22}$$

(2) $X_{\mathcal{L}}$ is a SODE: $\mathcal{J}(X_{\mathcal{L}}) = \Delta$.

Therefore, these trajectories are the holonomic curves $\mathbf{c} \colon I \subset \mathbb{R} \longrightarrow TQ \times \mathbb{R}$, which are the solutions to the equations

$$i(\widetilde{\mathbf{c}})(\eta_{\mathcal{L}} \circ \mathbf{c}) = -E_{\mathcal{L}} \circ \mathbf{c}, \quad i(\widetilde{\mathbf{c}})(\mathrm{d}\eta_{\mathcal{L}} \circ \mathbf{c}) = \big(\mathrm{d}E_{\mathcal{L}} - (\mathcal{R}_{\mathcal{L}}(E_{\mathcal{L}})) \eta_{\mathcal{L}}\big) \circ \mathbf{c}. \tag{6.23}$$

Equations (6.23) are called the *(pre)contact **Euler–Lagrange equations for curves.*** Equations (6.22) are called the *(pre)contact **Lagrangian equations for vector fields,*** and a vector field solution to them (if it exists) is a *(pre)contact **Lagrangian dynamical vector field.*** In addition, if $X_{\mathcal{L}}$ is a SODE, then it is called a *(pre)contact **Euler–Lagrange vector field*** of the system.

Definition 6.10. Given a (pre)contact Lagrangian dynamical system $(TQ \times \mathbb{R}, \mathcal{L})$, the **Lagrangian problem** posed by this system consists of finding a SODE vector field $X_{\mathcal{L}} \in \mathfrak{X}(TQ \times \mathbb{R})$ solution to (6.22).

6.3.2 The contact Euler–Lagrange equations

First, using (6.21), a simple computation in coordinates shows that

$$\mathcal{R}_{\mathcal{L}}(E_{\mathcal{L}}) = -\frac{\partial \mathcal{L}}{\partial s}. \tag{6.24}$$

Taking this into account, in natural coordinates, for a holonomic curve $\mathbf{c}(t) = (q^i(t), \dot{q}^i(t), s(t))$ on $TQ \times \mathbb{R}$, Eq. (6.23) gives the **Herglotz–Euler–Lagrange equations**,

$$\frac{ds}{dt} = \mathcal{L}, \tag{6.25}$$

$$\frac{d}{dt}\left(\frac{\partial \mathcal{L}}{\partial v^i}\right) - \frac{\partial \mathcal{L}}{\partial q^i} = \frac{\partial^2 \mathcal{L}}{\partial v^j \partial v^i}\frac{d^2 q^j}{dt^2} + \frac{\partial^2 \mathcal{L}}{\partial q^j \partial v^i}\frac{dq^j}{dt} + \frac{\partial^2 L}{\partial s \partial v^i}\frac{ds}{dt} - \frac{\partial \mathcal{L}}{\partial q^i}$$

$$= \frac{\partial \mathcal{L}}{\partial s}\frac{\partial \mathcal{L}}{\partial v^i}. \tag{6.26}$$

Furthermore, for a vector field $X_{\mathcal{L}} = f^i\frac{\partial}{\partial q^i} + F^i\frac{\partial}{\partial v^i} + g\frac{\partial}{\partial s}$, Eq. (6.22) leads to

$$0 = \mathcal{L} + \frac{\partial \mathcal{L}}{\partial v^i}(f^i - v^i) - g, \tag{6.27}$$

$$0 = (f^j - v^j)\frac{\partial^2 \mathcal{L}}{\partial q^i \partial v^j} + \frac{\partial \mathcal{L}}{\partial q^i} - \frac{\partial^2 \mathcal{L}}{\partial s \partial v^i}g - \frac{\partial^2 \mathcal{L}}{\partial q^j \partial v^i}f^j$$

$$- \frac{\partial^2 \mathcal{L}}{\partial v^j \partial v^i}F^j + \frac{\partial \mathcal{L}}{\partial s}\frac{\partial \mathcal{L}}{\partial v^i}, \tag{6.28}$$

$$0 = (f^j - v^j)\frac{\partial^2 \mathcal{L}}{\partial v^i \partial v^j}, \tag{6.29}$$

$$0 = (f^j - v^j)\frac{\partial^2 \mathcal{L}}{\partial v^j \partial s}, \tag{6.30}$$

and then it is straightforward to prove:

Proposition 6.9. *If $\mathcal{L}$ is a regular Lagrangian, then $X_{\mathcal{L}}$ is a SODE and Eqs. (6.27) and (6.28) become*

$$g = \mathcal{L}, \tag{6.31}$$

$$\frac{\partial^2 \mathcal{L}}{\partial v^j \partial v^i}F^j + \frac{\partial^2 \mathcal{L}}{\partial q^j \partial v^i}v^j + \frac{\partial^2 \mathcal{L}}{\partial s \partial v^i}\mathcal{L} - \frac{\partial \mathcal{L}}{\partial q^i} = \frac{\partial \mathcal{L}}{\partial s}\frac{\partial \mathcal{L}}{\partial v^i}, \tag{6.32}$$

which, for the integral curves of $X_{\mathcal{L}}$, give Eqs. (6.25) and (6.26).

Proof. In fact, if $\mathcal{L}$ is a regular Lagrangian, Eq. (6.29) leads to $v^i = f^i$, which is the SODE condition for the vector field $X_{\mathcal{L}}$. Then, (6.30) holds identically, and (6.27) and (6.28) give Eqs. (6.31) and (6.32), and then, for the integral curves of X_L, we get Eqs. (6.25) and (6.26).   □

Thus, the local expression of this Euler–Lagrange vector field is

$$X_{\mathcal{L}} = v^i \frac{\partial}{\partial q^i} + W^{ik}\left(\frac{\partial \mathcal{L}}{\partial q^k} - \frac{\partial^2 \mathcal{L}}{\partial q^j \partial v^k} v^j - \mathcal{L}\frac{\partial^2 \mathcal{L}}{\partial s \partial v^k} + \frac{\partial \mathcal{L}}{\partial s}\frac{\partial \mathcal{L}}{\partial v^k}\right)\frac{\partial}{\partial v^i} + \mathcal{L}\frac{\partial}{\partial s}.$$
$$(6.33)$$

Remark 6.6. It is important to point out that the expression in coordinates (6.25) of the second Lagrangian equation (6.22) relates the variation of the "dissipation coordinate" s to the Lagrangian function, and from here, we can identify this coordinate with the Lagrangian action,

$$s = \int \mathcal{L}\,dt.$$

Remark 6.7. If $\mathcal{L}$ is a singular Lagrangian, but $(TQ \times \mathbb{R}, \eta_{\mathcal{L}})$ is a precontact manifold, although the Reeb vector fields are not uniquely defined, the Lagrangian equations (6.22) are independent from the Reeb vector field used (see [117]). Alternatively, the Reeb-independent Eq. (6.17) for the precontact Hamiltonian system $(TQ \times \mathbb{R}, \eta_{\mathcal{L}}, E_{\mathcal{L}})$ can be used instead. In any case, solutions to the Lagrangian equations are not necessarily SODE and the condition $\mathcal{J}(X_{\mathcal{L}}) = \Delta$ must be added to the Lagrangian equations. Furthermore, these equations are not necessarily compatible everywhere on $TQ \times \mathbb{R}$ and a constraint algorithm must be implemented to find a final constraint submanifold $S_f \hookrightarrow TQ \times \mathbb{R}$ (if it exists) where there are SODE vector fields $X_{\mathcal{L}} \in \mathfrak{X}(TQ \times \mathbb{R})$, tangent to S_f, which are solutions to the above equations on S_f [117].

6.3.3 Canonical Hamiltonian formalism

Consider a hyperregular Lagrangian system $(TQ \times \mathbb{R}, \mathcal{L})$ (the regular case is the same, taking $\mathfrak{F}\mathcal{L}(TQ \times \mathbb{R}) \subset T^*Q \times \mathbb{R}$ instead of $T^*Q \times \mathbb{R}$, or locally, at least). Then, $\mathfrak{F}\mathcal{L}$ is a diffeomorphism between the contact manifolds $(TQ \times \mathbb{R}, \eta_{\mathcal{L}})$ and the canonical contact manifold $(T^*Q \times \mathbb{R}, \eta)$, with

$$\mathfrak{F}\mathcal{L}^*\eta = \eta_{\mathcal{L}}, \quad \mathfrak{F}\mathcal{L}_*\mathcal{R}_{\mathcal{L}} = \mathcal{R}.$$

Furthermore, there exists a function $h \in C^\infty(T^*Q \times \mathbb{R})$ such that $\mathfrak{F}\mathcal{L}^* h = E_\mathcal{L}$; so we have the contact Hamiltonian system $(T^*Q \times \mathbb{R}, \eta, h)$ and the contact Hamiltonian equations (6.13) (or their equivalent expressions (6.15) or (6.17)), which read as

$$i(X_h)\eta = -h, \quad i(X_h)d\eta = dh - \mathcal{R}(h)\,\eta, \tag{6.34}$$

and have a unique solution X_h. The dynamical trajectories are the integral curves of this contact Hamiltonian vector field, and then, they are the solutions to Eq. (6.14), which read as

$$\frac{dx^i}{dt} = \frac{\partial h}{\partial p_i}, \quad \frac{dp_i}{dt} = -\left(\frac{\partial h}{\partial x^i} + p_i \frac{\partial h}{\partial s}\right), \quad \frac{ds}{dt} = p_i \frac{\partial h}{\partial p_i} - h. \tag{6.35}$$

Then, if $X_h \in \mathfrak{X}(T^*Q \times \mathbb{R})$ is the contact Hamiltonian vector field associated with h, we have $\mathfrak{F}\mathcal{L}_* X_\mathcal{L} = X_h$.

For singular Lagrangians, as in the above chapters, we have:

Definition 6.11. A singular Lagrangian function $\mathcal{L} \in C^\infty(TQ \times \mathbb{R})$ is called **almost-regular** if $\mathcal{P} := \mathfrak{F}\mathcal{L}(TQ \times \mathbb{R})$ is a closed submanifold of $T^*Q \times \mathbb{R}$ (the natural embedding is denoted $j_\mathcal{P}: \mathcal{P} \hookrightarrow T^*Q \times \mathbb{R}$), $\mathfrak{F}\mathcal{L}$ is a submersion onto its image, and the fibres $\mathfrak{F}\mathcal{L}^{-1}(FL(p))$, for every $p \in TQ \times \mathbb{R}$, are connected submanifolds of $TQ \times \mathbb{R}$.

In these cases, we have $(\mathcal{P}, \eta_\mathcal{P})$, where $\eta_\mathcal{P} = j_\mathcal{P}^* \eta \in \Omega^1(\mathcal{P})$. Furthermore, the Lagrangian energy function $E_\mathcal{L}$ is $\mathfrak{F}\mathcal{L}$-projectable, that is, there is a unique $h_\mathcal{P} \in C^\infty(\mathcal{P})$, which is $\mathfrak{F}\mathcal{L}$-related with $E_\mathcal{L}$. Therefore, if $(\mathcal{P}, \eta_\mathcal{P})$ is a contact or a precontact manifold, then $(\mathcal{P}, \eta_\mathcal{P}, h_\mathcal{P})$ is a (pre)contact Hamiltonian system whose (pre)contact Hamiltonian equations are (6.13), or (6.17), adapted to this situation. As in the Lagrangian formalism, these equations are not necessarily compatible everywhere on $\mathcal{P}$, and a constraint algorithm must be implemented to find a final constraint submanifold $P_f \hookrightarrow \mathcal{P}$ (if it exists) where there are vector fields $X_{h_\mathcal{P}} \in \mathfrak{X}(\mathcal{P})$, tangent to P_f, which are solutions to the precontact Hamiltonian equations on P_f. (See [117] for a detailed analysis on all these topics.)

6.4 Unified Lagrangian–Hamiltonian Formalism for Contact Systems

The Lagrangian–Hamiltonian unified formalism of contact Lagrangian systems has been developed in [114]. Next, we present its main features.

6.4.1 Extended precontact unified bundle: Unified formalism

Definition 6.12. Let Q be an n-dimensional differentiable manifold. The **extended precontact unified bundle** (or **extended precontact Pontryagin bundle**) is

$$\mathfrak{M} = TQ \times_Q T^*Q \times \mathbb{R},$$

and it is endowed with the natural projections

$$\rho_1 \colon \mathfrak{M} \longrightarrow TQ \times \mathbb{R}, \quad \rho_2 \colon \mathfrak{M} \longrightarrow T^*Q \times \mathbb{R}, \quad \rho_0 \colon \mathfrak{M} \longrightarrow Q \times \mathbb{R}, \quad s \colon \mathfrak{M} \longrightarrow \mathbb{R}.$$

Natural coordinates in $\mathfrak{M}$ are (q^i, v^i, p_i, s).

Definition 6.13. A curve $\boldsymbol{\sigma} \colon \mathbb{R} \to \mathfrak{M}$ is said to be **holonomic** in $\mathfrak{M}$ if $\rho_1 \circ \boldsymbol{\sigma} \colon \mathbb{R} \longrightarrow TQ \times \mathbb{R}$ is holonomic curve. A vector field $\Gamma \in \mathfrak{X}(\mathfrak{M})$ satisfies the **second-order condition** in $\mathfrak{M}$ (for short: it is a SODE in $\mathfrak{M}$) if its integral curves are holonomic in $\mathfrak{M}$.

In coordinates, a holonomic curve and a SODE in $\mathfrak{M}$ are expressed as

$$\boldsymbol{\sigma} = \left(\sigma_1^i(t), \frac{d\sigma_1^i}{dt}(t), \sigma_{2\,i}(t), \sigma_0(t) \right),$$

$$\Gamma = v^i \frac{\partial}{\partial q^i} + F^i \frac{\partial}{\partial v^i} + G_i \frac{\partial}{\partial p_i} + f \frac{\partial}{\partial s}.$$

Definition 6.14. The bundle $\mathfrak{M}$ is endowed with the following canonical structures:

(1) The **coupling function** in $\mathfrak{M}$ is the map $\mathfrak{C} \colon \mathfrak{M} \longrightarrow \mathbb{R}$ defined as follows: for every $w = (v_q, p_q, s) \in \mathfrak{M}$ where $q \in Q$, $p_q \in T^*Q$ and $v_q \in TQ$, then $\mathfrak{C}(w) := \langle p_q, v_q \rangle$.

(2) The **canonical precontact 1-form** on $\mathfrak{M}$ is the ρ_1-semibasic form

$$\eta_{\mathfrak{M}} := \rho_2^* \eta = ds - \pi_1^* \rho_2^* \pi_1^* \Theta \in \Omega^1(\mathfrak{M}),$$

where η is the canonical contact form on $T^*Q \times \mathbb{R}$ (and Θ is the canonical 1-form on T^*Q).

In natural coordinates of $\mathfrak{M}$, we have $\eta_{\mathfrak{M}} = ds - p_i \, dq^i$.

Definition 6.15. Given a Lagrangian function $\mathcal{L} \in C^\infty(TQ \times \mathbb{R})$, let $\mathfrak{L} = \rho_1^* \mathcal{L} \in C^\infty(\mathfrak{M})$. The **Hamiltonian function** is defined as

$$\mathbf{H} := \mathfrak{C} - \mathfrak{L} = p_i v^i - \mathcal{L}(q^j, v^j, s) \in C^\infty(\mathfrak{M}). \qquad (6.36)$$

Remark 6.8. Observe that $(\mathfrak{M}, \eta_{\mathfrak{M}})$ is a precontact manifold. As a consequence, the Reeb vector fields are not uniquely defined, and in natural coordinates of $\mathfrak{M}$, the vector fields $\mathcal{R} = \dfrac{\partial}{\partial z} + F^i \dfrac{\partial}{\partial v^i}$ for arbitrary functions F^i are the general solution to (6.2). Nevertheless, as we have pointed out, the dynamics obtained from the formalism are independent of the choice of the Reeb vector fields; therefore, as $\mathfrak{M} = TQ \times_Q T^*Q \times \mathbb{R}$ is a trivial bundle over $\mathbb{R}$, the canonical vector field $\dfrac{\partial}{\partial s}$ on $\mathbb{R}$ can be lifted canonically to a vector field in $\mathfrak{M}$, which can be taken as a representative of the family of these Reeb vector fields.

Thus, we have $(\mathfrak{M}, \eta_{\mathfrak{M}}, \mathbf{H})$ as a precontact Hamiltonian system, and then, we can pose the dynamic problem for this system, which consists of finding a Hamiltonian vector field that is a solution to the precontact Hamiltonian equations

$$i(X_{\mathbf{H}})d\eta_{\mathfrak{M}} = d\mathbf{H} - (\mathcal{R}(\mathbf{H}))\eta_{\mathfrak{M}}, \quad i(X_{\mathbf{H}})\eta_{\mathfrak{M}} = -\mathbf{H}. \qquad (6.37)$$

Then, the integral curves $\sigma : I \subset \mathbb{R} \longrightarrow \mathfrak{M}$ of $X_{\mathbf{H}}$ are the solutions to the equations

$$i(\tilde{\sigma})(d\eta_{\mathfrak{M}} \circ \sigma) = (d\mathbf{H} - (\mathcal{R}(\mathbf{H}))\eta_{\mathfrak{M}}) \circ \sigma, \quad i(\tilde{\sigma})(\eta_{\mathfrak{M}} \circ \sigma) = -\mathbf{H} \circ \sigma. \quad (6.38)$$

As we shall see next, these equations are not compatible on $\mathfrak{M}$, and we have to implement the constraint algorithm to find the final constraint submanifold of $\mathfrak{M}$ where there are consistent solutions to the equations. In fact, in natural coordinates of $\mathfrak{M}$, we have

$$d\mathbf{H} = v^i dp_i + \left(p_i - \frac{\partial \mathcal{L}}{\partial v^i} \right) dv^i - \frac{\partial \mathcal{L}}{\partial q^i} dq^i - \frac{\partial \mathcal{L}}{\partial s} ds,$$

and if the local expression of a vector field $X_{\mathbf{H}} \in \mathfrak{X}(\mathfrak{M})$ is

$$X_{\mathbf{H}} = f^i \frac{\partial}{\partial q^i} + F^i \frac{\partial}{\partial v^i} + G_i \frac{\partial}{\partial p_i} + f \frac{\partial}{\partial s},$$

we obtain

$$i(X_{\mathbf{H}})\eta_{\mathfrak{M}} = f - f^i p_i, \quad i(X_{\mathbf{H}})d\eta_{\mathfrak{M}} = f^i dp_i - G_i dq^i,$$

$$(\mathcal{R}(\mathbf{H}))\eta_{\mathfrak{M}} = -\frac{\partial \mathcal{L}}{\partial s}(ds - p_i dq^i).$$

Then, Eq. (6.37) gives

$$f = (f^i - v^i)\, p_i + \mathcal{L}, \tag{6.39}$$

$$f^i = v^i, \tag{6.40}$$

$$p_i = \frac{\partial \mathcal{L}}{\partial v^i}, \tag{6.41}$$

$$G_i = \frac{\partial \mathcal{L}}{\partial q^i} + p_i \frac{\partial \mathcal{L}}{\partial s}. \tag{6.42}$$

Equation (6.40) is the holonomy condition (i.e., $X_{\mathbf{H}}$ must be a SODE), which arise naturally from the unified formalism. The algebraic Eq. (6.41) are compatibility conditions defining the submanifold $\mathfrak{M}_0 \hookrightarrow \mathfrak{M}$, where the equations are compatible. This $\mathfrak{M}_0$ is the *first constraint submanifold* of the Hamiltonian precontact system $(\mathfrak{M}, \eta_{\mathfrak{M}}, \mathbf{H})$ and is the graph of $\mathfrak{F}\mathcal{L}$, that is,

$$\mathfrak{M}_0 = \{(v_q, \mathfrak{F}\mathcal{L}(v_q)) \in \mathfrak{M}, \text{ for } v_q \in TQ\}.$$

In this way, as usual, the unified formalism includes the definition of the Legendre map as a consequence of the constraint algorithm.

Thus, vector fields that are solutions to (6.37) are

$$X_{\mathbf{H}} = v^i \frac{\partial}{\partial q^i} + F^i \frac{\partial}{\partial v^i} + \left(\frac{\partial \mathcal{L}}{\partial q^i} + p_i \frac{\partial \mathcal{L}}{\partial z}\right) \frac{\partial}{\partial p_i} + \mathcal{L} \frac{\partial}{\partial s} \quad (\text{on } \mathfrak{M}_0).$$

The constraint algorithm continues by demanding that $X_{\mathbf{H}}$ must be tangent to $\mathfrak{M}_0$. As $\xi_j^1 = p_j - \frac{\partial \mathcal{L}}{\partial v^j} \in C^\infty(\mathfrak{M})$ are the constraints defining $\mathfrak{M}_0$, this condition is

$$X_{\mathbf{H}}\left(p_j - \frac{\partial \mathcal{L}}{\partial v^j}\right) = -\frac{\partial^2 \mathcal{L}}{\partial q^i \partial v^j} v^i - \frac{\partial^2 \mathcal{L}}{\partial v^i \partial v^j} F^i - \mathcal{L}\frac{\partial^2 \mathcal{L}}{\partial z \partial v^j} + \frac{\partial \mathcal{L}}{\partial q^j} + p_j \frac{\partial \mathcal{L}}{\partial s}$$

$$= 0 \quad (\text{on } \mathfrak{M}_0), \tag{6.43}$$

and there are two options:

- If $\mathcal{L}$ is a regular Lagrangian, these equations determine all the functions F^i. Then the solution is unique, and the algorithm ends.
- If $\mathcal{L}$ is singular, then these equations establish relationships among the arbitrary functions F^i, so some of them remain undetermined, and the solutions are not unique. Furthermore, new constraints $\xi_\mu^2 \in C^\infty(\mathfrak{M})$ could appear, and then we would define a new submanifold $\mathfrak{M}_1 \hookrightarrow \mathfrak{M}_0 \hookrightarrow \mathfrak{M}$. Then, the algorithm continues by demanding the tangency of $X_{\mathbf{H}}$ to $\mathfrak{M}_1$, until we obtain a final constraint submanifold $\mathfrak{M}_f$ (if it exists) where tangent solutions $X_{\mathbf{H}}$ exist.

If $\boldsymbol{\sigma}(t) = (q^i(t), v^i(t), p_i(t), s(t))$ is an integral curve of $X_{\mathbf{H}}$, we have $f^i = \dfrac{dq^i}{dt}$, $F^i = \dfrac{dv^i}{dt}$, $G_i = \dfrac{dp_i}{dt}$, $f = \dfrac{ds}{dt}$, and Eqs. (6.39), (6.40), (6.41), and (6.42) contribute to the coordinate expression of Eq. (6.38), in particular:

- From (6.40), we have $v^i = \dfrac{dq^i}{dt}$, that is, the holonomy condition.
- Using (6.40) again, Eq. (6.39) gives Eq. (6.25) again.
- Equation (6.42) reads

$$\frac{dp_i}{dt} = \frac{\partial \mathcal{L}}{\partial q^i} + p_i \frac{\partial \mathcal{L}}{\partial s} = -\left(\frac{\partial \mathbf{H}}{\partial q^i} + p_i \frac{\partial \mathbf{H}}{\partial s}\right),$$

which is the second group of Hamilton's equations (6.12). Then, using (6.41), these equations are (on $\mathfrak{M}_0$),

$$\frac{d}{dt}\left(\frac{\partial \mathcal{L}}{\partial v^i}\right) = \frac{\partial \mathcal{L}}{\partial q^i} + \frac{\partial \mathcal{L}}{\partial v^i}\frac{\partial \mathcal{L}}{\partial s},$$

which are the Herglotz–Euler-Lagrange equations (6.26). The first group of Hamilton's equations (6.12) arises from the definition of the Hamiltonian function (6.36) using the holonomy condition.

- Using (6.41) and (6.25), the tangency condition (6.43) again gives the Herglotz–Euler–Lagrange equations (6.26). If $\mathcal{L}$ is singular, these equations may be incompatible.

6.4.2   *Recovering the Lagrangian and Hamiltonian formalisms and equivalence*

Now, we study the equivalence of the unified formalism with the Lagrangian and Hamiltonian formalisms for the hyperregular case (the regular case is the same, at least locally). See [114] for details about the singular case.

First, observe that in this case, denoted by $\jmath_0 \colon \mathfrak{M}_0 \hookrightarrow \mathfrak{M}$ the natural embedding, we have

$$(\rho_1 \circ \jmath_0)(\mathfrak{M}_0) = TQ \times \mathbb{R}, \quad (\rho_2 \circ \jmath_0)(\mathfrak{M}_0) = T^*Q \times \mathbb{R}.$$

Furthermore, as $\mathfrak{M}_0$ is the graph of the Legendre map $\mathfrak{F}\mathcal{L}$, it is diffeomorphic to $TQ \times \mathbb{R}$, being the restricted projection $\rho_1 \circ \jmath_0$ of this diffeomorphism.

So, we have the diagram

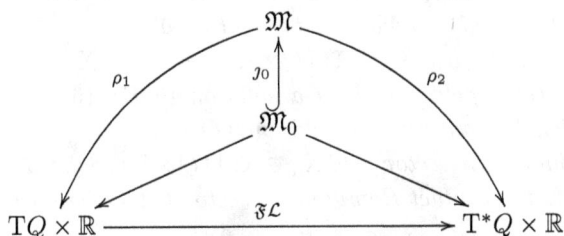

As in the other unified formalisms, the functions and differential forms on $\mathfrak{M}$, as well as vector fields on $\mathfrak{M}$ tangent to $\mathfrak{M}_0$, can be restricted to $\mathfrak{M}_0$. Then, they can be translated to the Lagrangian or the Hamiltonian side by using $\mathfrak{M}_0$ which is diffeomorphic to $TQ \times \mathbb{R}$, or projecting to $T^*Q \times \mathbb{R}$. In particular, a simple calculation in coordinates shows that

$$\rho_1^* E_L = \mathbf{H} = \rho_2^* \mathrm{h}.$$

Furthermore, every curve $\sigma \colon I \subseteq \mathbb{R} \longrightarrow \mathfrak{M}$ that takes values in $\mathfrak{M}_0$ can be split as $\sigma = (\sigma_L, \sigma_H)$, where $\sigma_L = \rho_1 \circ \sigma \colon I \subseteq \mathbb{R} \longrightarrow TQ \times \mathbb{R}$ and $\sigma_H = \mathfrak{F}\mathcal{L} \circ \sigma_L \colon I \subseteq \mathbb{R} \longrightarrow T^*Q \times \mathbb{R}$. Therefore, the results and discussion in the above section lead to

---

Theorem 6.3. *Let $\sigma \colon \mathbb{R} \longrightarrow \mathfrak{M}$, with $\mathrm{Im}\,(\sigma) \subset \mathfrak{M}_0$, be a curve solution to Eq. (6.38) on $\mathfrak{M}_0$. Then, σ_L is the lift of the curve $\mathbf{c} = \rho_0 \circ \sigma \colon \mathbb{R} \longrightarrow Q \times \mathbb{R}$ to $TQ \times \mathbb{R}$ (i.e., σ_L is a holonomic curve), and it is a solution to Eq. (6.23). Moreover, the curve $\sigma_H = \mathfrak{F}\mathcal{L} \circ \tilde{\mathbf{c}}$ is a solution to Eq. (6.8) on $T^*Q \times \mathbb{R}$.*

Conversely, for every curve $\mathbf{c} \colon \mathbb{R} \longrightarrow Q \times \mathbb{R}$ such that $\tilde{\mathbf{c}}$ is a solution to Eq. (6.23), we have that the curve $\sigma = (\tilde{\mathbf{c}}, \mathfrak{F}\mathcal{L} \circ \tilde{\mathbf{c}})$ is a solution to Eq. (6.38) on $\mathfrak{M}_0$, and the curve $\mathfrak{F}\mathcal{L} \circ \tilde{\mathbf{c}}$ is a solution to Eq. (6.8) on $T^*Q \times \mathbb{R}$.

The curves $\sigma \colon \mathbb{R} \longrightarrow \mathfrak{M}$, which are a solution to Eq. (6.38), are the integral curves of SODE vector fields $X_{\mathbf{H}} \in \mathfrak{X}(\mathfrak{M})$ which are a solution to (6.37) on $\mathfrak{M}_0$. The curves $\sigma_{\mathcal{L}} \colon \mathbb{R} \longrightarrow TQ \times \mathbb{R}$ are the integral curves of SODE vector fields $X_{\mathcal{L}} \in \mathfrak{X}(TQ \times \mathbb{R})$, and a solution to (6.23). Finally, the curves $\sigma_{\mathrm{h}} \colon \mathbb{R} \longrightarrow T^*Q \times \mathbb{R}$ are the integral curves of the contact Hamiltonian vector field $X_{\mathrm{h}} \in \mathfrak{X}(T^*Q \times \mathbb{R})$, and a solution to (6.34) on $T^*Q \times \mathbb{R}$. Then, a logical corollary of the above theorem is the following:

Theorem 6.4. *Let* $X_H \in \mathfrak{X}(\mathfrak{M})$ *be the contact Hamiltonian vector field solution to Eq.* (6.37) *on* $\mathfrak{M}_0$ *and tangent to* $\mathfrak{M}_0$.

Then the vector field $X_{\mathcal{L}} \in \mathfrak{X}(TQ \times \mathbb{R})$, *defined by* $X_{\mathcal{L}} \circ \rho_1 = T\rho_1 \circ X_H$, *is a* SODE *vector field, which is a solution to Eq.* (6.22), *that is, the contact Euler–Lagrange vector field, on* $TQ \times \mathbb{R}$.

Furthermore, the vector field $X_h \in \mathfrak{X}(T^*Q \times \mathbb{R})$, *defined by* $X_h \circ \rho_2 = T\rho_2 \circ X_H$, *is the contact Hamiltonian vector field solution to Eq.* (6.34) *on* $T^*Q \times \mathbb{R}$.

These results are analogous to those obtained for the unified formalism of nonautonomous dynamical systems.

Remark 6.9. For singular almost-regular Lagrangians, the equivalence between the constraint algorithms in the unified and in the Lagrangian formalism only holds when the second-order condition is imposed, since, unlike in the unified formalism, this condition does not hold in the Lagrangian case (see [281, 322]).

6.5   Symmetries of Contact Dynamical Systems

(See [118] and [169] for complementary approaches to these topics. See also [172] for a complete classification of symmetries of autonomous and nonautonomous contact systems and [8] for a reduction scheme.)

6.5.1   *Symmetries of contact Hamiltonian systems: Dissipated and conserved quantities*

Let (M, η, h) be a contact Hamiltonian system with Reeb vector field $\mathcal{R}$, and let X_h be the contact Hamiltonian vector field for this system, that is, the solution to the Hamilton equations (6.4).

As for other kinds of dynamical systems, a *dynamical symmetry* of a contact Hamiltonian system is a diffeomorphism $\Phi \colon M \longrightarrow M$ such that $\Phi_* X_h = X_h$. Similarly, an *infinitesimal dynamical symmetry* of a contact Hamiltonian system is a vector field $Y \in \mathfrak{X}(M)$ whose local flux is a dynamical symmetry, that is, $L(Y)X_h = [Y, X_h] = 0$. This means that dynamical symmetries map solutions into solutions.

Nevertheless, we are mainly interested in those symmetries that let the geometric structures be invariant.

Definition 6.16. A *contact Noether symmetry* of a contact Hamiltonian system (M, η, h) is a diffeomorphism $\Phi \colon M \to M$ such that

$$\Phi^* \eta = \eta, \quad \Phi^* h = h.$$

An **infinitesimal contact Noether symmetry** of a contact Hamiltonian system (M, η, h) is a vector field $Y \in \mathfrak{X}(M)$ whose local flux is a contact Noether symmetry, that is,

$$\mathrm{L}(Y)\eta = 0, \quad \mathrm{L}(Y)h = 0.$$

Proposition 6.10. *Every (infinitesimal) contact Noether symmetry preserves the Reeb vector field, that is, $\Phi_* \mathcal{R} = \mathcal{R}$ (or $[Y, \mathcal{R}] = 0$).*

Proof. We obtain

$$i(\Phi_*^{-1}\mathcal{R})(\Phi^* \mathrm{d}\eta) = \Phi^*(i(\mathcal{R})\mathrm{d}\eta) = 0,$$
$$i(\Phi_*^{-1}\mathcal{R})(\Phi^* \eta) = \Phi^*(i(\mathcal{R})\eta) = 1.$$

Since $\Phi^* \eta = \eta$ and the Reeb vector field are unique we get $\Phi_*^{-1}\mathcal{R} = \mathcal{R}$, or equivalently, $\Phi_* \mathcal{R} = \mathcal{R}$ from these equalities.

If $Y \in \mathfrak{X}(M)$ is an infinitesimal Noether symmetry, we have

$$\begin{aligned}
\mathrm{L}([Y, \mathcal{R}])\eta &= \mathrm{L}(Y)\,\mathrm{L}(\mathcal{R})\eta - \mathrm{L}(\mathcal{R})\,\mathrm{L}(Y)\eta \\
&= \mathrm{L}(Y)\,i(\mathcal{R})\mathrm{d}\eta + \mathrm{d}\,i(\mathcal{R})\eta = 0, \qquad (6.44) \\
\mathrm{L}([Y, \mathcal{R}])\mathrm{d}\eta &= \mathrm{d}\,\mathrm{L}([Y, \mathcal{R}])\eta = 0,
\end{aligned}$$

so $[Y, \mathcal{R}] = \mathrm{L}(Y)\mathcal{R} \in \ker \eta \cap \ker \mathrm{d}\eta = \{0\}$. $\qquad \square$

As in the dynamical systems studied in the previous chapters, we have:

Proposition 6.11. *Every (infinitesimal) contact Noether symmetry is a (infinitesimal) dynamical symmetry.*

Proof. In fact, using the fact that Φ is a contact Noether symmetry, and that by the above proposition $\Phi_* \mathcal{R} = \mathcal{R}$, we obtain:

$$\begin{aligned}
i(\Phi_*^{-1}X_h)\mathrm{d}\eta &= i(\Phi_*^{-1}X_h)(\Phi^* \mathrm{d}\eta) = \Phi^*(i(X_h)\mathrm{d}\eta) \\
&= \Phi^*(\mathrm{d}h - (\mathcal{R}(h))\,\eta) = \mathrm{d}h - (\mathcal{R}(h))\,\eta, \\
i(\Phi_*^{-1}X_h)\eta &= i(\Phi_*^{-1}X_h)(\Phi^* \eta) = \Phi^*(i(X_h)\eta) = \Phi^*(-h) = -h.
\end{aligned}$$

Then $\Phi_*^{-1}X_h$ is a solution to the dynamical equation (6.34), and as the solution is unique, we conclude that $\Phi_*X_h = X_h$, and hence Φ is a symmetry.

For the infinitesimal case, we have

$$i([Y, X_h])\eta = L(Y)\,i(X_h)\eta - i(X_h)\,L(Y)\eta = -L(Y)h = 0,$$
$$i([Y, X_h])d\eta = d\,i([Y, X_h])\eta = 0,$$

so $[Y, X_h] \in \ker \eta \cap \ker d\eta = \{0\}$, and this is a dynamical symmetry. $\square$

Now, the result stated in Proposition 6.6, $L(X_h)h = -(\mathcal{R}(h))\,h$ allows us to have:

Definition 6.17. A ***dissipated quantity*** of the contact Hamiltonian system (M, η, h) is a function $F \in C^\infty(M)$ satisfying

$$L(X_h)F = -(\mathcal{R}(h))\,F. \tag{6.45}$$

A ***conserved quantity*** of the contact Hamiltonian system (M, η, h) is a function $G \in C^\infty(M)$ satisfying

$$L(X_h)G = 0.$$

For contact Hamiltonian systems, as we are dealing with dissipative systems, symmetries are associated with dissipated quantities:

Theorem 6.5 (Dissipation Theorem). *If $Y \in \mathfrak{X}(M)$ is an infinitesimal dynamical symmetry, then the function $F = -i(Y)\eta$ is a dissipated quantity.*

Proof. In fact, Proposition 6.11 says that $[Y, X_h] = 0$; then,

$$\begin{aligned}
L(X_h)F &= -L(X_h)\,i(Y)\eta = -i(Y)\,L(X_h)\eta - i(L(X_h)Y)\,\eta \\
&= (\mathcal{R}(h))\,i(Y)\eta + i([Y, X_h])\eta \\
&= -(\mathcal{R}(h))\,F + i([Y, X_h])\eta = -(\mathcal{R}(h))\,F,
\end{aligned}$$

where we have applied $L(X_h)\eta = -\mathcal{R}(h))\,\eta$ (by (6.15)). $\square$

Remark 6.10. In particular, the Hamiltonian vector field X_h is trivially a symmetry, since $[X_h, X_h] = 0$. Then, Equation (6.16) establishes that its associated dissipated quantity is the energy, $h = -i(X_h)\eta$.

These are "non-conservation theorems" that account for the non-conservation of these quantities associated with the symmetries. In particular, every dissipated quantity changes with the same rate, $-\mathcal{R}(h)$, which suggests that the quotient of two dissipated quantities should be a conserved quantity. Indeed:

Proposition 6.12.

(1) If F_1 and F_2 are dissipated quantities and $F_2 \neq 0$, then $\dfrac{F_1}{F_2}$ is a conserved quantity.

(2) If F is a dissipated quantity and G is a conserved quantity, then FG is a dissipated quantity.

Proof. In fact, we have

$$L(X_h)\left(\frac{F_1}{F_2}\right) = \frac{1}{F_2}L(X_h)F_1 - \frac{F_1}{F_2^2}L(X_h)F_2$$

$$= -\frac{1}{F_2}(\mathcal{R}(h))F_1 + \frac{F_1}{F_2^2}(\mathcal{R}(h))F_2 = 0.$$

$$L(X_h)(FG) = G\, L(X_h)F + F\, L(X_h)G = -(\mathcal{R}(h))\,FG. \qquad \square$$

A straightforward consequence of this proposition is that symmetries can also have associated conserved quantities:

Corollary 6.1. *For every infinitesimal symmetry $Y \in \mathfrak{X}(M)$, if $h \neq 0$, then $-\dfrac{i(Y)\eta}{h}$ is a conserved quantity.*

Contact symmetries also allow us to generate new dissipated quantities from another one.

Proposition 6.13. *If $\Phi \colon M \to M$ is a contact Noether symmetry and $F \in C^\infty(M)$ is a dissipated quantity, then so is $\Phi^* F$. Similarly, if $Y \in \mathfrak{X}(M)$ is an infinitesimal contact Noether symmetry, then $L(Y)F$ is a dissipated quantity.*

Proof. In fact, we have

$$L(X_h)(\Phi^* F) = \Phi^* L(\Phi_* X_h)F = \Phi^* L(X_h)F$$

$$= \Phi^*(-\mathcal{R}(h))F = -(\mathcal{R}(h))(\Phi^* F).$$

For the infinitesimal case, considering that Y is a contact infinitesimal $L(X_h)F = 0$, that $[Y, X_h] = X_h$ (Proposition 6.11), and that $[Y, \mathcal{R}] = 0$ (Proposition 6.10), we have

$$
\begin{aligned}
L(X_h) L(Y)F &= L([X_h, Y])F - L(Y) L(X_h)F + - L(Y)\big((\mathcal{R}(h))\,F\big) \\
&= - L(X_h)F - F\, L(Y)(\mathcal{R}(h)) - (\mathcal{R}(h))(L(Y)F) \\
&= -F\, L([Y, \mathcal{R}])h - F\, L(\mathcal{R}) L(Y)h - (\mathcal{R}(h))(L(Y)F) \\
&= -(\mathcal{R}(h))(L(Y)F).
\end{aligned}
$$

$\square$

6.5.2   Symmetries for contact Lagrangian and canonical Hamiltonian systems

Next, we particularise the results on symmetries and dissipated quantities to the case of Lagrangian dissipative systems and their canonical Hamiltonian formalism.

Consider a regular contact Lagrangian system $(TQ \times \mathbb{R}, \mathcal{L})$, with a Reeb vector field $\mathcal{R}_{\mathcal{L}}$, and let $X_{\mathcal{L}}$ be the contact Euler–Lagrange vector field for this system, that is, the solution to the Lagrangian equations (6.22).

The definitions and results about symmetries and dissipated quantities stated in the preceding section hold for the contact system $(TQ \times \mathbb{R}, \eta_{\mathcal{L}}, E_{\mathcal{L}})$. In particular, the dissipation theorem states that $- i(Y)\eta_{\mathcal{L}}$ is a dissipated quantity, and for every infinitesimal dynamical symmetry $Y \in \mathfrak{X}(TQ \times \mathbb{R})$, the energy dissipation theorem states that

$$
L(X_{\mathcal{L}})E_{\mathcal{L}} = -(\mathcal{R}_{\mathcal{L}}(E_{\mathcal{L}}))\, E_{\mathcal{L}}. \tag{6.46}
$$

Furthermore, if $\varphi \colon Q \longrightarrow Q$ is a diffeomorphism, we construct the *canonical lift* of φ to $TQ \times \mathbb{R}$ as the diffeomorphism $\Phi := (T\varphi, \mathrm{Id}_{\mathbb{R}}) \colon TQ \times \mathbb{R} \longrightarrow TQ \times \mathbb{R}$. These kinds of diffeomorphisms Φ are usually called *natural transformations* on $TQ \times \mathbb{R}$. Similarly, for every vector field $Z \in \mathfrak{X}(Q)$, we can define its *complete lift* to $TQ \times \mathbb{R}$ as the vector field $Y \in \mathfrak{X}(TQ \times \mathbb{R})$ whose local flux is the canonical lift of the local flux of Z to $TQ \times \mathbb{R}$. The infinitesimal transformation generated by Y is called *infinitesimal natural transformation* on $TQ \times \mathbb{R}$.

Then, considering the definitions of the canonical endomorphism $\mathcal{J}$ and the Liouville vector field Δ in $TQ \times \mathbb{R}$, it can be proven that canonical lifts of diffeomorphisms and vector fields from Q to $TQ \times \mathbb{R}$ preserve these canonical structures. Furthermore, if these lifts leave the Lagrangian function invariant, they also preserve the Reeb vector field $\mathcal{R}_{\mathcal{L}}$. Therefore, as a logical consequence, we obtain the following relationship:

Proposition 6.14. *If* $\Phi \in \mathrm{Diff}(\mathrm{T}Q \times \mathbb{R})$ *(respectively,* $Y \in \mathfrak{X}(\mathrm{T}Q \times \mathbb{R})$*) is a canonical lift to* $\mathrm{T}Q \times \mathbb{R}$ *of a diffeomorphism* $\varphi \in \mathrm{Diff}(Q)$ *(of a vector field* $Z \in \mathfrak{X}(Q)$*) that leaves the Lagrangian invariant, then it is a (infinitesimal) contact Noether symmetry; that is,*

$$\Phi^* \eta_{\mathcal{L}} = \eta_{\mathcal{L}}, \ \ \Phi^* E_{\mathcal{L}} = E_{\mathcal{L}} \qquad (\text{resp. } \mathrm{L}(Y)\eta_{\mathcal{L}} = 0, \ \mathrm{L}(Y)E_{\mathcal{L}} = 0 \,).$$

As a consequence, it is an (infinitesimal) dynamical symmetry.

As a direct consequence, if $\dfrac{\partial \mathcal{L}}{\partial q^i} = 0$, then $\dfrac{\partial}{\partial q^i}$ is an infinitesimal contact Noether symmetry, and its associated dissipated quantity is the momentum $\dfrac{\partial \mathcal{L}}{\partial v^i}$; that is,

$$\mathrm{L}(X_{\mathcal{L}})\left(\frac{\partial \mathcal{L}}{\partial v^i}\right) = -(\mathcal{R}_{\mathcal{L}}(E_{\mathcal{L}}))\frac{\partial \mathcal{L}}{\partial v^i} = \frac{\partial \mathcal{L}}{\partial s}\frac{\partial \mathcal{L}}{\partial v^i}.$$

(See also [175] for a similar description.)

For the canonical Hamiltonian formalism, consider the canonical contact Hamiltonian system $(\mathrm{T}^*Q \times \mathbb{R}, \boldsymbol{\eta}, \mathrm{h})$. As in the Lagrangian formalism, if $\varphi \colon Q \longrightarrow Q$ is a diffeomorphism, we can construct the *canonical lift of* φ to $\mathrm{T}^*Q \times \mathbb{R}$, which is the diffeomorphism $\Phi := (\mathrm{T}^*\varphi, \mathrm{Id}_{\mathbb{R}}) \colon \mathrm{T}^*Q \times \mathbb{R} \longrightarrow \mathrm{T}^*Q \times \mathbb{R}$ and is called a *natural transformation* on $\mathrm{T}^*Q \times \mathbb{R}$. In the same way, for every vector field $Z \in \mathfrak{X}(Q)$, we define its *complete lift* to $\mathrm{T}^*Q \times \mathbb{R}$ as the vector field $Y \in \mathrm{X}(\mathrm{T}^*Q \times \mathbb{R})$ whose local flux is the canonical lift of the local flux of Z to $\mathrm{T}^*Q \times \mathbb{R}$. It is called a *natural infinitesimal transformation* on $\mathrm{T}^*Q \times \mathbb{R}$.

The canonical forms Θ and $\Omega = -\mathrm{d}\Theta$ in T^*Q and their extensions to $\mathrm{T}^*Q \times \mathbb{R}$ are invariant under the action of canonical lifts of diffeomorphisms and vector fields from Q to T^*Q and $\mathrm{T}^*Q \times \mathbb{R}$. Then, considering the definition of the contact form $\boldsymbol{\eta}$ in $\mathrm{T}^*Q \times \mathbb{R}$, we have:

Proposition 6.15. *If* $\Phi \in \mathrm{Diff}(\mathrm{T}^*Q \times \mathbb{R})$ $(Y \in \mathfrak{X}(\mathrm{T}^*Q \times \mathbb{R}))$ *is a canonical lift to* $\mathrm{T}^*Q \times \mathbb{R}$ *of a diffeomorphism* $\varphi \in \mathrm{Diff}(Q)$ *(of* $Z \in \mathfrak{X}(Q)$*), then:*

(1) $\Phi^*\boldsymbol{\eta} = \boldsymbol{\eta}$ $(\mathrm{L}(Y)\boldsymbol{\eta} = 0)$.
(2) If, in addition, $\Phi^*\mathrm{h} = \mathrm{h}$ $(\mathrm{L}(Y)\mathrm{h} = 0)$, *then it is a (infinitesimal) contact Noether symmetry.*

In particular, we have the following:

Theorem 6.6 (Momentum Dissipation Theorem). *If* $\dfrac{\partial h}{\partial q^i} = 0$,

then $\dfrac{\partial}{\partial q^i}$ *is an infinitesimal contact Noether symmetry, and its associated dissipated quantity is the corresponding momentum p_i; that is,* $L(X_h)p_i = -(\mathcal{R}(h))\, p_i$.

Proof. A simple computation in coordinates shows that $L\left(\dfrac{\partial}{\partial q^i}\right)\eta = 0$ and $L\left(\dfrac{\partial}{\partial q^i}\right) = 0$. Therefore, $\dfrac{\partial}{\partial q^i}$ is a contact Noether symmetry, and hence, a dynamical symmetry. The other results are consequences of the dissipation theorem. $\square$

6.6 Examples

Finally, we consider the systems studied as examples in the previous chapters, incorporating a standard dissipation term that accounts for dissipative forces proportional to velocity.

6.6.1 *The damped harmonic oscillator*

Consider a harmonic oscillator in a medium with friction. As in Section 3.7.1, the configuration bundle of the system is $Q = \mathbb{R}$, but for the contact formulation, we take the manifold $Q \times \mathbb{R}$ with coordinates (t, q).

6.6.1.1 *Lagrangian formalism*

The Lagrangian description of the one-dimensional damped harmonic oscillator is done in the phase space $TQ \times \mathbb{R} \simeq \mathbb{R}^2 \times \mathbb{R}$. It is described by the hyperregular contact Lagrangian function,

$$\mathcal{L} = \frac{1}{2}mv^2 - \frac{1}{2}kq^2 - \gamma s,$$

where $\gamma \in \mathbb{R}^+$ is the dissipation parameter.

We have the contact Lagrangian form,

$$\eta_{\mathcal{L}} = \mathrm{d}s - v\mathrm{d}q,$$

and the energy Lagrangian function is

$$E_{\mathcal{L}} = \frac{1}{2}mv^2 + \frac{1}{2}kq^2 + \gamma s.$$

For $X_{\mathcal{L}} = f\dfrac{\partial}{\partial q} + F\dfrac{\partial}{\partial v} + g\dfrac{\partial}{\partial s} \in \mathfrak{X}(TQ \times \mathbb{R})$, the contact Lagrangian equations (6.31) and (6.32) are

$$g = \frac{1}{2}mv^2 - \frac{1}{2}kq^2 - \gamma s, \quad f = v, \quad mF = -kq - \gamma mv,$$

whose solution is the SODE vector field,

$$X_{\mathcal{L}} = v\frac{\partial}{\partial q} - \left(\frac{k}{m}q + \gamma v\right)\frac{\partial}{\partial v} + \left(\frac{1}{2}mv^2 - \frac{1}{2}kq^2 - \gamma s\right)\frac{\partial}{\partial s}.$$

Its integral curves are the solutions to Eqs. (6.25) and (6.26), which for this system give the Herglotz–Euler–Lagrange equations

$$\frac{ds}{dt} = \frac{1}{2}m\left(\frac{dq}{dt}\right)^2 - \frac{1}{2}kq^2 - \gamma s \qquad \frac{d^2q}{dt^2} = -\frac{k}{m}q - \gamma\frac{dq}{dt}, \qquad (6.47)$$

and the last equation corresponds to the well-known dynamical equation of the damped harmonic oscillator.

The dissipation of the energy is given by (6.16), which in the Lagrangian formalism reads

$$L(X_{\mathcal{L}})E_{\mathcal{L}} = -\gamma\left(\frac{1}{2}mv^2 + \frac{1}{2}kq^2 + \gamma s\right).$$

6.6.1.2 *Hamiltonian formalism*

For the canonical Hamiltonian formalism of the system, $T^*Q \times \mathbb{R} \simeq \mathbb{R}^2 \times \mathbb{R}$, and the Legendre map is

$$\mathfrak{F}\mathcal{L}^*q = q, \quad \mathfrak{F}\mathcal{L}^*p = mv, \quad \mathfrak{F}\mathcal{L}^*s = s.$$

The canonical contact form is $\eta = ds - p\,dq$, and the canonical Hamiltonian function reads

$$h = \frac{p^2}{2m^2} + \frac{1}{2}kq^2 + \gamma s.$$

For $X_h = f\dfrac{\partial}{\partial q} + F\dfrac{\partial}{\partial p} + g\dfrac{\partial}{\partial s}$, the contact Hamiltonian equation (6.34) gives

$$f = \frac{p}{m}, \quad F = -kq - \gamma p, \quad g = \frac{p^2}{2m^2} - \frac{1}{2}kq^2 - \gamma s,$$

and its solution is the Hamiltonian vector field

$$X_h = \frac{p}{m}\frac{\partial}{\partial q} - (kq + \gamma p)\frac{\partial}{\partial p} + \left(\frac{p^2}{2m^2} - \frac{1}{2}kq^2 - \gamma s\right)\frac{\partial}{\partial s},$$

whose integral curves are the solutions to Eq. (6.35), which for this system gives the Hamilton equations

$$\frac{dq}{dt} = \frac{p}{m}, \quad \frac{dp}{dt} = -kq - \gamma p, \quad \frac{ds}{dt} = \frac{p^2}{2m^2} - \frac{1}{2}kq^2 - \gamma s. \tag{6.48}$$

Finally, the dissipation of the energy is given by (6.46),

$$L(X_h)h = -\gamma\left(\frac{p^2}{2m^2} + \frac{1}{2}kq^2 + \gamma s\right).$$

6.6.1.3   *Unified Lagrangian–Hamiltonian formalism*

The extended unified bundle is $\mathfrak{M} = TQ \times_Q T^*Q \times \mathbb{R} \simeq \mathbb{R}^4$ with coordinates (q, v, p, s). Then we have the precontact Hamiltonian system $(\mathfrak{M}, \eta_{\mathfrak{M}}, \mathbf{H})$, where the canonical contact form is

$$\eta_{\mathfrak{M}} = ds - p\,dq,$$

and the Hamiltonian function is

$$\mathbf{H} = pv - \frac{1}{2}mv^2 + \frac{1}{2}kq^2 + \gamma s,$$

and we can take $\mathcal{R} = \dfrac{\partial}{\partial s}$ as the Reeb vector field. For this system, the compatibility condition for Eq. (6.37) leads to the submanifold $\mathfrak{M}_0 \hookrightarrow \mathfrak{M}$, defined by

$$\mathfrak{M}_0 = \{(q, v, p, s) \in \mathfrak{M} \mid p - mv = 0\},$$

and the Hamiltonian vector field solution to (6.37) on $\mathfrak{M}_0$ is

$$X_{\mathbf{H}}|_{\mathfrak{M}_0} = v\frac{\partial}{\partial q} + F\frac{\partial}{\partial v} - (kq + \gamma p)\frac{\partial}{\partial p} + \left(\frac{p^2}{2m^2} - \frac{1}{2}kq^2 - \gamma s\right)\frac{\partial}{\partial s}.$$

The tangency condition for $X_{\mathbf{H}}$ on $\mathfrak{M}_0$ gives $F = -\dfrac{k}{m}q - \gamma\dfrac{p}{m}$, and thus finally,

$$X_{\mathbf{H}}|_{\mathfrak{M}_0} = v\frac{\partial}{\partial q} - \left(\frac{k}{m}q + \gamma\frac{p}{m}\right)\frac{\partial}{\partial v} - (kq + \gamma p)\frac{\partial}{\partial p} + \left(\frac{p^2}{2m^2} - \frac{1}{2}kq^2 - \gamma s\right)\frac{\partial}{\partial s}.$$

Therefore, the integral curves of $X_{\mathbf{H}}$ are the solutions to

$$\frac{dq}{dt} = v, \quad m\frac{dv}{dt} = -mkq - \gamma p, \quad \frac{dp}{dt} = -kq - \gamma p, \quad \frac{ds}{dt} = \frac{p^2}{2m^2} - \frac{1}{2}kq^2 - \gamma s.$$

The first two equations of this system are equivalent to the second Herglotz–Euler–Lagrange equation (6.47), and using the constraint $p = mv$ (the Legendre map), the first and third equations are the first pair of the Hamilton equations (6.48).

6.6.2 *The Kepler problem with friction*

Consider the motion of a massive particle (of mass m) under the action of Newtonian central force in a stellar media with friction. As in the previous cases, the motion is on a plane; hence, $Q = \mathbb{R}^2$, and we take (r, ϕ, s) as coordinates in $Q \times \mathbb{R}$, (with the origin of r at the centre of the force).

6.6.2.1 *Lagrangian formalism*

The Lagrangian formalism takes place in $TQ \times \mathbb{R} \simeq \mathbb{R}^5$, with local coordinates $(r, \phi, v_r, v_\phi, s)$. The contact Lagrangian function that describes the dynamics is

$$\mathcal{L} = \frac{1}{2}m(v_r^2 + r^2 v_\phi^2) - \frac{K}{r} - \gamma s, \quad K \neq 0 \,,$$

which is regular as in the above situations. The energy Lagrangian function and the contact Lagrangian form are

$$E_\mathcal{L} = \frac{1}{2}m(v_r^2 + r^2 v_\phi^2) + \frac{K}{r} + \gamma s,$$

$$\eta_\mathcal{L} = ds - m(v_r \, dr + r^2 v_\phi \, d\phi) \,,$$

and the Lagrangian is regular.

For $X_\mathcal{L} = f_r \dfrac{\partial}{\partial r} + f_\phi \dfrac{\partial}{\partial \phi} + g_r \dfrac{\partial}{\partial v_r} + g_\phi \dfrac{\partial}{\partial v_\phi} + g \dfrac{\partial}{\partial s} \in \mathfrak{X}(\mathrm{T}^*Q \times \mathbb{R})$, the contact Lagrangian equations (6.31) and (6.32) give

$$g = \frac{1}{2}m(v_r^2 + r^2 v_\phi^2) - \frac{K}{r} - \gamma s, \quad f_r = v_r, \quad f_\phi = v_\phi,$$

$$mg_r = 2mrv_\phi f_\phi - mrv_\phi^2 + \frac{K}{r^2} - \gamma m v_r, \quad g_\phi = -\frac{2v_\phi f_r}{r} - \gamma r^2 v_\phi,$$

and the contact Euler–Lagrange vector field is

$$X_\mathcal{L} = v_r \frac{\partial}{\partial r} + v_\phi \frac{\partial}{\partial \phi} + \left(rv_\phi^2 + \frac{K}{mr^2} - \gamma v_r\right)\frac{\partial}{\partial v_r} + \left(-\frac{2v_\phi v_r}{r} - \gamma v_\phi\right)\frac{\partial}{\partial v_\phi}$$

$$+ \left(\frac{1}{2}m(v_r^2 + r^2 v_\phi^2) - \frac{K}{r} - \gamma s\right)\frac{\partial}{\partial s}.$$

Then, its integral curves are the solutions to Eqs. (6.25) and (6.26), which read as

$$\frac{ds}{dt} = \frac{1}{2}m\left(\left(\frac{dr}{dt}\right)^2 + r^2\left(\frac{d\phi}{dt}\right)^2\right) - \frac{K}{r} - \gamma s,$$

$$m\frac{d^2 r}{dt^2} = mr\left(\frac{d\phi}{dt}\right)^2 + \frac{K}{r^2} - m\gamma v_r, \quad \frac{d^2\phi}{dt^2} = -\frac{2}{r}\frac{d\phi}{dt}\frac{dr}{dt} - \gamma\frac{d\phi}{dt}, \quad (6.49)$$

and are the Herglotz–Euler–Lagrange equations for this system.

Similar to the above example, the dissipation of the energy reads

$$L(X_{\mathcal{L}})E_{\mathcal{L}} = -\gamma\left(\frac{1}{2}m(v_r^2 + r^2 v_\phi^2) + \frac{K}{r} + \gamma s\right).$$

Furthermore, there exists a Lagrangian contact Noether symmetry that is generated by the vector field $Y = \dfrac{\partial}{\partial\phi}$, and in fact, considering (3.26) and (3.27), we obtain

$$L(Y)\eta_{\mathcal{L}} = L\left(\frac{\partial}{\partial\phi}\right)(ds - m(v_r\,dr + r^2 v_\phi\,d\phi))$$

$$= L\left(\frac{\partial}{\partial\phi}\right)ds - L\left(\frac{\partial}{\partial\phi}\right)(m(v_r\,dr + r^2 v_\phi\,d\phi)) = 0,$$

$$L(Y)E_{\mathcal{L}} = L\left(\frac{\partial}{\partial\phi}\right)\left(\frac{1}{2}m(v_r^2 + r^2 v_\phi^2) + \frac{K}{r} + \gamma s\right) = 0,$$

and the corresponding dissipated quantity is the angular momentum map

$$F = -i\left(\frac{\partial}{\partial\phi}\right)\eta_{\mathcal{L}} = mr^2 v_\phi.$$

Finally, using Corollary 6.1, we obtain $\dfrac{F}{E_{\mathcal{L}}}$ as a conserved quantity associated with Y.

6.6.2.2 *Hamiltonian formalism*

For the Hamiltonian formalism, $T^*Q \times \mathbb{R} \simeq \mathbb{R}^5$, with local coordinates $(r, \phi, p_r, p_\phi, s)$. The Legendre transformation is

$$\mathfrak{F}\mathcal{L}^* r = r,\quad \mathfrak{F}\mathcal{L}^*\phi = \phi,\quad \mathfrak{F}\mathcal{L}^* p_r = mv_r,\quad \mathfrak{F}\mathcal{L}^* p_\phi = mr^2 v_\phi,\quad \mathfrak{F}\mathcal{L}^* s = s,$$

which is a diffeomorphism, and the Lagrangian is hyperregular. The canonical Hamiltonian function and the canonical contact form are

$$h = \frac{p_r^2}{2m} + \frac{p_\phi^2}{2mr^2} + \frac{K}{r} + \gamma s,$$

$$\eta = ds - p_r\,dr - p_\phi\,d\phi.$$

For $X_h = F_r\dfrac{\partial}{\partial r} + F_\phi\dfrac{\partial}{\partial\phi} + G_r\dfrac{\partial}{\partial p_r} + G_\phi\dfrac{\partial}{\partial p_\phi} + g\dfrac{\partial}{\partial s} \in \mathfrak{X}(T^*Q \times \mathbb{R})$, the contact Hamiltonian equation (6.34) leads to

$$F_r = \frac{p_r}{m},\quad F_\phi = \frac{p_\phi}{mr^2},\quad G_r = \frac{p_\phi^2}{mr^3} + \frac{K}{r^2} - \gamma p_r,$$

$$G_\phi = -\gamma p_\phi,\quad g = \frac{p_r^2}{2m} + \frac{p_\phi^2}{2mr^2} - \frac{K}{r} - \gamma s,$$

and then, the contact Hamiltonian vector field is

$$X_h = \frac{p_r}{m}\frac{\partial}{\partial r} + \frac{p_\phi}{mr^2}\frac{\partial}{\partial \phi} + \left(\frac{p_\phi^2}{mr^3} + \frac{K}{r^2} - \gamma p_r\right)\frac{\partial}{\partial p_r} - \gamma p_\phi \frac{\partial}{\partial p_\phi}$$

$$+ \left(\frac{p_r^2}{2m} + \frac{p_\phi^2}{2mr^2} - \frac{K}{r} - \gamma s\right)\frac{\partial}{\partial s}.$$

Hence, its integral curves are the solutions to Eq. (6.35), which are

$$m\frac{dr}{dt} = p_r, \; mr^2\frac{d\phi}{dt} = p_\phi, \; \frac{dp_r}{dt} = \frac{p_\phi^2}{mr^3} + \frac{K}{r^2} - \gamma p_r, \; \frac{dp_\phi}{dt} = -\gamma p_\phi, \quad (6.50)$$

$$\frac{ds}{dt} = \frac{p_r^2}{2m} + \frac{p_\phi^2}{2mr^2} - \frac{K}{r} - \gamma s,$$

and are the Hamiltonian equations for this system.

As it is usual, using the Legendre map, one can easily check that $\mathfrak{F}\mathcal{L}_* X_{\mathcal{L}} = X_h$.

Similar to the above example, the dissipation of the energy reads

$$L(X_h)h = -\gamma\left(\frac{p_r^2}{2m} + \frac{p_\phi^2}{2mr^2} + \frac{K}{r} + \gamma s\right).$$

In addition, we have a Hamiltonian contact Noether symmetry, which is again the vector field $Y = \dfrac{\partial}{\partial \phi}$, since

$$L(Y)\eta = L\left(\frac{\partial}{\partial \phi}\right)(ds - p_r\, dr - p_\phi\, d\phi) = 0,$$

$$L(Y)h = L\left(\frac{\partial}{\partial \phi}\right)\left(\frac{p_r^2}{2m} + \frac{p_\phi^2}{2mr^2} + \frac{K}{r} + \gamma s\right) = 0.$$

The associated dissipated quantity is again the angular momentum

$$F = -i\left(\frac{\partial}{\partial \phi}\right)\eta = p_\phi,$$

and the corresponding dissipation law is given directly by the last Hamilton equation in (6.50). As above, $\dfrac{F}{h}$ is a conserved quantity associated with Y.

6.6.2.3 *Unified Lagrangian–Hamiltonian formalism*

In the precontact extended unified bundle $\mathfrak{M} = TQ \times_Q T^*Q \times \mathbb{R} \simeq \mathbb{R}^7$, with coordinates $(r, \phi, v_r, v_\phi, p_r, p_\phi, s)$, the canonical precontact form and the Hamiltonian function are

$$\eta_{\mathfrak{M}} = ds - p_r\, dr - p_\phi\, d\phi,$$

$$\mathbf{H} = p_r v_r + p_\phi v_\phi - \frac{1}{2}m(v_r^2 + r^2 v_\phi^2) + \frac{K}{r} + \gamma s,$$

and we can take $\mathcal{R} = \dfrac{\partial}{\partial s}$ as the Reeb vector field.

For $X_H = f_r \dfrac{\partial}{\partial r} + f_\phi \dfrac{\partial}{\partial \phi} + g_r \dfrac{\partial}{\partial v_r} + g_\phi \dfrac{\partial}{\partial v_\phi} + G_r \dfrac{\partial}{\partial p_r} + G_\phi \dfrac{\partial}{\partial p_\phi} + g \dfrac{\partial}{\partial s}$,

Eq. (6.37) lead to

$$f_r = v_r, \ f_\phi = v_\phi, \ G_r = \frac{K}{r^2} + mrv_\phi^2 - \gamma p_r, \ G_\phi = -\gamma p_\phi \,,$$

$$p_r = mv_r, \ p_\phi = mr^2 v_\phi, \ g = \frac{p_r^2}{2m} + \frac{p_\phi^2}{2mr^2} - \frac{K}{r} - \gamma s \ . \quad (6.51)$$

The first two equations in (6.51) are constraints defining the submanifold $\mathfrak{M}_0 \hookrightarrow \mathfrak{M}$, which give the Legendre map. Thus, the Hamiltonian vector field is

$$X_H|_{\mathfrak{M}_0} = v_r \frac{\partial}{\partial r} + v_\phi \frac{\partial}{\partial \phi} + g_r \frac{\partial}{\partial v_r} + g_\phi \frac{\partial}{\partial v_\phi}$$

$$+ \left(\frac{K}{r^2} + mrv_\phi^2 - \gamma p_r \right) \frac{\partial}{\partial p_r} - \gamma p_\phi \frac{\partial}{\partial p_\phi}$$

$$+ \left(\frac{p_r^2}{2m} + \frac{p_\phi^2}{2mr^2} - \frac{K}{r} - \gamma s \right) \frac{\partial}{\partial s},$$

and the tangency condition of X_H on $\mathfrak{M}_0$ finally leads to,

$$X_H|_{\mathfrak{M}_0} = v_r \frac{\partial}{\partial r} + v_\phi \frac{\partial}{\partial \phi} + \left(rv_\phi^2 + \frac{K}{mr^2} - \gamma \frac{p_r}{m} \right) \frac{\partial}{\partial v_r}$$

$$- \left(\frac{2v_r v_\phi}{r} + \gamma \frac{p_\phi}{m} \right) \frac{\partial}{\partial v_\phi} + \left(\frac{K}{r^2} + mrv_\phi^2 - \gamma p_r \right) \frac{\partial}{\partial p_r}$$

$$- \gamma p_\phi \frac{\partial}{\partial p_\phi} + \left(\frac{p_r^2}{2m} + \frac{p_\phi^2}{2mr^2} - \frac{K}{r} - \gamma s \right) \frac{\partial}{\partial s}.$$

Therefore, the integral curves of X_H, on $\mathfrak{M}_0$, are the solutions to

$$\frac{dr}{dt} = v_r, \ \frac{d\phi}{dt} = v_\phi, \ \frac{dv_r}{dt} = \frac{K}{mr^2} + rv_\phi^2 - \gamma \frac{p_r}{m}, \ \frac{dv_\phi}{dt} = -\frac{2v_r v_\phi}{r} - \gamma \frac{p_\phi}{m}, \quad (6.52)$$

$$\frac{dp_r}{dt} = \frac{K}{r^2} + mrv_\phi^2 - \gamma p_r, \ \frac{dp_\phi}{dt} = -\gamma p_\phi, \ \frac{ds}{dt} = \frac{p_r^2}{2m} + \frac{p_\phi^2}{2mr^2} - \frac{K}{r} - \gamma s. \quad (6.53)$$

Equations (6.52) are equivalent to the Herglotz–Euler–Lagrange equation (6.49), and using the constraints $p_r = mv_r$ and $p_\phi = mr^2 v_\phi$, that is, the Legendre map, the first and second equations in (6.52) and (6.53) are the Hamiltonian equations (6.50) of the system.

6.7 Exercises

Exercise 6.1

Prove Theorem 6.1.

Exercise 6.2

Deduce the local expressions (6.9), (6.10), and (6.11).

Exercise 6.3

Prove Proposition 6.8.

Exercise 6.4

Prove that for the Lagrangian $\mathcal{L} = \sum_{i=1}^{n} v^i s \in C^\infty(TQ \times \mathbb{R})$ introduced in Remark 6.5, the associated form $\eta_\mathcal{L}$ does not fulfill condition (6.1) and has no Reeb vector field associated with it.

Exercise 6.5

Prove that the contact Lagrangian equations (6.22) can be written as

$$i(X_\mathcal{L})\mathrm{d}s = \mathcal{L}, \quad i(X_\mathcal{L})\mathrm{d}\eta_\mathcal{L} = \mathrm{d}E_\mathcal{L} + \frac{\partial \mathcal{L}}{\partial s}\eta_\mathcal{L}.$$

(*Hint:* Use (6.24).)

Exercise 6.6

Prove that if F and G are dissipated quantities, then $\frac{1}{2}FG$ is a dissipated quantity.

    Study the Lagrangian, Hamiltonian, and unified formalisms of the following dynamical systems with dissipation, and discuss the existence of dissipated and conserved quantities in each case.

Exercise 6.7

A particle of mass m moving in a vertical plane with coordinates (x, y) (where y denotes the height), with friction, and submitted to the action of constant gravity, g, which is described by the following contact Lagrangian function:

$$\mathcal{L} = \frac{1}{2}m(v_x^2 + v_y^2) - mgy - \gamma s; \quad \gamma \in \mathbb{R}^+.$$

(*Solution: See [169].*)

Exercise 6.8 (Parachute problem).

A particle of mass m is falling vertically in a fluid under the action of constant gravity, g, with friction force proportional to the square of the velocity, and whose contact Lagrangian function is

$$\mathcal{L} = \frac{1}{2}mv_y^2 - \frac{mg}{2\gamma}\left(e^{2\gamma y} - 1\right) + 2\gamma v_y s; \quad \gamma \in \mathbb{R}^+,$$

where the coordinate y denotes the height. Using this information, deduce the *parachute equation*, $\ddot{y} - \gamma \dot{y}^2 + g = 0$.
(*Solution:* See [169].)

Exercise 6.9 (RLC circuit).
An RLC electric circuit can be described by the contact Lagrangian function

$$\mathcal{L}(I, v_I, s) = \frac{1}{2} L v_I^2 - \frac{1}{2C} I^2 - Rs; \quad R, L, C \in \mathbb{R}^+.$$

Using this information, obtain the usual RLC equation, $L\ddot{I} + R\dot{I} + \frac{1}{C}I = 0$.
(*Solution:* See [93].)

Appendix A

Additional Contents

A.1 Tangent and Cotangent Bundles

(For a detailed account of these subjects and the proof of the results, see, for example, $[1, 11, 97, 228, 241, 290, 295, 330, 357]$.)

A.1.1 *The tangent bundle of a manifold: Canonical lifts*

Let Q be a differentiable manifold and $q \in Q$. Every differentiable curve $\gamma \colon (-\epsilon, \epsilon) \subset \mathbb{R} \longrightarrow Q$ passing through q, that is, $\gamma(0) = q$, induces a derivation D_γ in $C^\infty(Q)$ in the following way:

$$D_\gamma \colon C^\infty(Q) \longrightarrow \mathbb{R}$$
$$f \longmapsto \lim_{t \mapsto 0} \frac{(f \circ \gamma)(t) - (f \circ \gamma)(0)}{t} \; .$$

In the set of such differentiable curves, we can define the following equivalence relationship:

$$\gamma_1 \sim \gamma_2 \iff D_{\gamma_1} = D_{\gamma_2}.$$

Definition A.1.

(1) A **tangent vector** to Q at q is every equivalence class defined by this relationship.

(2) The **tangent space** of Q at q, denoted by $T_q Q$, is the vector space of all the tangent vectors to Q at q.

(3) The **tangent bundle** of Q is defined as $TQ := \bigcup_{q \in Q} T_q Q$. We denote $\tau_Q \colon TQ \longrightarrow Q$ its natural projection.

Every point $p \in TQ$ is a couple $(q, v) \equiv v_q$, where $q = \tau_Q(p) \in Q$ and $v \in T_q Q$ (it is a tangent vector).

Proposition A.1. *The tangent bundle* TQ *is a* $2n$-*dimensional differentiable manifold whose differentiable structure is inherited from* Q. *In addition, the natural projection* τ_Q *is a submersion.*

Proof. If $\mathcal{A} = \{(U_\alpha; \phi_\alpha)\}$, with $\phi_\alpha \equiv (x^1, \ldots, x^n)$, is an atlas of local charts on Q, then the induced atlas on T^*Q is $T\mathcal{A} = \{(\tau_Q^{-1}U_\alpha; \psi_\alpha)\}$, with the coordinate functions ψ_α defined as follows:

$$\psi_\alpha : \tau_Q^{-1}(U_\alpha) \longrightarrow \phi_\alpha(U_\alpha) \times \mathbb{R}^n \subset \mathbb{R}^{2n}$$
$$p = (q, v) \mapsto (q^i(\text{p}), v^i(\text{p}))\quad;$$

where $i = 1, \ldots, n$, and

(1) $q^i(\text{p}) = (x^i \circ \tau_Q)(\text{p})$ [1].

(2) If $v = \lambda^j \dfrac{\partial}{\partial x^j}\Big|_q$, then $v^i(p) = \lambda^i = v(x^i)$, that is, $v^i(p)$ are the components of the tangent vector $v \in T_qQ$ in the basis $\left\{ \dfrac{\partial}{\partial x^i}\Big|_q \right\}$.

It is clear that $TQ = \bigcup_\alpha \tau_Q^{-1}(U_\alpha)$ and ψ_α are diffeomorphisms. Then, it is straightforward to prove that $T\mathcal{A}$ endows TQ with the structure of a differentiable manifold. If $(U; q^i)$ and $(\bar{U}; \bar{q}^i)$ are two local charts in Q such that $U \cap \bar{U} \neq \varnothing$ and $\bar{q}^j = \varphi^j(q^i)$ in $U \cap \bar{U}$, then for the induced charts $(\tau_Q^{-1}(U); q^i, v^i)$ and $(\tau_Q^{-1}\bar{U}); \bar{q}^i, \bar{v}^i)$ in TQ, we have, on $TU \cap T\bar{U}$, the relationship between the coordinates v^i and $\bar{v}^i$ as $\bar{v}^j = \dfrac{\partial \varphi^j}{\partial q^i} v^i$, since

$$v = v^i \frac{\partial}{\partial q^i}\Big|_q = v^i \frac{\partial \varphi^j}{\partial q^i}\frac{\partial}{\partial \bar{q}^j}\Big|_q = \bar{v}^j \frac{\partial}{\partial \bar{q}^j}\Big|_q,$$

and hence $\bar{v}^j = \dfrac{\partial \varphi^j}{\partial q^i} v^i$.

Considering this local description of the tangent bundle, it is evident that $\dim TQ = 2n$.

Finally, it is straightforward to prove that τ_Q is an surjective submersion, since the canonical projection $\tau_Q \colon TQ \twoheadrightarrow Q$ is a surjective map given, in natural coordinates, by $\tau_Q(q^i, v^i) = q^i$, and hence its tangent map $T\tau_Q \colon TTQ \twoheadrightarrow TQ$ is defined as follows: if $(q, v) \in TQ$ and $X \in T_{(q,v)}(TQ)$ we have $X = \lambda^i \dfrac{\partial}{\partial q^i}\Big|_{(q,v)} + \mu^i \dfrac{\partial}{\partial v^i}\Big|_{(q,v)}$, and

$$[T_{(q,v)}\tau_Q(X)](q^j) = X(q^j \circ \tau_Q) = X(q^j) = \lambda^j;$$

---

[1] It is normal to use q^i as the coordinates in both the base manifold Q and the tangent bundle TQ, and we will do so throughout the text.

therefore,

$$(T\tau_Q)((q,v),X) = (T\tau_Q)\left((q,v),\lambda^i\frac{\partial}{\partial q^i}\Big|_{(q,v)} + \mu^i\frac{\partial}{\partial v^i}\Big|_{(q,v)}\right)$$

$$= \left(q,\lambda^i\frac{\partial}{\partial q^i}\Big|_q\right),$$

and the associated matrix is

$$(I_{n\times n}, 0_{n\times n}) = \begin{pmatrix} 1 \dots 0 & 0 \dots 0 \\ \vdots & \vdots \vdots & \vdots \\ 0 \dots 1 & 0 \dots 0 \end{pmatrix}. \tag{A.1}$$

In this way, we conclude that τ_Q is a surjective submersion. $\qquad\square$

Definition A.2. The above charts of the tangent bundle are called **natural charts**, and their coordinates are **natural coordinates** (q^i are the **base coordinates** and v^i the **fibre coordinates**).

Observe that the coordinate change in TQ, from (q^i, v^i) to $(\bar{q}^i, \bar{v}^i)$, has the Jacobian matrix

$$\begin{pmatrix} \left(\dfrac{\partial\varphi^j}{\partial q^i}\right) & 0 \\ \left(\dfrac{\partial^2\varphi^j}{\partial q^k\partial q^i}\right)v^i & \left(\dfrac{\partial\varphi^j}{\partial q^i}\right) \end{pmatrix},$$

and considering that its determinant is positive at every point, we conclude:

Corollary A.1. *TQ is an orientable manifold.*

The tangent bundle of a manifold is an example of a *vector bundle*. This structure is defined as follows:

Definition A.3. A **vector bundle** is a triple (E, B, π), where E, B are differentiable manifolds (with $\dim B = m$, $\dim E = m+n$) and $\pi\colon E \longrightarrow B$ is a surjective submersion such that for every $p \in B$, there exists a local chart (U, ϕ), $p \in U$, and a diffeomorphism $\psi\colon \pi^{-1}(U) \longrightarrow \phi(U) \times \mathbb{R}^n$ satisfying

(1) If $\pi_1\colon \phi(U) \times \mathbb{R}^n \longrightarrow \phi(U)$ is the natural projection, then $\pi_1 \circ \psi = \pi$.
(2) $E_p = \pi^{-1}(p)$ is a vector space, and if $\pi_2\colon \phi(U) \times \mathbb{R}^n \longrightarrow \mathbb{R}^n$ is the

natural projection, then the maps

$$\psi_p : E_p \longrightarrow \mathbb{R}^n$$
$$v \longmapsto (\pi_2 \circ \psi)(p, v)$$

are vector space morphisms.

E is called the **total manifold** of the vector bundle, B is the **base manifold**, π is the **projection of the bundle**, $\mathbb{R}^n$ is the **typical fibre**, n is the **rank of the bundle**, and for every $p \in B$, E_p is the **fibre** over p. The pair (U, ϕ) is said to be a **trivialising open set**, and ψ is the associated **coordinate map**.

Observe that the family $(\pi^{-1}(U), \psi)$ is a differentiable atlas of E, which is said to be **adapted** to the projection $\pi : E \longrightarrow B$.

There is a natural way to lift diffeomorphisms, curves, and vector fields on a manifold to its tangent bundle.

Definition A.4. Let Q be a differentiable manifold and a diffeomorphism

$$\varphi : Q \longrightarrow Q$$
$$q \longmapsto \varphi(q)\,.$$

The **canonical lift** of φ to TQ is the diffeomorphism

$$T\varphi : TQ \longrightarrow TQ$$
$$(q, v) \longmapsto (q, T_q\varphi(v))\,.$$

This definition also holds for every map $\varphi \colon Q \longrightarrow Q$.
The following properties follow logically from the definition:

Proposition A.2. *For every* $\varphi, \phi \in \mathrm{Diff}\, Q$,

(1) $\varphi^{-1} \circ \tau_Q \circ T\varphi = \tau_Q$.
(2) $T(\varphi \circ \phi) = T\varphi \circ T\phi$.

Using the definition of canonical lift of diffeomorphisms, we get:

Definition A.5. Let $Z \in \mathfrak{X}(Q)$ be a vector field. The **total, complete**, or **canonical lift** of Z to TQ is the vector field $Z^C \in \mathfrak{X}(TQ)$, whose local uniparametric groups of diffeomorphisms are the canonical lifts $\{TF_t\}$ of the local uniparametric groups of diffeomorphisms $\{F_t\}$ of Z.

As a direct consequence of the definition and the theorem of existence and unicity of local uniparametric groups of diffeomorphisms, we obtain the following result:

Proposition A.3. *Let $Z \in \mathfrak{X}(Q)$. Then Z^C is τ_Q-projectable and $\tau_{Q*}Z^C = Z$, that is, $\mathrm{T}\tau_Q(Z^C_{(q,v)}) = Z_x$, for every $(q,v) \in \mathrm{T}Q$.*

Local expression: In a chart of coordinates $(U; q^i)$ of Q, if

$$Z|_U = f^i(q^j)\frac{\partial}{\partial q^i},$$

then we have, in the induced chart of natural coordinates $(\tau_Q^{-1}(U); q^i, v^i)$ of $\mathrm{T}Q$,

$$Z^C|_{\tau_Q^{-1}(U)} = f^i(q^j)\frac{\partial}{\partial q^i} + v^k\frac{\partial f^i}{\partial q^k}(q^j)\frac{\partial}{\partial v^i}.$$

To prove it, remember that if $(q, u) \in \tau_Q^{-1}(U)$, then

$$Z^C(q, u) = \frac{d}{dt}\Big|_{t=0}\mathrm{T}F_t(q, u),$$

where $\mathrm{T}F_t(q^i, v^i) = \left(F_t(q^i), \dfrac{\partial F_t}{\partial q^i}v^i\right)$, and as a consequence,

$$\frac{d}{dt}\Big|_{t=0}\mathrm{T}F_t(q, u) = \left(\frac{d}{dt}\Big|_{t=0}F_t(q), \frac{d}{dt}\Big|_{t=0}\left(\frac{\partial F_t}{\partial q^i}v^i\right)(q, u)\right)$$

$$= \left(f^i(q), v^j\frac{\partial f^i}{\partial q^j}(q^k)\right),$$

and the result follows.

Let $\gamma\colon (a, b) \subseteq \mathbb{R} \longrightarrow Q$ be a curve. If $t_0 \in (a, b)$ and $x = \gamma(t_0)$, then $\dot\gamma(t_0)$ is the tangent vector to the curve at the point $\gamma(t_0)$, that is, $\dot\gamma(t_0) \in \mathrm{T}_{\gamma(t_0)}Q$. As $\dot\gamma(t)$ is well defined for every $t \in (a, b)$, we can have:

Definition A.6. The **canonical lift** of a curve γ to $\mathrm{T}Q$ is the curve

$$\tilde\gamma\colon (a, b) \subset \mathbb{R} \longrightarrow \mathrm{T}Q$$
$$t \mapsto (\gamma(t), \dot\gamma(t))\,,$$

which is defined as

$$(\gamma(t_0), \dot\gamma(t_0)) = \mathrm{T}_{t_0}\gamma\frac{d}{dt}\,,$$

that is, for every $f: Q \longrightarrow \mathbb{R}$, then

$$(\gamma(t_0), \dot{\gamma}(t_0))f = (\mathrm{T}_{t_0}\gamma \frac{d}{dt})f = \frac{d}{dt}\Big|_{t_0}(f \circ \gamma) = \lim_{h \to 0} \frac{f(\gamma(t_0 + h)) - f(\gamma(t_0))}{h}.$$

Observe that $\tau_Q \circ \tilde{\gamma} = \gamma$.

Local expression: If (q^i) are local coordinates of Q in a neighbourhood of $\gamma(t_0)$, then $\gamma = (\gamma^1, \ldots, \gamma^n)$, with $\gamma^i = q^i \circ \gamma$. If (q^i, v^i) are the induced natural coordinates in $\mathrm{T}Q$, then $\tilde{\gamma}$ is given by $\tilde{\gamma} = (\gamma^1, \ldots, \gamma^n, \dot{\gamma}^1, \ldots, \dot{\gamma}^n)$, where $\dot{\gamma}^i = \dfrac{d\gamma^i}{dt}$.

The tangent vector to $\tilde{\gamma}$ at $\tilde{\gamma}(t_0)$ is $\mathrm{T}_{t_0}\tilde{\gamma}\dfrac{d}{dt}$, and if $f: \mathrm{T}Q \longrightarrow \mathbb{R}$, we have

$$\left(\mathrm{T}_{t_0}\tilde{\gamma}\frac{d}{dt}\right)f = \frac{d}{dt}\Big|_{t_0} f \circ \tilde{\gamma},$$

and hence,

$$\left(\mathrm{T}_{t_0}\tilde{\gamma}\frac{d}{dt}\right)q^i = \frac{d}{dt}\Big|_{t_0} q^i \circ \tilde{\gamma} = \dot{\gamma}^i(t_0),$$

$$\left(\mathrm{T}_{t_0}\tilde{\gamma}\frac{d}{dt}\right)v^i = \frac{d}{dt}\Big|_{t_0} v^i \circ \tilde{\gamma} = \frac{d}{dt}\Big|_{t_0} \dot{\gamma}^i = \ddot{\gamma}^i(t_0),$$

that is

$$\mathrm{T}_{t_0}\tilde{\gamma}\left(\frac{d}{dt}\right) = \dot{\gamma}^i(t_0)\frac{\partial}{\partial q^i}\Big|_{\tilde{\gamma}(t_0)} + \ddot{\gamma}^i(t_0)\frac{\partial}{\partial v^i}\Big|_{\tilde{\gamma}(t_0)}.$$

Remark A.1. Remember that given a vector field $X \in \mathfrak{X}(Q)$, a curve $\gamma: (a, b) \subseteq \mathbb{R} \longrightarrow Q$ is an *integral curve* of X at $\gamma(t) \in Q$, for $t \in (a, b)$, if

$$X(\gamma(t)) = \dot{\gamma}(t).$$

Considering this, if $\beta \in \Omega^k(Q)$, the contraction of β with X, denoted by $i(X)\beta$, allows us to define a new contraction $\left(i(\tilde{\gamma})(\beta \circ \gamma)\right)$ in a natural way as follows:

$$i(\tilde{\gamma}(t))\left(\beta(\gamma(t))\right) := i\left(\dot{\gamma}(t)\right)(\beta|_{\gamma(t)}), \quad \text{for } t \in (a, b). \qquad (A.2)$$

A.1.2 *The cotangent bundle of a manifold: Canonical lifts*

Definition A.7. Let Q be a differentiable manifold. The **cotangent bundle** of Q is the dual bundle T^*Q of the tangent bundle $\mathrm{T}Q$, that is, $T^*Q := \bigcup_{q \in Q} T^*_q Q$.

We denote that $\pi_Q \colon T^*Q \longrightarrow Q$ is the natural projection.

Thus, every point $p \in T^*Q$ is a pair $(q, \xi) \equiv \xi_q$, where $q = \pi_Q(p) \in Q$ and $\xi \in T^*_q Q$ (it is a linear form on $T_q Q$).

Proposition A.4. *Let Q be an n-dimensional differentiable manifold. The cotangent bundle T^*Q is a $2n$-dimensional differentiable manifold, whose differentiable structure is induced by one Q in a natural way. Moreover, the natural projection π_Q is a submersion.*

Proof. If $\mathcal{A} = \{(U_\alpha; \phi_\alpha = (x^i)\}$ is an atlas of local charts in Q, then the induced atlas $T^*\mathcal{A} = \{(T^*U_\alpha; \eta_\alpha = (q^i, p_i)\}$ is obtained as follows:

Given one of its elements $(U; x^i)$, we take

(1) $T^*U := \pi_Q^{-1}(U)$.

(2) For every $p \in T^*Q$, let $q^i(p) := x^i \circ \pi_Q(p).$[2]

(3) Let $p \in T^*Q$ with $p = (q, \xi)$. Then $p_i(p) := \left\langle \xi, \dfrac{\partial}{\partial q^i}\Big|_q \right\rangle$; that is, $p_i(p)$ are the components of the linear form ξ in the natural basis $dq^i\,|_q$ of $T^*_q M$: $\xi = p_i(p)\,dq^i\,|_q$.

If (q^i) and $(\bar{q}^i)$ are two local systems of coordinates in $U \subset Q$, and $\bar{q}^j = \phi^j(q^i)$, then the relationship between (p_i) and $(\bar{p}_i)$ is the following

$$p_i(q, \xi) = \left\langle \frac{\partial}{\partial q^i}, \xi \right\rangle = \left\langle \frac{\partial \phi^j}{\partial q^i}\Big|_q \frac{\partial}{\partial \bar{q}^j}, \xi \right\rangle = \frac{\partial \phi^j}{\partial q^i}\Big|_q \left\langle \frac{\partial}{\partial \bar{q}^j}, \xi \right\rangle = \frac{\partial \phi^j}{\partial q^i}\Big|_q \bar{p}_j(q, \xi),$$

that is, $p_i = \dfrac{\partial \phi^j}{\partial q^i} \bar{p}_j$. Then, the changes of coordinates in T^*Q, from (q^i, p_i) to $(\bar{q}^i, \bar{p}_i)$, have a Jacobian matrix

$$\begin{pmatrix} \left(\dfrac{\partial \phi^j}{\partial q^i}\right) & 0 \\ \left(\dfrac{\partial^2 \phi^j}{\partial q^k \partial q^i}\right)^{-1} p_i & \left(\dfrac{\partial \phi^j}{\partial q^i}\right)^{-1} \end{pmatrix}.$$

From this local description of the cotangent bundle, it is evident that its dimension is $2n$, as we know from it being the dual bundle of TQ. $\qquad\square$

---

[2] As noted above, we use q^i for the coordinates q^i and x^i.

Definition A.8. The above coordinate charts are called **natural charts** of the cotangent bundle, and their elements are **natural coordinates** (q^i are the **base coordinates** and p_i the **fibre coordinates**).

Observe that the determinant of the last matrix above is the unity, and this is the same for any change of natural coordinates, so the manifold T^*Q is also an orientable manifold.

Similar to the tangent bundle, there are natural (canonical) ways to lift diffeomorphisms and vector fields from a differentiable manifold Q to its cotangent bundle T^*Q.

Definition A.9. Consider a diffeomorphism

$$\varphi : Q \longrightarrow Q$$
$$x \mapsto \varphi(x) .$$

The **canonical lift** of φ to T^*Q is the diffeomorphism

$$T^*\varphi : \quad T^*Q \quad \longrightarrow \quad T^*Q$$
$$(\varphi(x), \xi) \mapsto (x, T^*\varphi(\xi)) ,$$

where $T^*\varphi(\xi)$ is defined by duality as follows: if

$$T\varphi : TQ \quad \longrightarrow \quad TQ$$
$$(q, v) \mapsto (x, T\varphi(v)) .$$

Then, for every $v \in T_qQ$,

$$(T^*\varphi(\xi))(v) := \xi(T\varphi(v)).$$

The following properties follow logically from the definition:

Proposition A.5. *If $\varphi, \phi \colon Q \longrightarrow Q$ are diffeomorphisms, then:*

(1) $\varphi \circ \pi_Q \circ T^*\varphi = \pi_Q$.

(2) $T^*(\varphi \circ \phi) = T^*\phi \circ T^*\varphi$.

Definition A.10. Consider a vector field $Z \in \mathfrak{X}(Q)$. The **canonical lift** of Z to T^*Q is the vector field $Z^* \in \mathfrak{X}(T^*Q)$ whose local uniparametric groups of diffeomorphisms are the canonical lifts T^*F_t of the local uniparametric groups of diffeomorphisms F_t of Z.

As a logical consequence of the definition and the theorem of existence and unicity of uniparametric groups of diffeomorphisms, we have:

> **Proposition A.6.** *If $Z \in \mathfrak{X}(Q)$, then Z^* is π_Q-projectable, and for every* $\mathrm{p} \equiv (q, \xi) \in \mathrm{T}^*Q$, *we have* $\mathrm{T}\pi_Q(Z_\mathrm{p}^*) = Z_q$.

The coordinate expression of the canonical lift of a vector field to T^*Q is given in (3.3).

A.2 Lie Groups and Lie Algebras

(For more information about definitions and properties of Lie groups and Lie algebras, see, for instance, [11, 108, 290, 336, 357].)

> **Definition A.11.** A (finite-dimensional) **Lie group** is a (finite-dimensional) differentiable manifold G such that:
>
> (1) G is a group.
> (2) The following group operations are smooth:
>
> $$\begin{array}{ccc} G \times G & \longrightarrow & G \\ (g_1, g_2) & \longmapsto & g_1 g_2 \end{array} \quad , \quad \begin{array}{ccc} G & \longrightarrow & G \\ g & \longmapsto & g^{-1} \end{array}.$$

> **Definition A.12.** Let G be a Lie group. A subset $H \subset G$ is a **Lie subgroup** of G if:
>
> (1) H is a subgroup of G and
> (2) H has a differentiable structure with respect to which it is a submanifold of G and is a Lie group.

Given a Lie group G and an element $g \in G$, we can define the diffeomorphisms

$$\begin{array}{ccc} L_g : G & \longrightarrow & G \\ g' & \longmapsto & gg' \end{array} \quad , \quad \begin{array}{ccc} R_g : G & \longrightarrow & G \\ g' & \longmapsto & g'g \end{array}$$

which are the **left** and **right translations**, respectively, and satisfy the following properties:

(1) For every $g, g' \in G$; $\quad L_g \circ L_{g'} = L_g L_{g'}, \quad R_g \circ R_{g'} = R_g R_{g'}$.
(2) For every $g \in G$; $\quad L_g^{-1} = L_{g^{-1}}, \quad R_g^{-1} = R_{g^{-1}}$.
(3) For every $g, g' \in G$, $\quad L_g \circ R_{g'} = R_{g'} \circ L_g$.

These translations also induce translations on the set of vector fields $\mathfrak{X}(G)$, which are denoted by L_{g_*} and R_{g_*}, respectively:

Definition A.13. A vector field $X \in \mathfrak{X}(G)$ is **left** (or **right**) **invariant** if, for every $g \in G$,

$$L_{g_*}X = X \quad (or \ R_{g_*}X = X).$$

The sets of these vector fields are denoted $\mathfrak{X}_L(G)$ and $\mathfrak{X}_R(G)$, respectively.

Proposition A.7. *The sets $\mathfrak{X}_L(G)$ and $\mathfrak{X}_R(G)$ are Lie subalgebras of $\mathfrak{X}(G)$.*

Proof. The sum of invariant vector fields is also invariant, and the same thing holds for the Lie bracket, since, for every $g \in G$ and $X_1, X_2 \in \mathfrak{X}_L(G)$ (or $X_1, X_2 \in \mathfrak{X}_R(G)$),

$$L_{g_*}[X_1, X_2] = [L_{g_*}X_1, L_{g_*}X_2] = [X_1, X_2]. \qquad \square$$

Definition A.14. The set $\mathfrak{X}_L(G)$ is the **Lie algebra** of G and is denoted by $\mathbf{g}$.

Proposition A.8. *If $e \in G$ is the neutral element of G, then there is a canonical vector space isomorphism between $\mathfrak{X}_L(G)$ and $\mathrm{T}_e G$.*

Proof. To every $X \in \mathfrak{X}_L(G)$ corresponds a vector of $\mathrm{T}_e G$ by means of the isomorphism

$$\rho_1 \colon \mathfrak{X}_L(G) \longrightarrow \mathrm{T}_e G$$
$$X \longmapsto X(e) := X_e \, .$$

Conversely, from every $X_e \in \mathrm{T}_e G$, we can obtain a unique vector field $X \in \mathfrak{X}_L(G)$ using the map

$$X \colon G \longrightarrow \mathrm{T}G$$
$$g \longmapsto (g, \mathrm{T}L_g X_e) \, ,$$

which is a vector field since, for every $f \in \mathrm{C}^\infty(G)$, the map $X(f)\colon G \longrightarrow \mathbb{R}$ is smooth, and hence, so is X. Moreover, X is left-invariant, as you can see considering its construction from X_e. The map giving X from X_e is denoted $\rho_2 \colon \mathrm{T}_e G \longrightarrow \mathfrak{X}_L(G)$. $\qquad \square$

Remark A.2.

(1) Thus, we identify $\mathbf{g}$ with T_eG, and the Lie bracket between two vectors of T_eG is defined as

$$[X_e, X'_e] := [\rho_2 X_e, \rho_2 X'_e](e).$$

Then, if $\{v_i\}$ is a basis of T_eG, there exists $c_{ij}^k \in \mathbb{R}$ such that

$$[v_i, v_j] = c_{ij}^k v_k,$$

which are called **structure constants** of the group G (related to the chosen basis of $\mathbf{g}$). It can be proven that $c_{ij}^k = 0$ (for all i, j, k) if, and only if, G is locally isomorphic to $\mathbb{R}^n$, that is, G is an *Abelian Lie group*.

(2) Clearly, we can take $\mathfrak{X}_R(G)$ instead of $\mathfrak{X}_L(G)$ to define $\mathbf{g}$. Then, the identification with T_eG would be established in a similar way, although the Lie algebra structure so obtained in T_eG is the opposite one to the above. Nevertheless, both structures are isomorphic since the map $-\mathrm{Id}_{T_eG}$ exchanges them.

Definition A.15. Let $\mathbf{g}$ be a Lie algebra. If

$$\mathbf{g}' := [\mathbf{g}, \mathbf{g}], \ \mathbf{g}'' := [\mathbf{g}', \mathbf{g}'], \ \ldots, \ \mathbf{g}^k := [\mathbf{g}^{k-1}, \mathbf{g}^{k-1}],$$

we have the following sequence of ideals:

$$\mathbf{g} \supseteq \mathbf{g}' \supseteq \mathbf{g}'' \supseteq \cdots \supseteq \mathbf{g}^k \supseteq \cdots.$$

Then, $\mathbf{g}^k$ is called the *k***th-derived algebra** of $\mathbf{g}$.

Observe that if $\mathbf{g}^k = 0$, then $\mathbf{g}^{k+1} = 0$. Then we have:

Definition A.16. Let $\mathbf{g}$ be a Lie algebra.

(1) $\mathbf{g}$ is **Abelian** if $\mathbf{g}' = 0$.
(2) $\mathbf{g}$ is **solvable** if $\mathbf{g}^k = 0$, for some $k > 0$.
(3) Consider now the sequence $\mathbf{g} = \mathbf{g}_1 \supseteq \mathbf{g}_2 \supseteq \cdots \supseteq \mathbf{g}_k \supseteq \ldots$, where

$$\mathbf{g}_1 := \mathbf{g}, \ \mathbf{g}_2 := [\mathbf{g}, \mathbf{g}_1], \ \ldots, \ \mathbf{g}_k := [\mathbf{g}, \mathbf{g}_{k-1}];$$

then, $\mathbf{g}$ is **nilpotent** if $\mathbf{g}_k = \{0\}$ for some $k > 0$.
(4) $\mathbf{g}$ is **simple** if it has no ideals other than $\{0\}$ and $\mathbf{g}$.
(5) $\mathbf{g}$ is **semisimple** if its largest solvable ideal is $\{0\}$, that is, if $\mathbf{g}^k = \mathbf{g}$, for every k.

Then, we can prove that

$$\mathbf{g} \text{ abelian } \Rightarrow \mathbf{g} \text{ nilpotent } \Rightarrow \mathbf{g} \text{ solvable,}$$

and that every non-Abelian simple Lie algebra is semisimple.

Definition A.17. Let $\mathbf{g}$ be a Lie algebra. The **centre** of $\mathbf{g}$ is the set

$$\{X \in \mathbf{g} \mid [X, Y] = 0, Y \in \mathbf{g}\},$$

which is a commutative ideal of $\mathbf{g}$.

Definition A.18. A p-form $\beta \in \Omega^p(G)$ is **left** (or **right**) **invariant** if, for every $g \in G$,

$$L_g^* \beta = \beta \quad (or \ R_g^* \beta = \beta).$$

The set of the left-invariant 1-forms is the dual vector space of $\mathfrak{X}_L(G) := \mathbf{g}$ and is denoted by $\mathbf{g}^*$.

Proposition A.9. $\mathbf{g}^*$ *is isomorphic to* $\mathrm{T}_e^* G$.

Proof. It is clear from the definition. $\qquad\square$

As a consequence, from a covector $\beta_e \in \mathrm{T}_e^* G$, we can obtain another one at every point $g \in G$ doing $L_{g^{-1}}^* \beta_e := \beta_g$. Then, the map

$$\begin{aligned} B : G &\longrightarrow \mathrm{T}^* G \\ g &\longmapsto (g, \beta_g) \end{aligned}$$

is a left-invariant 1-form. Therefore, it is straightforward to prove the following properties:

Proposition A.10.

(1) *For every* $g \in G$, $\beta \in \mathbf{g}^*$, *and* $X, Y \in \mathbf{g}$, *we have*

    (a) $d\beta$ *is left-invariant.*

    (b) $\langle X_g, \beta_g \rangle = \langle X_e, \beta_e \rangle$, *that is,* $\langle X, \beta \rangle = ctn.$

    (c) $d\beta(X, Y) = -\langle [X, Y], \beta \rangle.$

(2) *If* $\{v_i\}$ *is a basis of* $\mathrm{T}_e G$ *and* $\{\gamma^i\}$ *is the dual basis in* $\mathrm{T}_e^* G$, *then*

    (a) $d\gamma^k(v_i, v_j) = -c_{ij}^l \langle v_l, \gamma^k \rangle = -c_{ij}^k.$

(b) The following equations hold:

$$d\gamma^k = -\frac{1}{2}c^k_{ij}\gamma^i \wedge \gamma^j.$$

*The numbers $c^k_{ij} \in \mathbb{R}$ are called the **Maurer–Cartan structure equations** of the group G (related to the chosen basis of $\mathbf{g}^*$).*

Definition A.19. The **canonical** or **Maurer–Cartan form** of the Lie group G is the 1-form $\omega \in \Omega^1(G, \mathbf{g})$, with values on $\mathbf{g}$, defined as follows: For every $g \in G$ and $X \in \mathfrak{X}(G)$,

$$\omega_g(X_g) := TL_g^{-1}X_g.$$

Clearly, ω is left-invariant.

In the definition, observe that $X \in \mathfrak{X}(G)$. In particular, if $X \in \mathbf{g}$, (i.e., X is left-invariant), then we have $TL_g^{-1}X_g = X_e$. Then an alternative definition of ω is as follows:

Definition A.20. The **Maurer–Cartan form** of the Lie group G is the only left-invariant 1-form $\omega \in \Omega^1(G, \mathbf{g})$, with values on $\mathbf{g}$, such that $\omega(X) = X$, for every $X \in \mathbf{g}$, that is, it gives the identity one on $\mathbf{g}$.

Definition A.21. The **uniparametric subgroups** of G are the integral curves passing through e of the left (or right) invariant vector fields, that is, the maps

$$\begin{aligned} \alpha : \mathbb{R} &\longrightarrow \quad G \\ t &\longmapsto \quad g_t := \alpha(t) \end{aligned},$$

such that, if $X \in \mathfrak{X}_L(G)$, then:

(1) $X_{g_{t_0}} = \dfrac{d}{dt}\alpha(t)\Big|_{t_0}$ and

(2) $X_e = \dfrac{d}{dt}\alpha(t)\Big|_{0}.$

These uniparametric subgroups are complete, and there exists a bijective correspondence between $\mathbf{g}$ and the set of uniparametric subgroups of G. The map that implements this correspondence is defined as follows:

Definition A.22. The **exponential map** is defined as,

$$\exp : T_e G \longrightarrow G$$
$$tX_e \longmapsto g_t \, ,$$

for $t \in \mathbb{R}$, and such that $\exp(0) = e$, that is, if α is the uniparametric subgroup corresponding to $\rho_2 X_e$, then:

(1) $\exp(X_e) := \alpha(1)$,
(2) $\exp(0) := e$.

Remark A.3.

(1) Note that the exponential map maps a point of the line tX_e on $T_e G$ into a point of the integral curve of the vector field $\rho_2 X_e$ on G.
(2) Considering that $\mathbb{R}$ is a Lie group with respect to the sum and that $\alpha : \mathbb{R} \longrightarrow G$ is a Lie group homomorphism, we have

$$\exp(t_1 X_e)\exp(t_2 X_e) = g_{t_1} g_{t_2} = \alpha(t_1)\alpha(t_2) = \alpha(t_1 + t_2)$$
$$= g_{(t_1+t_2)} = \exp((t_1 + t_2)X_e).$$

Furthermore,

$$\frac{d}{dt}\exp(tX_e)\Big|_{t_0} = \frac{d}{dt}\alpha(t)\Big|_{t_0} = X_{g_{t_0}},$$

and in particular,

$$\frac{d}{dt}\exp(tX_e)\Big|_{t=0} = X_e.$$

All these properties justify the name given to this map. (Another justification is that if V is a vector space, and we consider $G = \mathrm{Aut}(V)$, then $\exp : \mathrm{End}(V) \longrightarrow \mathrm{Aut}(V)$, and for every $a \in \mathrm{End}(V)$, we obtain

$$\exp(a) = e^a := 1 + a + a^2/2! + a^3/3! + \ldots + a^n/n! + \ldots,$$

where 1 is the identity and $a^n := a \circ \ldots \circ a$ (n times)).

Observe that the last two equalities are natural from the definition, since $\alpha(t) = \exp(tX_e)$ is the integral curve of X. Hence, the flux of X is

$$\tau : \mathbb{R} \times G \longrightarrow G$$
$$(t, g) \longmapsto g\exp(tX_e) \, ,$$

that is, $\tau_t(g) = g\exp(tX_e) = R_{\exp(tX_e)}g$.

(3) The exponential map allows us to locally construct the group G from the algebra $\mathbf{g}$.

Bibliography

[1] R. Abraham and J. E. Marsden, *Foundations of Mechanics*, 2nd edn., Benjamin–Cummings, New York (1978).

[2] R. Abraham, J. E. Marsden, and T. S. Ratiu, *Manifolds, Tensor Analysis, and Applications*, 2nd edn. *American Math. Sci.*, vol 75, Springer, New York (1988). https://doi.org/10.1007/978-1-4612-1029-0.

[3] B. Aebischer, M. Borer, M. Kälin, C. Leuenberger, and H. M. Reimann, *Symplectic Geometry. Progress in Math.*, vol 124. Birkhäuser, Basel (2012). https://doi.org/10.1007/978-3-0348-7512-7.

[4] C. Albert, Le théorème de réduction de Marsden–Weinstein en géometrie cosymplectique et de contact, *J. Geom. Phys.* **6**, 4, pp. 627–649 (1989). https://doi.org/10.1016/0393-0440(89)90029-6.

[5] V. Aldaya and J. A. de Azcárraga, Variational principles on Rth-order jets of fiber bundles in field theory, *J. Math. Phys.* **19**, 9, pp. 1869–1875 (1978). https://doi.org/10.1063/1.523904.

[6] V. Aldaya and J. A. de Azcárraga, Vector bundles, rth-order Noether invariants and canonical symmetries in Lagrangian field theory, *J. Math. Phys.* **19**, 9, pp. 1876–1880 (1978). https://doi.org/10.1063/1.523905.

[7] V. Aldaya and J. A. de Azcárraga, Geometric formulation of classical mechanics and field theory, *Riv. Nuovo Cim.* **3** pp. 1–66 (1980). https://doi.org/10.1007/BF02906204.

[8] A. Anahory, L. Colombo, M. de León, J. C. Marrero, D. Martín de Diego, and E. Padrón, Reduction by symmetries of contact mechanical systems on Lie groups, *SIAM J. Appl. Algebra and Geom.* **8**, 4, pp. 821–845 (2024) https://doi.org/10.1137/23M1616935.

[9] V. Apostolov, D. Calderbank, P. Gauduchon, and E. Legendre, Toric contact geometry in arbitrary codimension, *Int. Math. Res. Not. IMRN* **2020**, 8, pp. 2436–2467. https://doi.org/10.1093/imrn/rny021.

[10] J. M. Arms, R. H. Cushman, and M. J. Gotay, A universal reduction procedure for Hamiltonian group actions, in *The Geometry of Hamiltonian Systems, M.S.R.I. Series*, vol. 22, T. Ratiu (ed.), Springer-Verlag, New York (1991). https://doi.org/10.1007/978-1-4613-9725-0_4.

[11] V. I. Arnold, *Mathematical Methods of Classical Mechanics, Graduate Texts in Math.* vol. 60, Springer-Verlag, New York (1989).

[12] D. Arovas, *Lecture Notes on Classical Mechanics*, Lecture notes for UCSD PHY 110A/B and PHY 200A, Create Space Indep. Pub. Platform (2014).

[13] M. Asorey, J. F. Cariñena, and L. A. Ibort, Generalized canonical transformations for time-dependent systems, *J. Math. Phys.* **24**, 12, pp. 2745–2750 (1983). https://doi.org/10.1063/1.525672.

[14] P. Balseiro, J. C. Marrero, D. Martín de Diego, and E. Padrón, A unified framework for mechanics: Hamilton–Jacobi equation and applications, *Nonlinearity* **23**, 8, pp. 1887–1918 (2010). https://doi.org/10.1088/0951-7715/23/8/006.

[15] G. Bande and A. Hadjar, Contact pairs, *Tohoku Math. J.* **57**, 2, pp. 247–260 (2005). https://doi.org/10.2748/tmj/1119888338.

[16] A. Banyaga, On isomorphic classical diffeomorphism groups I, *Proc. Am. Math. Soc.* **98**, 1, pp. 113–118 (1986). https://doi.org/10.2307/2045779.

[17] A. Banyaga, On isomorphic classical diffeomorphism groups II, *J. Diff. Geom.* **28**, 1, pp. 23–35 (1988). https://10.4310/jdg/1214442158.

[18] A. Banyaga, The structure of classical diffeomorphism groups, *Mathematics and its Applications*, vol. 400, Kluwer Academic Publishers Group, Dordrecht, pp. 113–118 (1997). https://doi.org/10.1007/978-1-4757-6800-8.

[19] A. Banyaga, The geometry surrounding the Arnold-Liouville theorem, in *Advances in Geometry, Progress in Mathematics*, vol. 172, J. L. Brylinski, R. Brylinski, V. Nistor, B. Tsygan, and P. Xu (eds.), Birkhäuser, Boston, MA (1999). https://doi.org/10.1007/978-1-4612-1770-1_3.

[20] A. Banyaga and D. F. Houenou, *A Brief Introduction to Symplectic and Contact Manifolds, Nankai Tracts in Math.*, vol. 15, World Scientific (2016). https://doi.org/10.1142/9667.

[21] M. Barbero-Liñán, M. de León, and D. Martín de Diego, Lagrangian submanifolds and Hamilton–Jacobi equation, *Monatshefte für Mathematik* **171**, 3–4, pp. 269–290 (2013). https://doi.org/10.3934/jgm.2012.4.207.

[22] M. Barbero-Liñán, M. de León, D. Martín de Diego, J. C. Marrero, and M. C. Muñoz-Lecanda, Kinematic reduction and the Hamilton–Jacobi equation, *J. Geom. Mech.* **4**, 3, pp. 207–237 (2012). https://doi.org/10.3934/jgm.2012.4.207.

[23] M. Barbero-Liñán, M. Delgado-Téllez, and D. Martín de Diego, A geometric framework for discrete Hamilton–Jacobi equation, in *Procs. XX Int. Fall Workshop on Geom. Phys.* (Madrid, Spain), *AIP Conference Procs.*, vol. 1460, pp. 164–168 (2012). https://doi.org/10.1063/1.4733374.

[24] M. Barbero-Liñán, A. Echeverría-Enríquez, D. Martín de Diego, M. C. Muñoz-Lecanda, and N. Román–Roy, Skinner–Rusk unified formalism for optimal control problems and applications, *J. Phys. A: Math. Theor.* **40**, 40, pp. 12071–12093 (2007). https://doi.org/10.1088/1751-8113/40/40/005.

[25] M. Barbero-Liñán, A. Echeverría-Enríquez, D. Martín de Diego, M. C. Muñoz-Lecanda, and N. Román-Roy, Unified formalism for non-autonomous mechanical systems, *J. Math. Phys.* **49**, 6, pp. 062902 (2008). https://doi.org/10.1063/1.2929668.

[26] L. M. Bates, F. Fassò, and N. Sansonetto, The Hamilton–Jacobi equation, integrability, and nonholonomic systems, *J. Geom. Mech.* **6**, 4, pp. 441–449 (2014). https://doi.org/10.3934/jgm.2014.6.441.

[27] L. M. Bates and J. Sniatycki, Nonholonomic reduction, *Rep. Math. Phys.* **32**, 1, pp. 99–115 (1993). https://doi.org/10.1016/0034-4877(93)90073-N.

[28] C. Batlle, J. Gomis, J. M. Pons, and N. Román-Roy, Equivalence between the Lagrangian and Hamiltonian formalism for constrained systems, *J. Math. Phys.* **27**, 12, pp. 2953–2962 (1986). https://doi.org/10.1063/1.527274.

[29] C. Batlle, J. Gomis, J. M. Pons, and N. Román-Roy, Lagrangian and Hamiltonian constraints, *Lett. Math. Phys.* **13**, 1, pp. 17–23 (1987). https://doi.org/10.1007/BF00570763.

[30] S. Benenti and W. M. Tulczyjew, The geometrical meaning and globalization of the the Hamilton–Jacobi method, in *Differential Geometric Methods in Mathematical Physics*, Proc. Conf. Aix-en-Provence/Salamanca 1979, *Lect. Notes in Math.*, Vol. 836, Springer, Berlin, pp. 9–21 (1980). https://doi.org/10.1007/BFb0089724.

[31] G. Blankenstein and A. J. Van der Schaft, Optimal control and implicit Hamiltonian systems, in *Nonlinear Control in the Year 2000*, A. Isidori, F. Lamnabhi-Lagarrigue, W. Respondek (eds). Springer–Verlag, LNCIS 258, pp. 185–206 (2000). https://doi.org/10.1007/BFb0110216.

[32] R. J. Blattner, Quantization and representation theory, *Proc. Symp. Pure Math.* **26**, pp. 147–165 (1973).

[33] D. Bleeker, *Gauge Theory and Variational Principles*, Addison-Wesley, Reading, MA (1981).

[34] G. A. Bliss, *Lectures on the Calculus of Variations*, University of Chicago Press, Chicago (1980).

[35] A. Bloch and P. Crouch, Optimal control, optimization and analytical mechanics, in *Mathematical Control Theory*, pp. 265–321, L. Bailieul and J. C. Willems (eds.), Springer-Verlag (1999). https://doi.org/10.1007/978-1-4612-1416-8_8.

[36] A. M. Bloch, P. S. Krishnaprasad, J. E. Marsden, and R. M. Murray, Nonholonomic mechanical systems with symmetry, *Arch. Rational Mech. Anal.* **136**, 1, pp. 21–99 (1996). https://doi.org/10.1007/BF02199365.

[37] P. Bolle, Une condition de contact pour les sous-variétés coisotropes d'une variété symplectique, *C. R. Math. Acad. Sci. Sér. 1* **1**, pp. 83–86 (1996).

[38] A. Bravetti, Contact Hamiltonian dynamics: The concept and its use, *Entropy* **19**, 10, pp. 535–546 (2017). https://doi.org/10.3390/e19100535.

[39] A. Bravetti, Contact geometry and thermodynamics, *Int. J. Geom. Methods Mod. Phys.* **16**, supp01, p. 1940003 (2019). https://doi.org/10.1142/S0219887819400036.

[40] A. Bravetti, H. Cruz, and D. Tapias, Contact Hamiltonian mechanics. *Ann. Phys.* **376**, pp. 17–39 (2017). https://doi.org/10.1016/j.aop.2016.11.003.

[41] A. Bravetti, M. de León, J. C. Marrero, and E. Padrón, Invariant measures for contact Hamiltonian systems: Symplectic sandwiches with contact bread, *J. Phys. A: Math. Gen.* **53**, 45, p. 455205 (2020). https://doi.org/10.1088/1751-8121/abbaaa.

[42] A. J. Bruce, K. Grabowska, and J. Grabowski, Higher order mechanics on graded bundles, *J. Phys. A: Math. Theor.* **48**, 20, p. 205203 (2015). https://doi.org/10.1088/1751-8113/48/20/205203.

[43] A.J. Bruce, K. Grabowska, and J. Grabowski, Remarks on contact and Jacobi geometry, *Symm. Integ. Geom. Meth. Appl. (SIGMA)*, **13**, p. 059 (2017). https://doi.org/10.3842/SIGMA.2017.059.

[44] R. Bryant, S. Chern, H. Gardner, P. Griffith, and H. Goldschmidt, *Exterior Differential Systems*, MSRI Book Series, vol. 18, Springer, New York (1991). https://doi.org/10.1007/978-1-4613-9714-4.

[45] F. Bullo and A. D. Lewis, *Geometric Control of Mechanical Systems*, Springer-Verlag, New York/Heidelberg/Berlin (2004). https://doi.org/10.1007/978-1-4899-7276-7.

[46] O. Calin and D. H. Chang, *Geometric Mechanics on Riemannian Manifolds*, Birkhäuser, Boston (2005). https://doi.org/10.1007/b138771.

[47] M. G. Calkin, *Lagrangian and Hamiltonian Mechanics*, World Scientific, Singapore (1996). https://doi.org/10.1142/3111.

[48] C. M. Campos, M. de León, D. Martín de Diego, and M. Vaquero, Hamilton–Jacobi theory in Cauchy data space, *Rep. Math. Phys.* **76**, 3, pp. 359–387 (2015). https://doi.org/10.1016/S0034-4877(15)30038-0.

[49] C. M. Campos, M. de León, D. Martín de Diego, and J. Vankerschaver, Unambiguous formalism for higher order Lagrangian field theories. *J. Phys. A: Math. Theor.* **42**, 47, pp. 475207 (2009). https://doi.org/10.1088/1751-8113/42/47/475207.

[50] A. Cannas da Silva, *Lectures on Symplectic Geometry*, Springer–Verlag, Heildelberg (2008). https://doi.org/10.1007/978-3-540-45330-7.

[51] F. Cantrijn, J. F. Cariñena, M. Crampin, and L. A. Ibort, Reduction of degenerate Lagrangian systems, *J. Geom. Phys.* **3**, 3, pp. 353–400 (1986). https://doi.org/10.1016/0393-0440(86)90014-8.

[52] F. Cantrijn, J. Cortés, and S. Martínez, Skinner-Rusk approach to time-dependent mechanics, *Phys. Lett. A* **300**, 2–3, pp. 250–258 (2002). https://doi.org/10.1016/S0375-9601(02)00777-6.

[53] F. Cantrijn, M. Crampin, and W. Sarlet, Higher-order differential equations and higher-order Lagrangian mechanics, in *Math. Proc. Cambridge Phil. Soc.*, vol. 99, 3, pp. 565–587 (1986). https://doi.org/10.1017/S0305004100064501.

[54] F. Cantrijn, M. de León, and E. A. Lacomba, Gradient vector fields on cosymplectic manifolds, *J. Phys. A: Math. Gen.* **25**, 1, pp. 175–188 (1986). https://doi.org/10.1088/0305-4470/25/1/022.

[55] F. Cantrijn, M. de León, J. C. Marrero, and D. Martín de Diego, Reduction of nonholonomic mechanical systems with symmetries, *Rep. Math. Phys.* **42**, 1–2, pp. 25–45 (1998). https://doi.org/10.1016/S0034-4877(98)80003-7.

[56] F. Cantrijn and J. Vankerschaver, The Skinner–Rusk approach for vakonomic and nonholonomic field theories, in *Diff. Geom. Meth. in Mech. and Field Theor.*, Academic Press, pp. 1–14 (2007).

[57] B. Cappelletti-Montano, A. de Nicola, and I. Yudin, A survey on cosymplectic geometry, *Rev. Math. Phys.* **25**, 10, pp.1343002 (2013). https://doi.org/10.1142/S0129055X13430022.

[58] G. Caratù, G. Marmo, A. Simoni, B. Vitale, and F. Zaccaria, Lagrangian and Hamiltonian formalisms: An analysis of classical mechanics on tangent and cotangent bundles, *Nuov Cim B* **31**, pp. 152–172 (1976). https://doi.org/10.1007/BF02730325.

[59] R. Cardona, E. Miranda, D. Peralta-Salas, and F. Presas, Universality of Euler flows and flexibility of Reeb embeddings, *Adv. in Math.* **428**, 1, pp. 109142 (2023). https://doi.org/10.1016/j.aim.2023.109142.

[60] J. F. Cariñena, Theory of singular Lagrangians, *Fortschr. Phys.* **38**, 9, pp. 641–679 (1990). https://doi.org/10.1002/prop.2190380902.

[61] J. F. Cariñena, J. Clemente-Gallardo, and G. Marmo, Reduction procedures in classical and quantum mechanics, *Int. J. Geom. Meth. Mod. Phys.* **4**, 8, pp.1363–1403 (2007). https://doi.org/10.1142/S0219887807002594.

[62] J. F. Cariñena and J. Fernández-Nuñez, Geometric theory of time-dependent singular Lagrangians, *Fortschr. Phys.* **41**, 6, pp. 517–552 (1993). https://doi.org/10.1002/prop.2190410603.

[63] J. F. Cariñena, J. Gomis, L. A. Ibort, and N. Román-Roy, Canonical transformations theory for presymplectic systems, *J. Math. Phys.* **26**, 8, pp. 1961–1969 (1985). https://doi.org/10.1063/1.526864.

[64] J. F. Cariñena, J. Gomis, L. A. Ibort, and N. Román-Roy, Applications of the canonical transformations theory for presymplectic systems, *Nuovo Cim. B* **98**, 2, pp. 172–196 (1987). https://doi.org/10.1007/BF02721479.

[65] J. F. Cariñena, X. Gràcia, G. Marmo, E. Martínez, M. C. Muñoz-Lecanda, and N. Román-Roy, Geometric Hamilton–Jacobi theory, *Int. J. Geom. Meth. Mod. Phys.* **3**, 7, pp. 1417–1458 (2006). https://doi.org/10.1142/S0219887806001764.

[66] J. F. Cariñena, X. Gràcia, G. Marmo, E. Martínez, M. C. Muñoz-Lecanda, and N. Román-Roy, Geometric Hamilton–Jacobi theory for nonholonomic dynamical systems, *Int. J. Geom. Meth. Mod. Phys.* **7**, 3, pp. 431–454 (2010). https://doi.org/10.1142/S0219887810004385.

[67] J. F. Cariñena, X. Gràcia, G. Marmo, E. Martínez, M. C. Muñoz-Lecanda, and N. Román-Roy, Structural aspects of Hamilton–Jacobi theory, *Int. J. Geom. Meth. Mod. Phys.* **13**, 2, p. 1650017 (2016). https://doi.org/10.1142/S0219887816500171.

[68] J. F. Cariñena and P. Guha, Nonstandard Hamiltonian structures of the Liénard equation and contact geometry, *Int. J. Geom. Meth. Mod. Phys.* **16**, supp 01, pp. 1940001 (2019). https://doi.org/10.1142/S0219887819400012.

[69] J. F. Cariñena, L. A. Ibort, G. Marmo, and G. Morandi, *Geometry from Dynamics, Classical and Quantum*, Springer, Dordrecht (2015). https://doi.org/10.1007/978-94-017-9220-2.

[70] J. F. Cariñena and C. López, The time evolution operator for singular Lagrangians, *Lett. Math. Phys.* **14**, 3, pp. 203–210 (1987). https://doi.org/10.1007/BF00416849.

[71] J. F. Cariñena and C. López, The time-evolution operator for higher-order singular Lagrangians, *Int. J. Mod. Phys. A* **7**, 11, pp. 2447–2468 (1992). https://doi.org/10.1142/S0217751X92001083.

[72] J. F. Cariñena, C. López, and N. Román-Roy, Geometric study of the connection between the Lagrangian and Hamiltonian constraints, *J. Geom. Phys.* **4**, 3, pp. 315–334 (1987). https://doi.org/10.1016/0393-0440(87)90017-9.

[73] J. F. Cariñena, C. López, and N. Román-Roy, Origin of the Lagrangian constraints and their relation with the Hamiltonian formulation, *J. Math. Phys.* **29**, 5, pp. 1143–1149 (1988). https://doi.org/10.1063/1.527955.

[74] J. F. Cariñena, G. Marmo, and M. F. Rañada, Non-symplectic symmetries and bi-Hamiltonian structures of the rational harmonic oscillator, *J. Phys. A: Math. Gen.* **35**, 47, p. L679 (2002). https://doi.org/10.1088/0305-4470/35/47/101.

[75] J. F. Cariñena and M. C. Muñoz-Lecanda, Geodesic and Newtonian vector fields and symmetries of mechanical systems, *Symmetry* **2023**, 15, p. 181 (2023). https://doi.org/10.3390/sym15010181.

[76] J. F. Cariñena, J. Nunes da Costa, and P. Santos, Reduction of Lie algebroid structures, *Int. J. Geom. Meth. Mod. Phys.* **2**, 5, pp. 965–991 (2005). https://doi.org/10.1142/S0219887805000909.

[77] J. F. Cariñena, J. Nunes da Costa, and P. Santos, Reduction of Lagrangian mechanics on Lie algebroids, *J. Geom. Phys.* **57**, 3, pp. 977–990 (2007). https://doi.org/10.1016/j.geomphys.2006.08.001.

[78] J. F. Cariñena, J. Nunes da Costa, and P. Santos, Quasi-coordinates from the point of view of Lie algebroid structures, *J. Geom. Phys.* **40**, 33, p. 10031 (2007). https://doi.org/10.1088/1751-8113/40/33/008.

[79] E. Cartan, *Leçons sur les Invariants Integraux*, Hermann, Paris (1922).

[80] M. Castrillón-López, P. L. García-Pérez, and T. S. Ratiu, Euler–Poincaré reduction on principal bundles, *Lett. Math. Phys.* **58**, 2, pp. 167–180 (2001). https://doi.org/10.1023/A:1013303320765.

[81] M. Castrillón-López, P. L. García-Pérez, and C. Rodrigo, Euler–Poincaré reduction in principal fibre bundles and the problem of Lagrange, *Diff. Geom. Appl.* **25**, 6, pp. 585–593 (2007). https://doi.org/10.1016/j.difgeo.2007.06.007.

[82] M. Castrillón-López and B. Langerock, Routh reduction for singular Lagrangians, *Int. J. Geom. Meth. Mod. Phys.* **7**, 8, pp. 1451–1489 (2010). https://doi.org/10.1142/S0219887810004907.

[83] M. Castrillón-López and J. E. Marsden, Some remarks on Lagrangian and Poisson reduction for field theories, *J. Geom. Phys.* **48**, 1, pp. 52–83 (2003). https://doi.org/10.1016/S0393-0440(03)00025-1.

[84] M. Castrillón-López and T. S. Ratiu, Reduction in principal bundles: Covariant Lagrange–Poincaré equations, *Comm. Math. Phys.* **236**, 2, pp. 223–250 (2003). https://doi.org/10.1007/s00220-003-0797-5.

[85] M. Castrillón-López, T. S. Ratiu, and S. Shkoller, Reduction in principal fiber bundles: Covariant Euler–Poincaré equations, *Proc. Amer. Math. Soc.* **128**, 7, pp. 2155–2164 (2000). https://doi.org/10.1090/S0002-9939-99-05304-6.

[86] H. Cendra, J. E. Marsden, and T. S. Ratiu, *Lagrangian Reduction by Stages*, *Mem. Amer. Math. Soc.*, vol. 152, 722 (2001). https://doi.org/10.1090/memo/0722.

[87] G. Chavchanidze, Non-Noether symmetries and their influence on phase space geometry, *J. Geom. Phys.* **48**, 2–3, pp. 190–202 (2003). https://doi.org/10.1016/S0393-0440(03)00040-8.

[88] G. Chavchanidze, Non-Noether symmetries in Hamiltonian dynamical systems, *Mem. Diff. Eqs. Math. Phys.* **36**, pp. 81–134 (2005).

[89] D. Chinea, M. de León, and J. C. Marrero, Locally conformal cosymplectic manifolds and time-dependent Hamiltonian systems, *Comment. Math. Univ. Carolin.* **32**, 2, pp. 383–387 (1991).

[90] D. Chinea, M. de León, and J.C. Marrero, The constraint algorithm for time-dependent Lagrangians, *J. Math. Phys.* **35**, 7, pp. 3410–3447 (1994). https://doi.org/10.1063/1.530476.

[91] Y. Choquet-Bruhat, C. Dewitt-Morette, and M. Dillard-Bleick, *Analysis, Manifolds and Physics*, 2nd edn., North-Holland Publishing Co., Amsterdam/New York (1982).

[92] D. Cline, *Variational Principles in Classical Mechanics*, 2nd edn., University of Rochester, Rochester, New York (2018).

[93] F. Ciaglia, H. Cruz, and G. Marmo, Contact manifolds and dissipation, classical and quantum. *Ann. Phys.* **398**, pp. 159–179 (2018). https://doi.org/10.1016/j.aop.2018.09.012.

[94] L. Colombo, D. Martín de Diego, and M. Zuccalli, Optimal control of underactuated mechanical systems: A geometric approach, *J. Math. Phys.* **51**, 8, pp. 083519 (2010). https://doi.org/10.1063/1.3456158.

[95] L. Colombo, M. de León, P. D. Prieto-Martínez, and N. Román-Roy, Geometric Hamilton–Jacobi theory for higher-order autonomous systems, *J. Phys. A: Math. Gen.* **47**, 3), p. 235203 (2014). https://doi.org/10.1088/1751-8113/47/23/235203.

[96] L. Colombo, M. de León, P. D. Prieto-Martínez, and N. Román-Roy, Unified formalism for the generalized kth-order Hamilton–Jacobi problem, *Int. J. Geom. Meth. Mod. Phys.* **11**, 9, p. 1460037 (2014). https://doi.org/10.1142/S0219887814600378.

[97] L. Conlon, *Differentiable Manifolds*, Birkhäuser, Boston (2001). https://doi.org/10.1007/978-0-8176-4767-4.

[98] G. C. Constantelos, On the Hamilton–Jacobi Theory with derivatives of higher order, *Nuovo Cim. B (11)* **84**, 1, pp. 91–101 (1984). https://doi.org/10.1007/BF02721650.

[99] E. Corinaldesi, *Classical Mechanics for Physics Graduate Students*, World Scientific Publishing Co., Singapore (1999). https://doi.org/10.1142/3926.

[100] J. Cortés, M. de León, D. Martín de Diego, and S. Martínez, Geometric description of vakonomic and nonholonomic dynamics. Comparison of solutions. *SIAM J. Control Opt.* **41**, 5, pp. 1389–1412 (2002). https://doi.org/10.1137/S036301290036817X.

[101] J. Cortés, S. Martínez, and F. Cantrijn, Skinner-Rusk approach to time-dependent mechanics, *Phys. Lett. A* **300**, 2-3, pp. 250–258 (2002). https://doi.org/10.1016/S0375-9601(02)00777-6.

[102] M. Crampin, Constants of the motion in Lagrangian mechanics, *Int. J. Theor. Phys.* **16**, 10, pp. 741–754 (1977). https://doi.org/10.1007/BF01807231.

[103] M. Crampin, On the differential geometry of the Euler–Lagrange equations and the inverse problem of Lagrangian dynamics, *J. Phys. A: Math. Gen.* **14**, 10, pp. 2567–2575 (1981). https://doi.org/10.1088/0305-4470/14/10/012.

[104] M. Crampin, A note on non-Noether constants of motion, *Phys. Lett. A* **95**, 5, pp. 209–212 (1983). https://doi.org/10.1016/0375-9601(83)90605-9.

[105] M. Crampin, Tangent bundle geometry for Lagrangian dynamics, *J. Phys. A: Math. Gen.* **16**, pp. 3755–3772 (1983). https://doi.org/10.1088/0305-4470/16/16/014.

[106] M. Crampin, *Jet Bundle Techniques in Analytical Mechanics*, *Quaderni del Consiglio Nazionale delle Ricerche*, Gruppo Nazionale de Fisica Matematica, Florence, vol. 47 (1995).

[107] M. Crampin and T. Mestdag, Reduction and reconstruction aspects of second-order dynamical systems with symmetry, *Acta Appl. Math.* **105**, pp. 241–266 (2009). https://doi.org/10.1007/s10440-008-9274-7.

[108] M. Crampin and F. A.E. Pirani, *Applicable Differential Geometry*, LMS Lecture Notes Ser., vol. 59, Cambridge University Press (1986). https://doi.org/10.1017/CBO9780511623905.

[109] M. Crampin, G. E. Prince, and G. Thompson, A geometrical version of the Helmholtz conditions in time-dependent Lagrangian dynamics, *J. Phys. A: Math. Gen.* **17**, 7, pp. 1437–1447 (1984). https://doi.org/10.1088/0305-4470/17/7/011.

[110] G. Darboux, Sur le problème de Pfaff, *Bull. Sci. Math. Astr. (Ser. 2)* **6**, 1, pp. 14–36 (1984).

[111] U. N. de Almeida, *Contact Anosov Actions with Smooth Invariant Bundles*, Ph.D. Thesis, Universidade de São Paulo (2018). https://doi.org/10.11606/T.55.2018.tde-01112018-110622.

[112] M. de León, J. Gaset, X. Gràcia, M. C. Muñoz-Lecanda, and X. Rivas, Time-dependent contact mechanics, *Monatsh. Math.* **201**, pp. 1149–1183 (2023). https://doi.org/10.1007/s00605-022-01767-1.

[113] M. de León, J. Gaset, M. Laínz, M. C. Muñoz-Lecanda, and N. Román-Roy, Higher-order contact mechanics. *Ann. Phys.* **425**, p. 168396 (2021). https://doi.org/10.1016/j.aop.2021.168396.

[114] M. de León, J. Gaset, M. Laínz-Valcázar, X. Rivas, and N. Román-Roy, Unified Lagrangian-Hamiltonian formalism for contact systems, *Fortsch. Phys.* **68**, 8, p. 2000045 (2020). https://doi.org/10.1002/prop. 202000045.

[115] M. de León, J. Gaset, M. C. Muñoz-Lecanda, X. Rivas, and N. Román-Roy, Multicontact formulation for non-conservative field theories, *J. Phys. A: Math. Theor.* **56**, 2, p. 025201 (2023). https://doi.org/10.1088/ 1751-8121/acb575.

[116] M. de León, V. M. Jiménez, and M. Laínz, Contact Hamiltonian and Lagrangian systems with nonholonomic constraints, *J. Geom. Mech.* **13**, 1, pp. 25–53 (2021). https://doi.org/10.3934/jgm.2021001.

[117] M. de León and M. Laínz-Valcázar, Singular Lagrangians and precontact Hamiltonian systems. *Int. J. Geom. Meth. Mod. Phys.* **16**, 10, p. 1950158 (2019). https://doi.org/10.1142/S0219887819501585.

[118] M. de León and M. Laínz-Valcázar, Infinitesimal symmetries in contact Hamiltonian systems, *J. Geom. Phys.* **153**, pp. 153651 (2020). https:// doi.org/10.1016/j.geomphys.2020.103651.

[119] M. de León, M. Laínz-Valcázar, A. López-Gordón, and X. Rivas, Hamilton–Jacobi theory and integrability for autonomous and non-autonomous contact systems, *J. Geom.Phys.* **187**, p. 104787 (2023). https://doi.org/10. 1016/j.geomphys.2023.104787.

[120] M. de León, M. Laínz-Valcázar, and A. Muñiz-Brea, The Hamilton–Jacobi theory for contact Hamiltonian systems, *Mathematics* **16**, 9, p. 1993 (2021). https://doi.org/10.3390/math9161993.

[121] M. de León, M. Laínz-Valcázar, and M. C. Muñoz-Lecanda, Optimal control, contact dynamics and Herglotz variational problem, *J Nonlinear Sci.* **33**, 9 (2023). https://doi.org/10.1007/s00332-022-09861-2.

[122] M. de León, M. Laínz-Valcázar, M. C. Muñoz-Lecanda, and N. Román-Roy, Constrained Lagrangian dissipative contact dynamics, *J. Math. Phys.* **62**, 12, pp. 122902 (2021). https://doi.org/10.1063/5.0071236.

[123] M. de León, J. Marín-Solano, and J. C. Marrero, The constraint algorithm in the jet formalism, *Diff. Geom. Appl.* **6**, 3, pp. 275–300 (1996). https: //doi.org/10.1016/0926-2245(96)82423-5.

[124] M. de León, J. Marín-Solano, J. C. Marrero, M. C. Muñoz-Lecanda, and N. Román-Roy, Singular Lagrangian systems on jet bundles, *Fortsch. Phys.* **50**, 2, pp. 105–169 (2002). https://doi.org/10.1002/1521-3978(200203) 50:2<105::AID-PROP105>3.0.CO;2-N.

[125] M. de León, J. C. Marrero, and D. Martín de Diego, Time-dependent constrained Hamiltonian systems and Dirac brackets, *J. Phys. A: Math. Gen.* **29**, 21, pp. 6843–6859 (1996). https://doi.org/10.1088/0305-4470/29/ 21/016.

[126] M. de León, J. C. Marrero, and D. Martín de Diego, A new geometrical setting for classical field theories, in *Classical and Quantum Integrability*, Banach Center Pub., vol. 59, pp. 189–209, Institute of Mathematics, Polish Academy of Sciences, Warsaw (2003). https://doi.org/10.4064/bc59-0-10.

[127] M. de León, J. C. Marrero, and D. Martín de Diego, Vakonomic mechanics versus nonholonomic mechanics: A unified geometric approach, *J. Geom. Phys.* **35**, 2–3, pp. 126–144 (2000). http://doi.org/10.1016/S0393-0440(00)00004-8.

[128] M. de León, J. C. Marrero, and D. Martín de Diego, A geometric Hamilton–Jacobi theory for classical field theories, in *Variations, Geometry and Physics*, pp. 129–140, Nova Science Publishers, New York (2009). http://hdl.handle.net/10261/4169.

[129] M. de León, J.C. Marrero, and D. Martín de Diego, Linear almost Poisson structures and Hamilton–Jacobi equation. Applications to nonholonomic mechanics, *J. Geom. Mech.* **2**, 2, pp. 159–198 (2010). https://doi.org/10.3934/jgm.2010.2.159.

[130] M. de León, J. C. Marrero, and E. Martínez, Lagrangian submanifolds and dynamics on Lie algebroids, *J. Phys. A: Math. Gen.* **38**, 24, p. R241 (2005). https://doi.org/10.1088/0305-4470/38/24/R01.

[131] M. de León and D. Martín de Diego, On the geometry of nonholonomic Lagrangian systems, *J. Math. Phys.* **37**, 7, pp. 3389–3414 (1996). https://doi.org/10.1063/1.531571.

[132] M. de León, D. Martín de Diego, J. C. Marrero, M. Salgado, and S. Vilariño, Hamilton–Jacobi theory in k-symplectic field theories, *Int. J. Geom. Meth. Mod. Phys.* **7**, 8, pp. 1491–1507 (2010). https://doi.org/10.1142/S0219887810004919.

[133] M. de León, D. Martín de Diego, and A. Santamaría-Merino, Symmetries in classical field theories, *Int. J. Geom. Meth. Mod. Phys.* **1**, 5, pp. 651–710 (2004). https://doi.org/10.1142/S0219887804000290.

[134] M. de León, D. Martín de Diego, and M. Vaquero, A Hamilton–Jacobi theory for singular Lagrangian systems in the Skinner and Rusk setting, *Int. J. Geom. Meth. Mod. Phys.* **9**, 8, p. 1250074 (2012). https://doi.org/10.1142/S0219887812500740.

[135] M. de León, D. Martín de Diego, and M. Vaquero, A Hamilton–Jacobi theory on Poisson manifolds, *J. Geom. Mech.* **6**, 1, pp. 121–140 (2014). https://doi.org/10.3934/jgm.2014.6.121.

[136] M. de León, M. H. Mello, and P. R. Rodrigues, Reduction of degenerate non-autonomous Lagrangian systems, in *Mathematical Aspects of Classical Field Theory*, Contemp. Math. vol. 132, pp. 275–305, M. J. Gotay, J. E. Marsden, V. Moncrief (eds.), American Mathematical Society, Seattle, WA (1992). http://dx.doi.org/10.1090/conm/132.

[137] M. de León, P. D. Prieto-Martínez, N. Román-Roy, and S. Vilariño, Hamilton–Jacobi theory in multisymplectic classical field theories, *J. Math. Phys.* **58**, 9, p. 092901 (2017). https://doi.org/10.1063/1.5004260.

[138] M. de León and P. R. Rodrigues, *Methods of Differential Geometry in Analytical Mechanics, North-Holland Math. Studies*, vol. 158, North-Holland Publishing Co., Amsterdam (1989).

[139] M. de León and P. R. Rodrigues, *Generalized Classical Mechanics and Field Theory, North-Holland Math. Studies*, vol. 112, Elsevier, Amsterdam (1985).

[140] M. de León and M. Saralegi, Cosymplectic reduction for singular momentum maps, *J. Phys. A: Math. Gen.* **26**, 19, pp. 1–11 (1993). https://doi.org/10.1088/0305-4470/26/19/032.

[141] M. de León and C. Sardón, Cosymplectic and contact structures to resolve time-dependent and dissipative Hamiltonian systems, *J. Phys. A: Math. Theor.* **50**, 25, pp. 255205 (2017). https://doi.org/10.1088/1751-8121/aa711d.

[142] M. de León and S. Vilariño, Hamilton–Jacobi theory in k-cosymplectic field theories, *Int. J. Geom. Meth. Mod. Phys.* **11**, 1, p. 1450007 (2014). https://doi.org/10.1142/S0219887814500078.

[143] J. de Lucas, X. Gràcia, X. Rivas, and N. Román-Roy, On Darboux theorems for geometric structures induced by closed forms, *Rev. Real Acad. Cienc. Exactas Fis. Nat. Ser. A-Mat* **118**, p. 131 (2024). https://doi.org/10.1007/s13398-024-01632-w.

[144] J. de Lucas, X. Gràcia, X. Rivas, N. Román-Roy, and S. Vilariño, Reduction and reconstruction of multisymplectic Lie systems, *J. Phys. A: Math. Theor.* **55**, 29, p. 295204 (2022). https://doi.org/10.1088/1751-8121/ac78ab.

[145] J. de Lucas and X. Rivas, Contact Lie systems: Theory and applications, *J. Phys. A: Math. Theor.* **56**, 33, p. 335203 (2023). https://doi.org/10.1088/1751-8121/ace0e7.

[146] J. de Lucas, X. Rivas, S. Vilariño, and B. M. Zawora, On k-polycosymplectic Marsden–Weinstein reductions, *J. Geom. Phys.* **191**, p. 104899 (2023). https://doi.org/10.1016/j.geomphys.2023.104899.

[147] A. de Nicola and W. M. Tulckzyjew, A variational formulation of analytical mechanics in an affine space, *Rep. Math. Phys.* **58**, 3, pp. 335–350 (2006). https://doi.org/10.1016/S0034-4877(07)00004-3.

[148] A. Deriglazov, *Classical Mechanics Hamiltonian and Lagrangian Formalism*, Springer-Verlag, Berlin/Heidelberg (2010). https://doi.org/10.1007/978-3-642-14037-2.

[149] P. A. M. Dirac, *Lectures on Quantum Mechanics*, Belfer Graduate School of Science, New York, Yeshiva University (1964).

[150] D. Dominici and J. Gomis, Poincaré–Cartan integral invariant and canonical transformations for singular Lagrangians, *J. Math. Phys.* **21**, 8, pp. 2124–2130 (1980). https://doi.org/10.1063/1.524721.

[151] D. Dominici, J. Gomis, G. Longhi, and J. M. Pons, Hamilton–Jacobi theory for constrained systems, *J. Math. Phys.* **25**, 8, pp. 2439–2460 (1984). https://doi.org/10.1063/1.526452.

[152] A. Echeverría-Enríquez, A. Ibort, M. C. Muñoz-Lecanda, and N. Román-Roy, Invariant forms and automorphisms of locally homogeneous multisymplectic manifolds, *J. Geom. Mech.* **4**, 4, pp. 397–419 (2012). https://doi.org/10.3934/jgm.2012.4.397.

[153] A. Echeverría-Enríquez, C. López, J. Marín–Solano, M. C. Muñoz-Lecanda, and N. Román–Roy, Lagrangian-Hamiltonian unified formalism for field theory, *J. Math. Phys.* **45**, 1, pp. 360–385 (2004). https://doi.org/10.1063/1.1628384.

[154] A. Echeverría-Enríquez, M. C. Muñoz-Lecanda, and N. Román-Roy, Geometrical setting of time-dependent regular systems. Alternative models, *Rev. Math. Phys.* **3**, 3, pp. 301–330 (1991). https://doi.org/10.1142/S0129055X91000114.

[155] A. Echeverría-Enríquez, M. C. Muñoz-Lecanda, and N. Román-Roy, Nonstandard connections in classical mechanics, *J. Phys. A: Math. Gen.*, **28**, 19, pp. 5553–5567 (1995). https://doi.org/10.1088/0305-4470/28/19/011.

[156] A. Echeverría-Enríquez, M. C. Muñoz-Lecanda, and N. Román-Roy, Reduction of presymplectic manifolds with symmetry, *Rev. Math. Phys.* **11**, 10, pp. 1209–1248 (1999). https://doi.org/10.1142/S0129055X99000386.

[157] A. Echeverría-Enríquez, M. C. Muñoz-Lecanda, N. Román-Roy, and C. Victoria, Mathematical foundations of geometric quantization, *Extracta Math.* **13**, 2, pp. 135–238 (1998).

[158] A. Echeverría-Enríquez, M. C. Muñoz-Lecanda, and N. Román-Roy, Remarks on multisymplectic reduction, *Rep. Math. Phys.* **81**, 3, pp. 415–424 (2018). https://doi.org/10.1016/S0034-4877(18)30057-0.

[159] L. Elsgolts, *Differential Equations and the Calculus of Variations*, Mir Publishers, Moscow (1971).

[160] G. Esposito, G. Marmo, and G. Sudarshan, *From Classical to Quantum Mechanics*, Cambridge University Press, Cambridge (2004). https://doi.org/10.1017/CBO9780511610929.

[161] R. P. Feynman, R. B. Leighton, and M. Sands, *Lectures on Physics*, vol. 1, 2nd edn., Basic Books, New York (2010).

[162] D. Finamore, Contact foliations and generalised Weinstein conjectures, arxiv.org: 2202.07622 [math.SG] (2022). https://doi.org/10.48550/arXiv.2202.07622.

[163] C. R. Galley, Classical mechanics of nonconservative systems, *Phys. Rev. Lett.* **110**, 17, p. 174301 (2013). https://doi.org/10.1103/PhysRevLett.110.174301.

[164] T. Gallissot, Les formes extérieures en Mécanique, *Ann. Inst. Fourier Grenoble* **4**, pp. 145–297 (1952). https://doi.org/10.5802/aif.49.

[165] F. Gantmacher, *Lectures in Analytical Mechanics*, Mir Publishers, Moscow (1970).

[166] P. L. García, Geometric quantization, *Mem. Real Acad. Cienc. Exact. Fis. Nat.* **11**, (1979).

[167] P. L. García and J. Muñoz-Masqué, Higher order analytical dynamics, in *Dynamical systems and partial differential equations*, pp. 19–47, University of Simon Bolivar, Equinoccio, Caracas (1986).

[168] J. Gaset, X. Gràcia, M. C. Muñoz-Lecanda, X. Rivas, and N. Román-Roy, A contact geometry framework for field theories with dissipation, *Ann. Phys.* **414**, p. 168092 (2020). https://doi.org/10.1016/j.aop.2020.168092.

[169] J. Gaset, X. Gràcia, M. C. Muñoz-Lecanda, X. Rivas, and N. Román-Roy, New contributions to the Hamiltonian and Lagrangian contact formalisms for dissipative mechanical systems and their symmetries, *Int. J. Geom. Meth. Mod. Phys.* **17**, 6, pp. 2050090 (2020). https://doi.org/10.1142/S0219887820500905.

[170] J. Gaset, X. Gràcia, M. C. Muñoz-Lecanda, X. Rivas, and N. Román-Roy, A k-contact Lagrangian formulation for nonconservative field theories, *Rep. Math. Phys.* **87**, 3, pp. 347–368 (2021). https://doi.org/10.1016/S0034-4877(21)00041-0.

[171] J. Gaset, M. Laínz, A. Mas, and X. Rivas, The Herglotz variational principle for dissipative field theories, arxiv.org: 2211.17058 [math-ph] (2022). https://doi.org/10.48550/arXiv.2211.17058.

[172] J. Gaset, A. López-Gordón, A. Mas, and X. Rivas, Symmetries, conservation and dissipation in time-dependent contact systems, *Fortschr. Phys.* **71**, p. 2300048 (2023). https://doi.org/10.1002/prop.202300048.

[173] I. M. Gelfand and S. V. Fomin, *Calculus of Variations*, Prentice-Hall, Englewood Cliffs, NJ (1963).

[174] H. Geiges, *An Introduction to Contact Topology*. Cambridge University Press, Cambridge (2008). https://doi.org/10.1017/CB09780511611438.

[175] B. Georgieva and R. Guenther, First Noether-type theorem for the generalized variational principle of Herglotz, *Topol. Methods Nonlinear Anal.* **20**, 2, pp. 261–273 (2002). https://doi.org/10.12775/TMNA.2002.036.

[176] A. Ghosh and C. Bhamidipati, Contact geometry and thermodynamics of black holes in AdS spacetimes, *Phys. Rev. D* **100**, p. 126020 (2019). https://doi.org/10.1103/PhysRevD.100.126020.

[177] R. Giachetti, Hamiltonian systems with symmetry: An introduction, *Riv. Nuovo Cim.* **4**, 12, pp. 1–63 (1981). https://doi.org/10.1007/BF02740644.

[178] C. Godbillon, *Géométrie Différentielle et Mécanique Analytique*, Hermann, Paris (1969).

[179] L. Godinho and J. Natário, *An Introduction to Riemannian Geometry (With Applications to Mechanics and Relativity)*, Springer (2014). https://doi.org/10.1007/978-3-319-08666-8.

[180] H. Goldstein, C. P. Poole, and J. L. Safko, *Classical Mechanics*, 3rd edn., Addison-Wesley Press, Inc., Cambridge, MA (2001).

[181] J. Gomis, J. Llosa, and N. Román-Roy, Lee Hwa Chung theorem for presymplectic manifolds. Canonical transformations for constrained systems, *J. Math. Phys.* **25**, 5, pp. 1348–1355 (1984). https://doi.org/10.1063/1.526303.

[182] M. Gosson, *Symplectic Geometry and Quantum Mechanics*, Birkhäuser, Basel (2009). https://doi.org/10.1007/3-7643-7575-2.

[183] M. J. Gotay, J. Isenberg, and J. E. Marsden, Momentum maps and classical relativistic fields, Part I: Covariant field theory, arxiv.org: 9801019 [physics] (2004). https://doi.org/10.48550/arXiv.physics/9801019. Momentum maps and classical relativistic fields, Part II: Canonical analysis of Field Theories, arxiv.org:0411032[math-ph] (2004). https://doi.org/10.48550/arXiv.math-ph/0411032.

[184] M. J. Gotay and J. M. Nester, Presymplectic Lagrangiany systems I: The constraint algorithm and the equivalence problem, *Ann. Inst. H. Poincaré A* **30**, pp. 129–142 (1979). https:::://www.numdam.org/item/AIHPA_1979_30_2_129_0/.

[185] M. J. Gotay, and J. M. Nester, Presymplectic Lagrangian systems II: The second order equation problem, *Ann. Inst. H. Poincaré A* **32**, pp. 1–13 (1980). https://www.numdam.org/item/AIHPA_1980_32_1_1_0/.

[186] M. J. Gotay, J. M. Nester, and G. Hinds, Presymplectic manifolds and the Dirac-Bergmann theory of constraints, *J. Math. Phys.* **27**, 11, pp. 2388–2399 (1978). https://doi.org/10.1063/1.523597.

[187] S. Goto, Contact geometric descriptions of vector fields on dually flat spaces and their applications in electric circuit models and nonequilibrium statistical mechanics, *J. Math. Phys.* **57**, 10, p. 102702 (2016). https://doi.org/10.1063/1.4964751.

[188] K. Grabowska and J. Grabowski, Dirac algebroids in Lagrangian and Hamiltonian mechanics, *J. Geom. Phys.* **61**, 11, pp. 2233–2253 (2011). https://doi.org/10.1016/j.geomphys.2011.06.018.

[189] K. Grabowska and J. Grabowski, A geometric approach to contact Hamiltonians and contact Hamilton–Jacobi theory, *J. Phys. A: Math. Theor.* **55**, 43, pp. 435204 (2022). https://doi.org/10.1016/10.1088/1751-8121/ac9adb.

[190] K. Grabowska and J. Grabowski, Reductions: Precontact versus presymplectic, *Annali di Matematica* **202**, pp. 2803–2839 (2023). https://doi.org/10.1007/s10231-023-01341-y.

[191] K. Grabowska and J. Grabowski, Contact geometric mechanics: The Tulczyjew triples, arxiv.org: 2209.03154 [math.SG] (2022). https://doi.org/10.48550/arXiv.2209.03154.

[192] K. Grabowska, J. Grabowski, and P. Urbanski, Geometrical mechanics on algebroids, *Int. J. Geom. Meth. Mod. Phys.* **3**, 3, pp. 559–575 (2006). https://doi.org/10.1142/S0219887806001259.

[193] J. Grabowski, G. Landi, G. Marmo, and G. Vilasi, Generalized reduction procedure: Symplectic and Poisson formalism, *Fortschr. Phys.* **42**, 5, pp. 393–427 (1994). https://doi.org/10.1002/prop.2190420502.

[194] X. Gràcia and R. Martín, Geometric aspects of time-dependent singular differential equations, *Int. J. Geom. Meth. Mod. Phys.* **2**, 4, pp. 597–618 (2005). https://doi.org/10.1142/S0219887805000697.

[195] X. Gràcia and J. M. Pons, On an evolution operator connecting Lagrangian and Hamiltonian formalisms, *Lett. Math. Phys.* **17**, pp. 175–180 (1989). https://doi.org/10.1007/BF00401582.

[196] X. Gràcia, and J. M. Pons, A generalized geometric framework for constrained systems, *Diff. Geom. Appl.* **2**, pp. 223–247 (1992). https://doi.org/10.1016/0926-2245(92)90012-C.

[197] X. Gràcia, J. M. Pons, and N. Román-Roy, Higher order Lagrangian systems: Geometric structures, dynamics and constraints, *J. Math. Phys.* **32**, 10, pp. 2744–2763 (1991). https://doi.org/10.1063/1.529066.

[198] X. Gràcia, J. M. Pons, and N. Román-Roy, Higher-order conditions for singular Lagrangian systems, *J. Phys. A: Math. Gen.* **25**, 7, pp. 1981–2004 (1992). https://doi.org/10.1088/0305-4470/25/7/037.

[199] W. Greub, S. Halpering, and S. Vanstone, *Connections, Curvature and Cohomology*, Academic Press, New York (1972).

[200] J. Grifone, Structure presque-tangente et conexions I, *Ann Inst. Fourier Grenoble* **22**, 1, pp. 287–334 (1972). https://doi.org/10.5802/aif.407.

[201] J. Grifone, Structure presque-tangente et conexions II, *Ann Inst. Fourier Grenoble* **22**, 3, pp. 291–338 (1972). https://doi.org/10.5802/aif.431.

[202] S. Grillo, J. C. Marrero, and E. Padrón, Extended Hamilton–Jacobi theory, symmetries and integrability by quadratures, *Mathematics* **9**, 12, p. 1357 (2021). https://doi.org/10.3390/math9121357.

[203] S. Grillo and E. Padrón, A Hamilton–Jacobi theory for general dynamical systems and integrability by quadratures in symplectic and Poisson manifolds, *J. Geom. Phys.* **10**, 1, pp. 101–129 (2016). https://doi.org/10.1016/j.geomphys.2016.07.010.

[204] S. Grillo and E. Padrón, Extended Hamilton–Jacobi theory, contact manifolds, and integrability by quadratures, *J. Math. Phys.* **61**, 1, pp. 012901 (2020). https://doi.org/10.1063/1.5133153.

[205] A. Guerra IV and N. Román-Roy, More insights into symmetries in multisymplectic field theories, *Symmetry* **15** 2, p. 390 (2023). https://doi.org/10.3390/sym15020390.

[206] A. Guerra IV and N. Román-Roy, Canonical lifts in multisymplectic De Donder–Weyl Hamiltonian field theories, *J. Phys. A.: Math. Theor.* **57**, 33, p. 335203 (2024). https://doi.org/10.1088/1751-8121/ad6654.

[207] V. Guillemin, V. Ginzburg, and Y. Karshon, *Moment maps, Cobordisms, and Hamiltonian Group Actions, Mathematical Surveys and Monographs Ser.*, vol. 98, American Mathematical Society (2002).

[208] V. Guillemin and S. Sternberg, *Geometric Asymptotics, Mathematical Surveys and Monographs*, vol. 14, American Mathematical Society (1977).

[209] V. Guillemin and S. Sternberg, *Symplectic Techniques in Physics*, Cambridge University Press, Cambridge (1990).

[210] I. Gutierrez-Sagredo, D. Iglesias-Ponte, J. C. Marrero and E. Padrón, Mechanical presymplectic structures and Marsden-Weinstein reduction of time-dependent Hamiltonian systems, *J. Geom. Phys*, **213**, p. 105492 (2025). https://doi.org/10.1016/j.geomphys.2025.105492.

[211] H. Hamoui and A. Lichnerowicz, Geometry of the dynamical systems with time-dependent constraints and time-dependent Hamiltonian: An approach towards quantization, *J. Math. Phys.* **25**, 4, pp. 923–934 (1984). https://doi.org/10.1063/1.526247.

[212] A. J. Hanson, T. Regge, and C. Teitelboim, *Constrained Hamiltonian systems*, Accademia Nazionale dei Lincei, Rome (1976). https://hdl.handle.net/2022/3108.

[213] G. Herglotz, Berührungstransformationen, Lectures at the University of Gottingen (1930).

[214] G. Herglotz, *Vorlesungen über die Mechanik der Kontinua, Teubner-Archiv zur Mathematik*, vol. 3, Teubner, Leipzig (1985).

[215] G. Herczeg and A. Waldron, Contact geometry and quantum mechanics, *Phys. Lett. B* **781**, 1, pp. 312–315 (2018). https://doi.org/10.1016/j.physletb.2018.04.008.

[216] R. Hermann, *Differential Geometry and the Calculus of Variations, Math. in Sci. and Engineering*, vol. 49, Academic Press, New York (1968).

[217] D. D. Holm. *Geometric Mechanics. Part I. Dynamics and Symmetry*, Imperial College Press, London (2008). https://doi.org/10.1142/p801.

[218] L. A. Ibort, The geometry of dynamics, *Extracta Math.* **11**, 1, pp. 80–105 (1996).

[219] L. A. Ibort, M. de León, J. C. Marrero, and D. Martín de Diego, Dirac brackets in constrained dynamics, *Forschr. Phys.* **47**, 5, pp. 459–492 (1999). https://doi.org/10.1002/(SICI)1521-3978(199906)47:5<459::AID-PROP459>3.0.CO;2-E.

[220] L. A. Ibort and J. Marín-Solano, A geometric classification of Lagrangian functions and the reduction of the evolution space, *J. Phys. A: Math. Gen.* **25**, 11, pp. 3353–3367 (1992). https://doi.org/10.1088/0305-4470/25/11/036.

[221] D. Iglesias-Ponte, M. de León, and D. Martín de Diego, Towards a Hamilton–Jacobi theory for nonholonomic mechanical systems, *J. Phys. A: Math. Theor.* **41**, 1, p. 015205 (2008). https://doi.org/10.1088/1751-8113/41/1/015205.

[222] J. V. José and E. J. Saletan, *Classical Dynamics. A Contemporary Approach*, Cambridge University Press, Cambridge (1998).

[223] K. Kamimura, Singular Lagrangians and constrained Hamiltonian systems, generalized canonical formalism, *Nuovo Cim. B* **69**, pp. 33–54 (1982). https://doi.org/10.1007/BF02888859.

[224] A. L. Kholodenko, *Applications of Contact Geometry and Topology in Physics*, World Scientific (2013). https://doi.org/10.1142/8514.

[225] T. Kibble and F. H. Berkshire, *Classical Mechanics*, 5th edn., World Scientific Publishing Company, London (2004). https://doi.org/10.1142/p310.

[226] A. A. Kirillov, Geometric quantization, in *Encyclopaedia of Mathematical Sciences, Vol 4: Dynamical Systems*, pp. 137–172, Springer, Berlin (1985). https://doi.org/10.1007/978-3-662-06793-2_2.

[227] J. Klein, Espaces variationelles et mécanique, *Ann. Inst. Fourier* **12**, pp. 1–124 (1962). https://www.numdam.org/item?id=AIF_1962_12_1_0.

[228] S. Kobayashi and K. Nomizu, *Foundations of Differential Geometry*, vol. 1, Wiley, New York (1996).

[229] Y. Kosmann-Schwarzbach, *The Noether Theorems. Invariance and Conservation Laws in the Twentieth Century*, Springer, New York (2011). https://doi.org/10.1007/978-0-387-87868-3.

[230] J. L. Koszul and Yi Ming Zou, *Introduction to Symplectic Geometry*, Springer, Singapore (2019). https://doi.org/10.1007/978-981-13-3987-5.

[231] O. Krupkova, Higher-order mechanical systems with constraints, *J. Math. Phys.* **41**, 8, pp. 5304–5324 (2000). https://doi.org/10.1063/1.533411.

[232] O. Krupková and A. Vondra, On some integration methods for connections on fibered manifolds, in *Differential Geometry and its Applications, Proc. Conf.*, pp. 89–101, Silesian University, Opava (1993).

[233] R. Kuwabara, Time-dependent mechanical symmetries and extended Hamiltonian systems, *Rep. Math. Phys.* **19**, pp. 27–38 (1984). https://doi.org/10.1016/0034-4877(84)90023-5.

[234] C. Lanczos, *The Variational Principles of Mechanics*, 4th edn., Dover, New York (1970).

[235] M. Laínz-Valcázar and M. de León, Contact Hamiltonian systems, *J. Math. Phys.* **60**, 10, p. 102902 (2019). https://doi.org/10.1063/1.5096475.

[236] C. Lanczos, *The Variational Principles of Mechanics*, 4th edn., Dover Publishers, New York (1986).

[237] L. D. Landau and E. M. Lifshits, *Mechanics*, 3rd edn., Elsevier (1976).

[238] S. Lang, *Differential and Riemannian Manifolds*, Springer-Verlag, New York (1995). https://doi.org/10.1007/978-1-4612-4182-9.

[239] B. Langerock, F. Cantrijn, and J. Vankerschaver, Routhian reduction for quasi-invariant Lagrangians, *J. Math. Phys.* **51**, 2, p. 022902 (2010). https://doi.org/10.1063/1.3277181.

[240] M. J. Lazo, J. Paiva, J. T. S. Amaral, and G. S. F. Frederico, An action principle for action-dependent Lagrangians: Toward an action principle to non-conservative systems, *J. Math. Phys.* **59**, 3, p. 032902 (2018). https://doi.org/10.1063/1.5019936.

[241] J. M. Lee. *Introduction to Smooth Manifolds*, Springer (2013). https://doi.org/10.1007/978-0-387-21752-9.

[242] J. M. Lee. *Introduction to Riemannian Manifolds*, Springer (2018). https://doi.org/10.1007/978-3-319-91755-9.

[243] Lee Hwa Chung, The universal integral invariants of Hamiltonian systems and applications to the theory of canonical transformations, *Proc. Roy. Soc.* **LXIIA**, pp. 237–246 (1947). https://doi.org/10.1017/S0080454100006646.

[244] M. Leok, T. Ohsawa, and D. Sosa, Hamilton–Jacobi theory for degenerate Lagrangian systems with holonomic and nonholonomic constraints, *J. Math. Phys.* **53**, 7, p. 072905 (2012). https://doi.org/10.1063/1.4736733.

[245] M. Leok and D. Sosa, Dirac structures and Hamilton–Jacobi theory for Lagrangian mechanics on Lie algebroids, *J. Geom. Mech.* **4**, 4, pp. 421–442 (2012). https://doi.org/10.3934/jgm.2012.4.421.

[246] M. Levi, *Classical Mechanics with Calculus of Variations and Optimal Control*, American Mathematical Society (2014).

[247] A. D. Lewis, The physical foundations of geometric mechanics, *J. Geom. Mech.* **9**, 4, pp. 487–574 (2017). https://doi.org/10.3934/jgm.2017019.

[248] A. D. Lewis, Nonholonomic and constrained variational mechanics, *J. Geom. Mech.* **12**, 2, pp. 165–308 (2020). https://doi.org/10.3934/jgm.2020013.

[249] P. Libermann and C. M. Marle, *Symplectic Geometry and Analytical Dynamics*, D. Reidel Publishing Co., Dordrecht (1987). https://doi.org/10.1007/978-94-009-3807-6.

[250] A. Lichnerowicz, Variétés symplectiques et dynamique associée à une sous-variété, *C. R. Acad. Sci. Paris* **280A**, pp. 523–527 (1975).

[251] Q. Liu, P. J. Torres, and C. Wang, Contact Hamiltonian dynamics: Variational principles, invariants, completeness and periodic behavior, *Ann. Phys.* **395**, pp. 26–44 (2018). https://doi.org/10.1016/j.aop.2018.04.035.

[252] J. Llosa and N. Román-Roy, Invariant forms and Hamiltonian systems: A geometrical setting, *Int. J. Theor. Phys.* **27**, 12, pp. 1533–1543 (1988). https://doi.org/10.1007/BF00669290.

[253] R. Loja-Fernandes , J. P. Ortega, and T. S. Ratiu, The momentum map in Poisson geometry, *Am. J. Math.* **131**, 5, pp. 1261–1310 (2009). https://doi.org/10.1353/ajm.0.0068.

[254] C. López, E. Martínez, and M. F. Rañada, Dynamical symmetries, non-Cartan symmetries and superintegrability of the n-dimensional harmonic oscillator, *J. Phys. A: Math. Gen.* **32**, 7, pp. 1241–1249 (1999). https://doi.org/10.1088/0305-4470/32/7/013.

[255] L. Mangiarotti and G. Sardanashvily, *Gauge Mechanics*, World Scientific, Singapore (1998). https://doi.org/10.1142/3905.

[256] C. M. Marle, Poisson manifolds in mechanics, in *Bifurcation Theory, Mechanics and Physics*, pp. 47–76, D. Reidel, Dordrecht (1983). https://doi.org/10.1007/978-94-009-7192-9_3.

[257] C. M. Marle, Reduction of constrained mechanical systems and stability of relative equilibria, *Comm. Math. Phys.* **174**, pp. 295–318 (1995). https://doi.org/10.1007/BF02099604.

[258] G. Marmo, G. Mendella, and W. M. Tulczyjew, Constrained Hamiltonian systems as implicit differential equations, *J. Phys. A: Math. Gen.* **30**, 1, pp. 277–293 (1997). https://doi.org/10.1088/0305-4470/30/1/020.

[259] G. Marmo, G. Morandi, and N. Mukunda, The Hamilton–Jacobi theory and the analogy between classical and quantum mechanics, *J. Geom. Mech.* **1**, 3, pp. 317–355 (2009). https://doi.org/10.3934/jgm.2009.1.317.

[260] G. Marmo, E. J. Saletan, A. Simoni, and B. Vitale, *Dynamical Systems, a Differential Geometric Approach to Symmetry and Reduction.* J. Wiley, New York (1985). https://doi.org/10.1112/blms/18.5.523.

[261] J. C. Marrero, N. Román-Roy, M. Salgado, and S. Vilariño, On a kind of Noether symmetries and conservation laws in k-cosymplectic field theory, *J. Math. Phys.* **52**, 2, p. 022901 (2011). https://doi.org/10.1063/1.3545969.

[262] J. C. Marrero, N. Román-Roy, M. Salgado, and S. Vilariño, Reduction of polysymplectic manifolds, *J. Phys. A: Math. Theor.* **48**, 5, p. 055206 (2015). https://doi.org/10.1088/1751-8113/48/5/055206.

[263] J. E. Marsden, G. Misiolek, J.P. Ortega, M. Perlmutter, and T. S. Ratiu, *Hamiltonian Reduction by Stages, Lecture Notes in Mathematics,* vol. 1913, Springer Berlin, Heidelberg (2007). https://doi.org/10.1007/978-3-540-72470-4.

[264] J. E. Marsden and T. S. Ratiu, Reduction of Poisson manifolds, *Lett. Math. Phys.* **11**, pp. 161–170 (1986). https://doi.org/10.1007/BF00398428.

[265] J. E. Marsden and T. S. Ratiu, *Introduction to Mechanics and Symmetry, Texts in Applied Mathematics,* vol. 17, Springer-Verlag, New York (1999). https://doi.org/10.1007/978-0-387-21792-5.

[266] J. E. Marsden and A. Weinstein, Reduction of symplectic manifolds with symmetry, *Rep. Math. Phys.* **5**, 1, pp. 121–130 (1974). https://doi.org/10.1016/0034-4877(74)90021-4.

[267] J. E. Marsden and A. Weinstein, Some comments on the history, theory, and applications of symplectic reduction, *Quantization of Singular Symplectic Quotiens,* pp. 1–20 N. Landsman, M. Pflaum, M. Schlichenmanier (eds.), Birkhauser, Boston (2001). https://doi.org/10.1007/978-3-0348-8364-1.

[268] J. E. Marsden, T. S. Ratiu, and A. Weinstein, Semi-direct products and reduction in mechanics, *Trans. Am. Math. Soc.* **281**, 1, pp. 147–177, (1984). https://doi.org/10.2307/1999527.

[269] J. E. Marsden and J. Scheurle, The reduced Euler–Lagrange equations, *Fields Inst. Comm.* **1**, pp. 139–164 (1993). http://dx.doi.org/10.1090/fic/001.

[270] E. Martínez, Lagrangian mechanics on Lie algebroids, *Acta App. Math.* **67**, pp. 295–320 (2001). https://doi.org/10.1023/A:1011965919259.

[271] E. Martínez, Reduction in optimal control theory, *Rep. Math. Phys.* **53**, 1, pp. 79–90 (2004). https://doi.org/10.1016/S0034-4877(04)90005-5.

[272] E. Martínez, T. Mestdag, and W. Sarlet, Lie algebroid structures and Lagrangian systems on affine bundles, *J. Geom. Phys.* **44**, 1, pp. 70–95 (2002). https://doi.org/10.1016/S0393-0440(02)00114-6.

[273] S. Martínez, J. Cortés, and M. de León, The geometrical theory of constraints applied to the dynamics of vakonomic mechanical systems: The vakonomic bracket, *J. Math. Phys.* **41**, 4, pp. 2090–2120 (2000). http://doi.org/10.1063/1.533229.

[274] S. Martínez, J. Cortés, and M. de León, Symmetries in vakonomic dynamics. Applications to optimal control, *J. Geom. Phys.* **38**, 3–4, pp. 343–365 (2001). https://doi.org/10.1016/S0393-0440(00)00069-3.

[275] E. Massa and S. Vignolo, A new geometrical framework for time-dependent Hamiltonian mechanics, *Extracta Math.* **18**, 1, pp. 107–118 (2003).

[276] M. R. Menzio and W. M. Tulczyjew, Infinitesimal symplectic relations and generalized Hamiltonian dynamics, *Ann. Inst. H. Poincaré A* **28**, 4, pp. 349–367 (1978). http::://www.numdam.org/item?id=AIHPA1978_28_4_349_0.

[277] E. Miranda, Integrable systems and group actions, *Centr. Eur. J. Math.* **12**, pp. 240–270 (2014). https://doi.org/10.2478/s11533-013-0333-6.

[278] B. Montano, Integral submanifolds of r-contact manifolds, *Demonstr. Math.* **41**, 1, pp. 189–202 (2008). https://doi.org/10.1515/dema-2013-0054.

[279] G. Morandi, C. Ferrario, G. Lo Vecchio, G. Marmo, and C. Rubano, The inverse problem in the calculus of variations and the geometry of the tangent bundle, *Phys. Rep.* **188**, 3–4, pp. 147–284 (1990). https://doi.org/10.1016/0370-1573(90)90137-Q.

[280] M. C. Muñoz-Lecanda, Hamiltonian systems with constraints: A geometric approach, *Int. J. Theor. Phys.* **28**, 11, pp. 1405–1417 (1989). https://doi.org/10.1007/BF00671858.

[281] M. C. Muñoz-Lecanda and N. Román-Roy, Lagrangian theory for presymplectic systems, *Ann. Inst. H. Poincaré A* **57**, 1, pp. 27–45 (1992). https://www.numdam.org/item/AIHPA_1992_57_1_27_0/.

[282] J. I. Neĭmark and N. A. Fufaev, *Dynamics of Nonholonomic Systems*, Translated from the 1967 Russian original by J. R. Barbour. Translations of Mathematical Monographs, vol. 33, American Mathematical Society, Providence, RI (1972). https://doi.org/10.1090/mmono/033.

[283] E. Noether, Invariante Variationsprobleme, *Nachrichten von der Gesellschaft der Wissenschaften zu Göttingen. Mathematisch-Physikalische Klasse* **1918**, pp. 235–257 (1918). http://eudml.org/doc/59024.

[284] J. Nunes da Costa and F. Petalidou, Reduction of Jacobi–Nijenhuis manifolds, *J. Geom. Phys.* **41**, 3, pp. 181–195 (2002). https://doi.org/10.1016/S0393-0440(01)00054-7.

[285] J. Nunes da Costa and F. Petalidou, Reduction of Jacobi–Nijenhuis manifolds via Dirac structures theory, *Diff. Geom. App.* **23**, 3, pp. 282–304 (2005). https://doi.org/10.1016/j.difgeo.2005.06.003.

[286] J. T. Oden and J. N. Reddy, *Variational Methods in Theoretical Mechanics*, Springer-Verlag, Berlin Heidelberg (1983). https://doi.org/10.1007/978-3-642-68811-9.

[287] T. Ohsawa, and A. M. Bloch, Nonholomic Hamilton–Jacobi equation and integrability, *J. Geom. Mech.* **1**, 4, pp. 461–481 (2011). https://doi.org/10.3934/jgm.2009.1.461.

[288] T. Ohsawa, A. M. Bloch, and M. Leok, Discrete Hamilton–Jacobi theory, *SIAM J. Control Optim.* **49**, 4, pp. 1829–1856 (2011). https://doi.org/10.1137/090776822.

[289] T. Ohsawa, O. E. Fernández, A. M. Bloch, and M. D.V. Zenkov, Nonholonomic Hamilton–Jacobi theory via Chaplygin hamiltonization *J. Geom. Phys.* **61**, 8, pp. 1263–1291 (2011). https://doi.org/10.1016/j.geomphys.2011.02.015.

[290] T. Okubo, *Differential Geometry, Monographs in Pure and Applied Mathematics*, vol. 112, Marcel Dekker Inc, New York (1987).

[291] W. M. Oliva, *Geometric Mechanics, Lecture Notes in Math.*, vol. 1798, Springer-Verlag, Berlin (2002).

[292] J. P. Ortega, The symplectic reduced spaces of a Poisson action, *Comptes Rendus Math.* **334**, 11, pp. 999–1004 (2002). https://doi.org/10.1016/ S1631-073X(02)02394-4.

[293] J. P. Ortega and T. S. Ratiu, The optimal momentum map, in *Geometry, Mechanics, and Dynamics*, pp. 329–362, P. Newton, P. Holmes, and A. Weinstein (eds.), Springer, New York (2002). https://doi.org/10. 1007/0-387-21791-6_11.

[294] J. P. Ortega and T. S. Ratiu, *Momentum maps and Hamiltonian reduction*, *Progress in Mathematics*, vol. 222, Birkhäuser, New York (2004). https: //doi.org/10.1007/978-1-4757-3811-7.

[295] Pham-Mau-Quan, *Introduction à la Géométrie des Variétés Différentiables*, Dunod, Paris (1969).

[296] P. D. Prieto-Martínez and N. Román-Roy, Lagrangian-Hamiltonian unified formalism for autonomous higher-order dynamical systems, *J. Phys. A: Math. Theor.* **44**, 38, p. 385203 (2011). https://doi.org/10.1088/ 1751-8113/44/38/385203.

[297] P. D. Prieto-Martínez and N. Román-Roy Unified formalism for higher-order non-autonomous dynamical systems, *J. Math. Phys.* **53**, 3, p. 032901 (2012). https://doi.org/10.1063/1.3692326.

[298] P. D. Prieto-Martínez and N. Román-Roy Higher-order mechanics. Variational principles and other topics, *J. Geom. Mech.* **5**, 4, pp. 493–510 (2015). https://doi.org/10.3934/jgm.2013.5.493.

[299] P. D. Prieto–Martínez, and N. Román–Roy, A new multisymplectic unified formalism for second order classical field theories, *J. Geom. Mech.* **7**, 2, pp. 203–253 (2015). https://doi.org/10.3934/jgm.2015.7.203.

[300] H. Ramirez, B. Maschke, and D. Sbarbaro, Partial stabilization of input-output contact systems on a Legendre submanifold, *IEEE Trans. Automat. Control* **62**, 3, pp. 1431–1437 (2017). https://doi.org/10.1109/ TAC.2016.2572403.

[301] M. F. Rañada, Extended Legendre tangent bundle formalism for time-dependent Lagrangian systems, *J. Math. Phys.* **32**, 2, pp. 500–505 (1991). https://doi.org/10.1063/1.529442.

[302] M. F. Rañada, Extended Legendre transformation approach to the time-dependent Hamiltonian formalism, *J. Phys. A: Math. Gen.* **25**, 14, pp. 4025–4035 (1992). https://doi.org/10.1088/0305-4470/25/14/017.

[303] T. S. Ratiu, R. Tudoran, L. Sbano, E. Sousa Dias, and G. Terra, A crash course in geometric mechanics, in *Geometric Mechanics and Symmetry*, *Lecture Note Ser.*, vol. 306, pp. 23–156, Cambridge University Press, Cambridge (2005). https://doi.org/10.1017/CBO9780511526367.003.

[304] M. Razavy, *Classical and Quantum Dissipative Systems*, Imperial College Press (2006). https://doi.org/10.1142/10391|April2017.

[305] A. M. Rey, N. Román-Roy, and M. Salgado, Günther's formalism in classical field theory: Skinner-Rusk approach and the evolution operator, *J. Math. Phys.* **46**, 5, p. 052901 (2005). https://doi.org/https: //doi.org/10.1063/1.1876872.

[306] A. M. Rey, N. Román-Roy, M. Salgado, and S. Vilariño, k-cosymplectic classical field theories: Tulckzyjew and Skinner-Rusk formulations, *Math. Phys. Anal. Geom.* **15**, pp. 1–35 (2011). https://doi.org/10.1007/s11040-012-9104-z.

[307] X. Rivas, Nonautonomous k-contact field theories, *J. Math. Phys.* **64**, 3, p. 033507 (2023). https://doi.org/10.1063/5.0131110.

[308] X. Rivas, M. Salgado, and S.Souto, Some contributions to k-contact Lagrangian field equations, symmetries and dissipation laws, *Rev. Math. Phys.* **36**, 8, p. 2450019 (2024). https://doi.org/10.1142/S0129055X24500193.

[309] X. Rivas and D. Torres, Lagrangian–Hamiltonian formalism for cocontact systems, *J. Geom. Mech.* **15**, 1, pp. 1–26 (2022). https://doi.org/10.3934/jgm.2023001.

[310] N. Román-Roy, A summary on symmetries and conserved quantities of autonomous Hamiltonian systems, *J. Geom. Phys.* **12**, 3, pp. 541–551 (2020). https://doi.org/10.3934/jgm.2020009.

[311] N. Román-Roy, An overview of the Hamilton–Jacobi theory: The classical and geometrical approaches and some extensions and applications, *Mathematics* **9**, 1, p. 85 (2021). https://doi.org/10.3390/math9010085.

[312] N. Román-Roy, M. Salgado, and S. Vilariño, Symmetries and conservation laws in the Günther k-symplectic formalism of field theory, *Rev. Math. Phys.* **19**, 10, pp. 1117–1147 (2007). https://doi.org/10.1142/S0129055X07003188.

[313] E. J. Saletan and A. H. Cromer, *Theoretical Mechanics*, Wiley & Sons, New York (1971).

[314] W. Sarlet and F. Cantrijn, Higher-order Noether symmetries and constants of the motion, *J. Phys. A: Math. Gen.* **14**, 2, pp. 479–492 (1981). https://doi.org/10.1088/0305-4470/14/2/023.

[315] W. Sarlet and F. Cantrijn, Generalizations of Noether's theorem in classical mechanics. *SIAM Rev.* **23**, 4, pp. 467–494 (1981). https://doi.org/10.1137/1023098.

[316] W. Sarlet, F. Cantrijn, and M. Crampin, A new look at second-order equations and Lagrangian mechanics, *J. Phys. A Math. Gen.* **17**, 10, pp. 1999–2009 (1984). https://doi.org/10.1088/0305-4470/17/10/012.

[317] D. J. Saunders, *The Geometry of Jet Bundles*, Lecture Notes Series **142**, Cambridge University Press, Cambridge, New York (1989). https://doi.org/10.1017/CBO9780511526411.

[318] F. Scheck, *Mechanics: From Newton Laws to Deterministic Chaos*, 4th edn., Springer-Verlag, Berlin (2005). https://doi.org/10.1007/978-3-642-05370-2.

[319] D. J. Simms and N. M. J. Woodhouse, *Lectures on Geometric Quantization*, *Lecture Notes in Physics*, vol. 53, Springer, New York (1976). https://doi.org/10.1007/3-540-07860-6.

[320] R. Sjamaar and E. Lerman, Stratified symplectic spaces and reduction, *Ann. Math.* **134**, 2, pp. 375–422 (1991). https://doi.org/10.2307/2944350.

[321] A. A. Simoes, M. de León, M. Laínz-Valcázar, and D. Martín de Diego, Contact geometry for simple thermodynamical systems with friction, *Proc. Royal Soc. A: Math. Phys. Eng. Sci.* **476**, 2241, p. 20200244 (2020). https://doi.org/10.1098/rspa.2020.0244.

[322] R. Skinner and R. Rusk, Generalized Hamiltonian dynamics I: Formulation on $T^*Q \otimes TQ$, *J. Math. Phys.* **24**, 11, pp. 2589–2594 (1983). https://dx.doi.org/10.1063/1.525654.

[323] R. Skinner and R. Rusk, Generalized Hamiltonian dynamics II: Gauge transformations, *J. Math. Phys.* **24**, 11, pp. 2595–2601 (1983). https://dx.doi.org/10.1063/1.525655.

[324] J. Śniatycki, Dirac brackets in geometric dynamics, *Ann. I.H.P. Phys. Théor.* **20**, 4, pp. 365–372 (1974). https://www.numdam.org/item?id=AIHPA_1974_20_4_365_0.

[325] J. Śniatycki, *Geometric Quantization and Quantum Mechanics*, Springer-Verlag, Berlin (1980). https://dx.doi.org/10.1007/978-1-4612-6066-0.

[326] A. Sommerfeld, *Mechanics: Lectures on Theoretical Physics*, vol. I, Academic Press, Inc., New York (1952).

[327] J. M. Souriau, *Structure of Dynamical Systems: A Symplectic View of Physics*, Springer Science (2012).

[328] M. Spivak, *A Comprehensive Introduction to Differential Geometry*, vol. II (1972).

[329] M. Spivak, *Physics for Mathematicians, Mechanics I*. Publish or Perish, Inc., Houston, TX (2010).

[330] S. Sternberg, *Lectures on Differential Geometry*, Prentice Hall, Englewood Cliffs, NJ (1964).

[331] J. Struckmeier, Hamiltonian dynamics on the symplectic extended phase space for autonomous and non-autonomous systems, *J. Phys. A: Math. Gen.* **38**, 6, pp. 1257–1278 (2005). https://doi.org/10.1088/0305-4470/38/6/006.

[332] E. C. G. Sudarshan and N. Mukunda, *Classical Dynamics: A Modern Perspective*, Wiley-Interscience, New York (1974). https://doi.org/10.1142/9751.

[333] L.Susskind, *Classical Mechanics: The Theoretical Minimum*, Basic Books, New York (2013).

[334] H. J. Sussmann, Symmetries and integrals of motion in optimal control, in *Geometry in Nonlinear Control and Differential Inclusions*, vol. 32, pp. 379–393, A. Fryszkowski, B. Jakubczyk, W. Respondek, T. Rzezuchowski (eds.), Banach Center Publications, Institute of Mathematics of the Polish Academy of Sciences, Warsaw, Poland (1995).

[335] A. Tomassini and L. Vezzoni, Contact Calabi–Yau manifolds and special Legendrian submanifolds, *Osaka J. Math.* **45**, 1, pp. 127–147 (2008).

[336] P. Tondeur, *Introduction to Lie Groups and Transformation Groups, Lect. Notes Math.*, vol. 7, Springer, Berlin (1965). https://doi.org/10.1007/BFb0097829.

[337] W. M. Tulczyjew, Hamiltonian systems, Lagrangian systems and the Legendre transformation, *Symp. Math.* **14**, 1, pp. 247–258 (1974).

[338] W. M. Tulczyjew, Les sous-variétés lagrangiennes et la dynamique hamil-tonienne, *C. R. Acad. Sci. Paris Sér. A-B* **283**, 1, pp. A15–A18 (1976).

[339] W. M. Tulczyjew, Les sous-variétés lagrangiennes et la dynamique lagrang-ienne, *C. R. Acad. Sci. Paris Sér. A-B* **283**, 8, pp. A675–A678 (1976).

[340] G. M. Tuynman, Geometric quantization, in *Proc. Sem. 1983-1985 Mathematical Structures in field theories*, vol. 1, CWI Syllabus (1985).

[341] A. J. Van der Schaft, Symmetries in optimal control, *SIAM J. Control and Optimization* **25**, 2, pp. 245–259 (1987). https://doi.org/10.1137/0325.

[342] A. J. Van der Schaft, Liouville geometry of classical thermodynamics, *J. Geom. Phys.* **170**, p. 104365 (2021). https://doi.org/10.1016/j.geomphys.2021.104365.

[343] J. Vankerschaver, Euler–Poincaré reduction for discrete field theories, *J. Math. Phys.* **48**, 3, p. 032902 (2007). https://doi.org/10.1063/1.2712419.

[344] A. M. Vershik, *Classical and Non-classical Dynamics with Constraints*, in *Lecture Notes in Math.*, vol. 1108, pp. 278–301, Springer-Verlag, Berlin (1984). https://doi.org/10.1007/BFb0099563.

[345] S. Vignolo, A new presymplectic framework for time-dependent Lagrangian systems: The constraint algorithm and the second-order differential equation problem, *J. Phys. A: Math. Gen.* **33**, 28, pp. 5117–5135 (2000). https://doi.org/10.1088/0305-4470/33/28/314.

[346] A. M. Vinogradov and I. Krasil'shchik, What is the Hamiltonian formalism?, *Russ. Math. Surveys* **30**, 1, pp. 177–202 (1975). https://doi.org/10.1070/RM1975V030N01ABEH001403.

[347] A. M. Vinogradov and B. A. Kupershrnidt, The structures of Hamiltonian mechanics, *Russ. Math. Surveys* **32**, 4, pp. 177–243 (1977). https://doi.org/10.1070/RM1977v032n04ABEH001642.

[348] L. Vitagliano, The Lagrangian-Hamiltonian formalism for higher order field theories, *J. Geom. Phys.* **60**, 6–8, pp. 857–873 (2010). https://doi.org/10.1016/j.geomphys.2010.02.003.

[349] L. Vitagliano, The Hamilton–Jacobi formalism for higher-order field theory, *Int. J. Geom. Meth. Mod. Phys.* **07**, pp. 1413–1436 (2010). https://doi.org/10.1142/S0219887810004889.

[350] L. Vitagliano, Hamilton–Jacobi diffieties, *J. Geom. Phys.* **61**, 10, pp. 1932–1949 (2011). https://doi.org/10.1016/j.geomphys.2011.05.003.

[351] L. Vitagliano, Geometric Hamilton–Jacobi field theory, *Int. J. Geom. Meth. Mod. Phys.* **9**, 2, p. 1260008 (2012). https://doi.org/10.1142/S0219887812600080.

[352] L. Vitagliano, Characteristics, bicharacteristics, and geometric singularities of solutions of PDEs, *Int. J. Geom. Meth. Mod. Phys.* **11**, 9, p. 1460039 (2014). https://doi.org/10.1142/S0219887814600391.

[353] L. Vitagliano, L_∞-algebras from multicontact geometry, *Diff. Geom. Appl.* **59**, pp. 147–165 (2015). https://doi.org/10.1016/j.difgeo.2015.01.006.

[354] M. Visinescu, Contact Hamiltonian systems and complete integrability, *AIP Conference Proceedings* **1916**, 1, p. 020002 (2017). https://doi.org/10.1063/1.5017422.

[355] H. Wang, Symmetric reduction and Hamilton–Jacobi Equation of rigid spacecraft with a rotor, *J. Geom. Sym. Phys.* **32**, pp. 87–111 (2013). https://doi.org/10.7546/jgsp-32-2013-87-111.

[356] H. Wang, Hamilton–Jacobi theorems for regular reducible Hamiltonian systems on a cotangent bundle, *J. Geom. Phys.* **119**, pp. 82–102 (2017). https://doi.org/10.1016/j.geomphys.2017.04.011.

[357] F. W. Warner, *Foundations on Differentiable Manifolds and Lie Groups*, Scott, Foresman and Co., Glenview (1971). https://doi.org/10.1007/978-1-4757-1799-0.

[358] A. Weinstein, *Lectures on Symplectic Manifolds*, *Reg. Conf. Ser. Math.*, vol. 29, American Mathematical Society, Providence, RI (1977).

[359] A. Weinstein, On the hypotheses of Rabinowitz' periodic orbit theorems, *J. Diff. Eqs.* **33**, 3, pp. 353–358 (1979). https://doi.org/10.1016/0022-0396(79)90070-6.

[360] A. Weinstein, Lagrangian mechanics and groupoids, *Fields Inst. Comm.* **7**, pp. 207–231 (1996).

[361] E. T. Whittaker, *A Treatise on the Analytical Dynamics of Particles and Rigid Bodies*, 4th edn., Cambridge University Press, Cambridge (1937).

[362] C. Willett, Contact reduction, *Trans. Amer. Math. Soc.* **354**, 10, pp. 4245–4260 (2002). https://doi.org/10.1090/S0002-9947-02-03045-3.

[363] N. M. J. Woodhouse, *Geometric Quantization* , 2nd edn., Clarendon Press, Oxford (1992).

[364] N. M. J. Woodhouse, *Introduction to Analytical Dynamics*, 2nd edn., *Springer Undergraduate Mathematics Series*, Springer Science & Business Media, Oxford (2009). https://doi.org/10.1007/978-1-84882-816-2.

Index